3 \ Third Edition

Cellular Pathology

RELATED BIOMEDICAL SCIENCE TITLES

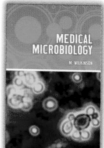

Cellular Pathology

An Introduction to Techniques and Applications

D.J. Cook

Formerly at *University of Portsmouth, School of Pharmacy and Biomedical Science, Portsmouth, UK*

and

P.J. Warren

University of Portsmouth, School of Pharmacy and Biomedical Science, Portsmouth, UK

Scion

© **Scion Publishing Limited, 2015**

ISBN 978 1 907904 35 6

First published 2015

Second edition published in 2006 by Scion Publishing (9781904842309)

First edition published in 1998 by Butterworth Heinemann (0759631112)

A CIP catalogue record for this book is available from the British Library.

Scion Publishing Limited

The Old Hayloft, Vantage Business Park, Bloxham Road, Banbury OX16 9UX, UK

www.scionpublishing.com

Important Note from the Publisher

The information contained within this book was obtained by Scion Publishing Ltd
from sources believed by us to be reliable. However, while every effort has been made
to ensure its accuracy, no responsibility for loss or injury whatsoever occasioned to
any person acting or refraining from action as a result of information contained herein
can be accepted by the authors or publishers.

Readers are reminded that medicine is a constantly evolving science and while the
authors and publishers have ensured that all dosages, applications and practices
are based on current indications, there may be specific practices which differ
between communities. You should always follow the guidelines laid down by the
manufacturers of specific products and the relevant authorities in the country in
which you are practising.

Although every effort has been made to ensure that all owners of copyright material
have been acknowledged in this publication, we would be pleased
to acknowledge in subsequent reprints or editions any omissions brought to
our attention.

Registered names, trademarks, etc. used in this book, even when not marked
as such, are not to be considered unprotected by law.

Typeset by Phoenix Photosetting, Chatham, Kent, UK

Printed in India

The photomicrograph on the cover illustrates picro-mallory staining on a section of skin x100 to show:
- Nuclei – dark blue/black
- Muscle/fibrin – red
- Collagen – blue
- Red blood cells and air shafts – yellow

Contents

Preface

The rewriting of this third edition was a chance to bring this student-oriented text up to date and firmly into the 21st century. For the rewriting, the writing team has increased to two authors. The collaboration and friendly critiques between the authors have been both beneficial and pleasurable.

The new text has been substantially changed, with a new chapter on the background pathological disorders, since investigation of these disorders constitutes the main task in biomedical histology laboratories. The separate chapter on autoradiography in the second edition is now incorporated into a fresh chapter on molecular biology to emphasize the development of DNA investigation as an important method in the diagnosis of disease.

The coverage of cytological methods has been expanded to update the coverage of liquid-based cytology, which has become the standard method, and also to expand the coverage of the use of cytology in different organ systems. This has resulted in the splitting of cytology into two chapters, the first covering gynaecological cytology and the second dealing with the other applications.

The chapter on microscopy has expanded coverage of confocal microscopy and the use of digital techniques in interpreting images. There has also been a significant update for the chapter on management, with more emphasis given to safety and quality control.

The changes in the text are now supported by a greater use of colour images and this helps readers to better visualize the topics covered in the text. Thanks are due to Behad Shambayati and Libby Soar for help with obtaining cytology images.

Rewriting has been challenging, interesting and pleasurable for us and we hope the readers of the book will have the same feelings about the work.

D.J. Cook
P.J. Warren
Portsmouth, July 2014

Preface to the second edition

It is nice to have the opportunity to produce a second edition of this book. Some things have changed but the popular structure of the book remains the same. Some of the changes are additions of new material, particularly in chapters 10, 12 and 17, whilst others are removal of older outdated material such as the references to stripping film in autoradiography where this particular product is, sadly, no longer produced. Chapter 10 on immunology techniques includes a range of newer techniques that have become popular since the first edition. Chapter 12 now introduces the changes to cytology screening in the UK where the use of fluid-based systems is replacing the traditional direct smearing and includes details of the Bethesda 2001 system for classifying cellular abnormalities. Some of the changes in chapter 17 relate to changes in legal requirements for safety in the laboratory. In the first edition safety was mentioned at many points in passing but there was no consideration of safety as a single topic. In this edition the individual references to safety remain embedded in the individual chapters but there is now a section purely on the organization and legal aspects of safety. Also changed in chapter 17 is a revised section mentioning the Human Tissue Act and introducing inter-laboratory quality control.

Other changes have occurred because of a change in publisher. The layout of the chapters has been improved and the learning objectives expanded. The contents list has been expanded significantly to help readers to navigate through the book more easily. The most important change is the inclusion of colour photographs in the book to help readers gain a better understanding of the subject in practice. So my thanks go to Jonathan Ray of Scion Publishing Ltd for encouraging the inclusion of the many colour photographs and also for his help in producing a new and better edition. My thanks also go to the copy-editor, Jane Hoyle, who checked my drafts and corrected my errors.

I am fortunate to be revising this edition at all. I began revising the material in the autumn of 2005 but then had a cycling accident at the end of November 2005. This fractured my skull and resulted in a subdural haematoma. The clot was removed at Southampton Neurological Centre and I was in Intensive Care on life support. Fortunately, I have recovered completely and so was able to complete the revisions in January 2006. I would like to thank the staff at Southampton Neurological Centre for their care and support. I would also like to thank Judy, my wife, and my children, Malcolm and Elaine, for their support and encouragement during my convalescence.

John Cook, Portsmouth, May 2006

The study of tissues and their diseases

Learning objectives

After studying this chapter you should confidently be able to:

- **Describe the major types of artefacts in sections**
  Loss of materials, gain of materials and distortion of tissue structure.

- **Describe why several different methods are needed in histology**
  Since tissues are altered, alternatives are needed to show which appearances are genuine and which are artefacts.

Ever since the microscope was invented somewhere around the beginning of the 17th century it has been used to magnify human cells and tissues. The great difficulty with the application of the microscope is that the tissues themselves are really quite featureless and uninteresting unless they are first prepared in some way to enhance the detail. This book is an attempt to explain the methods used in examining cells and to try to give a theoretical outline of the basis of the techniques.

1.1 What is cellular pathology?

I will begin with the words used to describe the topic itself. There are many words that have been used to describe the subject over the years and they all have some merit. The parent subject is **anatomy** (from the Greek meaning cutting up: *ana* = up and *tome* = to cut), which is the study of the organs and their shape, arrangement and structure. Anatomy is an ancient discipline. Ever since humans first started butchering animals, people must have realized that the body is made of different organs and that all of these organs have varying structures. Over many centuries, the study of anatomy catalogued and described the visible structure of the human body and laid the foundations of physiology by trying to explain the functions of each organ and how the organs interacted with each other (see *Box 1.1* for comments).

With the development of the microscope, the organs could be seen in finer detail and the discipline of **microscopic anatomy** was born. This is now normally called **histology**, the study of tissues and their structure. The term histology usually means the study of normal tissues, whilst the study of diseases in tissues is more correctly called **histopathology**, but fashions in science change and the more common term now is **cellular pathology**. This is a good name for the science since the basic structure of the human body is the cell and the basis of most human disease is also at the cellular level.

Tissues consist of cells and the study of the structure of cells is **cytology**, which can be used diagnostically. The prefix *cyto* can usually replace the prefix *histo* with the inference that the emphasis is on subcellular detail rather than on multicellular structures. Thus, we

> **Box 1.1 // Histology is complementary to other biological sciences**
>
> Histology does not exist in isolation but constantly interacts with other disciplines such as biochemistry and physiology. Many of the histochemical techniques used today are modifications of methods used originally in biochemistry. Biochemistry in turn depends on the information provided by histology to identify the sites where the biochemical events are occurring. The scientific disciplines are complementary. The main advantage of histology is its ability to localize materials and metabolic processes within the tissues and even within individual organelles in the cell. The main disadvantage of histology is the requirement that the materials being examined cannot be allowed to move and must be kept in their original site (*in situ*). The advantage and disadvantage are the same: the very thing that makes histology so useful also imposes its own problems. This is why histology needs the other disciplines and the other disciplines need histology. They are not competitors but allies in the scientific investigation of disease.

can have cytochemistry, immunocytochemistry, cytopathology and so on. In many cases the terms are simply interchangeable. Although this book is entitled *Cellular Pathology*, it is not limited to the diagnosis of disease and can be used equally as an introduction to normal histological techniques and/or research applications. There is no dividing line between the preparation of normal tissues and diseased tissues; the methods are the same, just the application is different.

1.2 Preparation of tissues

One recurring feature of the investigation of biological materials is the problem of preparing the tissues. Ideally we would like to study living tissues but this is not always possible. Any method that alters the tissues making them more readily studied must also risk the possibility that the tissues are no longer representative of the living tissue. The only way to solve this dilemma is to have several quite different ways of studying the same problem so that any errors produced by one method are quite different to the errors introduced by the other methods. Anything which remains the same, regardless of how the tissue was prepared, can safely be assumed to be a feature of the tissue and not of the method of preparing the tissue.

The more preparative techniques we have, the better our understanding

To make sure we have got the right answers, we therefore need many different methods. The methods may not all be equally easy and they may not all be equally good, but we need them to make sure the methods we are using are valid. Some techniques have been very important historically but have been superseded by better and more reliable techniques. However, their influence may have been immense and so they should still be remembered.

It is also important to realize that you cannot do everything with one method. The large numbers of methods developed in histology often relate to just one or two problems. These problems may well recur in a different way, so knowledge of a variety of techniques may enable you to see how to overcome a modern problem using an older method. Thus, this book considers not only the currently important techniques but also some older or less popular ones.

> **Box 1.2 // Artefacts**
>
> The term artefacts can also be spelt artifacts. They are both pronounced the same but I prefer the 'e' since it comes from the Latin *arte* (by art), although others disagree.
>
> The most bizarre artefact I have had was a complete small spider in a stained section. During the drying of the section, a student had used a hotplate and put the section face up on the hotplate. A spider must have fallen onto it and was killed and dried down. The student never noticed, stained the section and then when it was being examined under the microscope there in the middle of the kidney was a spider.

Preparing tissues alters them

All of the techniques used in histological preparation have their own particular problems and all produce significant changes in the tissues. These changes in the tissues are known as **artefacts**. Artefacts are changes in the tissue brought about by the techniques used; they are errors introduced during the processing of tissues. The artefacts that can be produced are as many and as varied as the techniques that produce them. It would take several pages just to list all of the artefacts that can occur, but they can be grouped into three major types. Within each type there will be a variety of causes and a range of individual errors, but if we can understand the types of errors it often helps to recognize the errors and sometimes helps us to know how to avoid them.

Materials can be lost from the tissue during section preparation

The loss of materials is one of the commonest types of artefact. Sometimes it is not crucial if many materials are lost provided the important materials, and in particular the substance we are looking for, are retained.

If we take paraffin wax-processed tissues as an example, the materials that are lost from the tissues include:

1. All of the small water-soluble molecules. These are lost when we attempt to preserve the tissue by immersing it in a fixative. The materials lost include simple sugars (glucose, fructose, etc.), ions (sodium, potassium, chlorides, sulphates, phosphates, etc.), vitamins (thiamine, riboflavin, etc.), cofactors (ADP, NAD, etc.) and many others.
2. All of the lipids and lipid-soluble materials that are extracted by the alcohols, clearing agents and hot waxes used in preparing wax sections. The membrane lipids, steroid hormones and fat-soluble vitamins are lost as well as the simple adipose fats.

Thus, it is no use preparing a paraffin wax section to try and identify where NAD is located within a cell because it will all have disappeared. In paraffin wax sections, just about the only things that are retained are the proteins, some nucleic acids and a few other molecules. What is perhaps more surprising is how much information about tissue structure and function can be learned from such a depleted material.

Materials can be gained by the tissue during section preparation

This usually involves chemical additions to the structures. Any material prepared using aldehydes as preservatives will have extra aldehydes that were not there in life. Aldehydes form addition compounds (see **Chapter 3**) with at least some of the amine groups in proteins. If the preservation is done with mercury compounds instead of aldehydes, there will be a black deposit of mercury within the tissues that was not there before the fixative was added.

These chemical additions are an inherent part of the technique but there are also additions that can be less predictable. For example, during the handling of the tissues in the laboratory, the blocks may be placed on filter paper or handled with rubber gloves. Fibres from the paper or glove powder from the gloves may be transferred accidentally to the specimen and appear in the final preparation. Handling slides or specimens with bare hands can transfer foreign cells shed from the skin into the preparation. These contaminants are artefacts since they were not there in the living body.

Box 1.3 // Reagents can be contaminated with microorganisms

Many histological reagents can be contaminated with living organisms. Dyes are often a good source of nutrients for bacteria and fungi which will live, grow and multiply in dye solutions. During the staining of tissues these may be transferred to the specimen and be thought to be part of the tissue. This can result in an incorrect diagnosis of infection when in reality the original tissue was completely free of microorganisms.

Structures can be altered or distorted during section preparation

The most common distortion is that tissues shrink during processing. Shrinkage in itself is not a serious problem provided everything alters to the same extent. As long as the shrinkage is regular, we can compare tissues with each other knowing that they will have been affected to the same degree. Unfortunately shrinkage is often irregular. Some materials shrink more than others and give a false impression. Take bone as an example. Bone consists of a very hard matrix containing a few scattered cells called osteocytes, which lie in small spaces in the bone called lacunae. In life, the whole of each lacuna is filled by the cytoplasm of the osteocytes. However, during processing, the soft tissue of the osteocyte shrinks much more than the hard matrix. As a result, in many sections of bone the osteocytes appear shrunken and a gap appears around the cells. This gap is an artefact of the processing.

Box 1.4 // Histology, science, art and craft

Histology is one area of science where there is a great deal of manual skill needed as well as scientific knowledge. The ability to cut sections is certainly crucial to obtaining acceptable results and a good microtomist needs a long apprenticeship. Staining requires skills in differentiation to get the optimal contrast in specimens. This has led to a certain amount of mystique in histology and the assertion that histology is more of an art than a science.

The results of histology are slides and these can certainly be aesthetically pleasing in an abstract way and are often used as illustrations and for covers of journals and magazines. But these images always reflect reality and are always based on scientific principles. Histologists do not actually create the results; they just bring out what is present in the tissues.

The other assertion is that histology is getting left behind. This again is not true. Histology may be involved with dyeing but it is not a dying art.

All sections have some artefacts

Whenever a section is seen, there will be artefacts and the recognition of these is an important part of histology. It is also important to be able to avoid certain artefacts and so histologists need a range of methods that have different properties so that they can always manage to get a section that is useful. It may have other artefacts but the section will avoid the specific artefacts that would make it useless for a particular purpose. Thus, wax sections do not contain lipids but they do not have any ice crystal damage. Frozen sections retain the lipids but may show ice crystal damage. You must choose your method to suit your requirements.

Fortunately histological workers over the history of the subject have been very clever in approaching the problems in different ways and so most of what we now acknowledge as the structure of cells and tissues stands on a very firm base. But whenever the boundaries of a technique are extended, it is always correct to ask 'Is this real or is this just what has been done to the tissues?'

1.3 Evolution of histology

The preparation of tissues has been improving constantly and it seems likely that it will continue to improve in future. The earliest microscopists simply used whole mounts of small specimens with little or no use of dyeing. The earliest sections used free-hand sectioning but this gives only poor results and it was the introduction of the technique of embedding tissues in a solid medium and the use of dyes to enhance contrast and visibility that made histology into an accurate and reliable method.

The embedding techniques have been many and varied. Using these differing techniques, much of the tissue structure still remains very similar. The appearances do change slightly, but by comparing techniques it is possible to determine the structure of the tissues as they would be inside the living organism. We can therefore also state with confidence that one technique will cause shrinkage, another will destroy some tissue components and yet another may alter the chemical nature of the tissues. Whenever a piece of tissue is to be investigated, the choice of the preparative technique will impose its own limitations and advantages. A single technique is never the best for every possible situation. You must know what you want to show and then choose the appropriate method. Thus, in this book there will be descriptions of several alternative techniques and a comparison of their advantages and problems.

Demonstration techniques are now more selective and specific

One of the major advances has been the improvement of demonstration methods. In the earliest years of micro-technique, the only method available to demonstrate structure was to use the few natural dyes that were available. These were often unreliable and the demonstrations were poor and the results did not keep well. The phenomenal increase in the number of dyes available following the synthesis of dyes (particularly from coal tar) made it possible to produce more selective methods and brighter colours. The use of many of these techniques is still an important part of micro-technique today. The development of the use of antibodies and histochemical techniques has allowed even more precision and it is now possible to identify and localize individual molecules and chemicals rather than just the vague staining of 'proteins'. Now there is a rapidly increasing use of nucleic acid techniques, which have unprecedented accuracy and can identify the actual genes and mutations of genes that are associated with many diseases. Where this will lead in

> **Box 1.5 // Histology is central to much modern research**
>
> Histology is still one of the most important research methods. A quick look in any major biomedical or biological journal will show that more than half of the papers will include some histological aspect. Histology can appeal to humans in a way that most scientific disciplines do not. Histology is essentially a visual science and human beings are highly visual, with the amount of information received and understood through vision being greater than all of the other senses combined. The display of a good histological preparation always appeals both aesthetically and intellectually since it is possible to see for oneself. No amount of tables or graphs can have quite the same impact as a good section.

the future is not absolutely clear, but histological micro-technique is certainly not a dead end with no significant further developments; rather, it is a lively and growing scientific discipline.

Histology is likely to change dramatically in the future

Histological techniques have not altered as much in the past century as those of other scientific disciplines such as biochemistry. A histologist from the late 19th century would probably adjust to working in a modern laboratory with little difficulty. He might need a little retraining in the advances that have taken place but at least he would understand the general concepts. Biochemistry, on the other hand, has changed dramatically and a 19th century biochemist would not recognize most of the instruments and techniques that are now routine. My feelings are that histological micro-technique will undergo a similar change in the coming years. It will be more important to understand the concepts behind the methods than to know the best recipe for a stain since many of the current techniques will become obsolete and will be replaced. Only the underlying principles are likely to remain the same and will be as important in 50 years' time as they will be in 5 years' time.

This book, unlike the majority of textbooks on histological technique, is not a practical book full of recipes and methods, but aims to be an introduction to the concepts of tissue preparation. This does not mean that I think that the detail of the techniques is unimportant. I believe that the details and tips present in the laboratory recipe books are often quite fascinating in themselves and also are essential to anyone intending to work in a histological laboratory or use histology in a regular and practical way. The detail is always important when you come to actually prepare a tissue section, but when you are just beginning to learn about the subject the detail may get in the way of a deeper understanding. When you want to do histology, then you will need a laboratory manual, but I hope that having read this book first, you will be able to follow the method with understanding and not just an unthinking adherence to the recipe.

The layout includes an outline of the learning objective at the beginning of each chapter and some Self-Assessment questions at the end. Answers to the questions are grouped together at the end of the book rather than immediately following the questions to discourage students from reading the answer before trying to answer it for themselves.

Also at the end of each chapter is some suggested reading and reference sources. For the first time this will now include internet websites (see ***Box 1.6***) as well as books and journals.

> **Box 1.6 // Websites**
>
> Websites can be a great place for information but can also be totally unreliable, either because their content is of dubious quality or because they suddenly become unavailable. I hope all the websites we have included will remain good quality and available but obviously this cannot be guaranteed.
>
> The best websites have some sort of peer review or other quality control. An example of this is Wikipedia which has sometimes had unreliable data on some pages but they do invite readers to correct any mistakes and report biased opinions. This has been effective on many occasions resulting in dubious pages being corrected or removed.
>
> Social networking sites such as Facebook and Twitter may be great for gossip but not for scientific information.
>
> When using websites remember that they are not all peer reviewed and may have a very biased point of view but distinguishing between good and bad sources of information is a great learning experience and a good skill to acquire.

This book is not only intended for histologists but may also be useful for people who are not intending to become histologists but who may work alongside histologists and will need to talk to histologists and make use of histology but without needing to do all the preparation themselves. You may not need to know the minute-to-minute steps in preparing tissues, but if you are going to understand histology and histologists, then you need to understand their problems and their language.

Like all scientific disciplines, histology is absolutely enthralling to its disciples but can often seem boring and routine to non-histologists. I hope to try and bridge the gap between these extremes and show the fascination but without the tedious detail, to show the scope that histology can cover. If in the process I can make you think that histological methodology is worthwhile learning, then I will feel that the book has succeeded.

Suggested further reading

Alberts, B., Johnson, A., Lewis, J., Raff, M. and Roberts, K. (2008) *Molecular Biology of the Cell*, 5th edn. Garland Science.

Bancroft, J., Layton, C. and Suvarna, S. (2012) Chapter 5: The gross room, surgical cut-up. In: *Bancroft's Theory and Practice of Histological Technique*, 7th edn. Elsevier Health Science.

Orchard, G. and Nation, B. (2011) *Fundamentals of Biomedical Science: Histopathology*. Oxford University Press.

Wallington, E.A. (1979) Artifacts in tissue sections. *Medical Laboratory Sciences*, **36**: 3–61.

Young, B., Stewart, W. and O'Dowd, B. (2014) *Wheater's Functional Histology: a colour atlas*, 6th edn. Churchill Livingstone.

Suggested websites

http://library.med.utah.edu/WebPath/webpath.html
www.histology.leeds.ac.uk/
www.medicalschoolpathology.com/ShotgunHistology.htm

Self-assessment questions

1. Outline what is meant by an artefact.
2. List the major types of artefact and give one example of each type.
3. Two sections are prepared by different methods to show the location of glycogen. The first section shows glycogen in hepatocytes (liver cells) evenly spread throughout the cells. The second shows glycogen present in the same cells but concentrated at one end of the cell.

   Can we conclude that hepatocytes contain glycogen?

   Can we conclude that glycogen is evenly spread in the hepatocytes?
4. Why is there no single method of tissue preparation that everyone agrees is the best?

Learning objectives

After studying this chapter you should confidently be able to:

- **Outline the limitations of the definitions of health, the concept of disease and the causes of disease**
  Health includes an absence of disease but also includes other factors of wellbeing. Disease is a deviation from health which prompts medical intervention to eliminate or minimise distress.

- **Explain the process of acute inflammation**
  Acute inflammation is characterized by pain, redness, heat and swelling in the affected area. These changes are due to changes in the vascular system resulting in increased blood flow, loss of fluid into tissue and emigration of white blood cells.

- **Describe the differences between acute and chronic inflammation**
  In chronic inflammation the classic signs of inflammation are reduced or absent, blood flow is less abnormal and the proportions of different types of white blood cells are altered. Acute inflammation is relatively brief and is usually resolved in a few days. Chronic inflammation can persist for years and usually requires treatment to give resolution.

- **Explain the differences between different forms of healing**
  Recovery is when tissue returns to normal with no loss or replacement of tissue. Regeneration is replacement of lost tissue with tissue that is visually and functionally identical to the lost tissue. Repair is when lost tissue is replaced with scar tissue which is visually and functionally different to the original tissue.

- **Outline the meaning of atrophy and name several types of atrophy**
  Atrophy is the shrinkage of tissues either by reduction in cell size or cell number.

- **Distinguish between hypertrophy and hyperplasia and explain the meaning of dysplasia and metaplasia**
  Hypertrophy is enlargement of the tissues due to an increase in cell size and is associated with increased demands on the cells. Hyperplasia is an increase in cell numbers and is associated with irritation of the tissues. Hyperplasia is sometimes associated with malignant change whilst hypertrophy is not associated with tumours. Dysplasia and metaplasia represent early premalignant changes.

- **Compare benign and malignant tumours**
  Benign tumours are usually slow growing with relatively normal cell and tissue structure. Benign tumours show only local growth and are rarely a significant risk to general health. Malignant tumours may grow very rapidly and have abnormal cell and tissue structure with changes in chromosomes being common. Malignant cells not only differ from normal but show cellular pleomorphism within a tumour. Growth can be invasive and metastatic resulting in significant health risks, including a risk of death if untreated.

- **Indicate the significance of invasion and metastasis in malignancy**
  Invasion is the growth of the tumour into adjacent tissues and often extensive local spread. In metastasis cells travel in body fluids to distant sites where they establish secondary tumours. Both invasion and metastasis make effective treatment more difficult.

- **List the features of cells which indicate malignancy**
  Variation in size of nucleus and cytoplasm, increased nucleocytoplasmic ratio, irregular nuclear border, prominent nucleoli, increased rates of cell division and altered staining.

- **Briefly describe the process of necrosis**
  Necrosis is the death of cells following loss of control, especially of energy usage. The cell becomes unable to maintain homeostasis and ionic and water balance disruption results in swelling of cells and organelles. The eventual rupture of lysosomes releases catabolic enzymes which break down the cell and tissue structure.

- **Explain the meaning and significance of apoptosis**
  Apoptosis is the controlled death of cells by specific cellular pathways. The cells are destroyed quickly without inducing inflammation. Apoptosis is found in both health and disease and is an important feature of embryonic development and disease resistance.

The distinction between health and disease is not always easy. Some definitions of health involve the absence of disease and some definitions of disease give it as the absence of health (the word itself reflects this as it was originally dis-ease (discomfort)). Thus the definitions often become circular. The Oxford English Dictionary defines disease as "a disorder of structure or function in a human, animal, or plant, especially one that produces specific symptoms or that affects a specific location and is not simply a direct result of physical injury". The World Health Organization defines health as follows: "Health is a state of complete physical, mental and social well-being and not merely the absence of disease or infirmity". This ideal state is rarely, if ever, achieved and is only an aspirational target.

Most people have significant pathologies most of the time yet do not consult the doctor or appear in hospital. In order to present to the medical profession the patient must feel that his/her life is in some way inadequate or difficult and can be improved by medical intervention. The prompt to seek help can be something nasty (e.g. pain or discomfort), alarming (e.g. blood in the urine or growth of a mole), disabling or embarrassing.

It is only after patients have entered the medical system and had a provisional diagnosis which suggests that surgery is indicated that specimens will arrive in the histopathology laboratory. The effective treatment of a disease needs an accurate diagnosis, for instance the treatment of toxic food poisoning and infective food poisoning are quite different, but the symptoms can be identical. Histological investigations can help to provide such a diagnosis, but the investigations are limited to the recognition of structural changes or to the detection of chemical changes in the tissue. Not all diseases give such histologically recognisable features and so cannot be diagnosed by histology laboratories.

Many of the changes that occur in disease are not directly caused by the agent inducing the disease, but occur because of the tissue's responses to the causation. This often helps because there are only a limited number of changes that a tissue can undergo, so recognition of these general changes is relatively easy. Changes can be very broadly grouped into three categories:

1. degenerative – in which cells are damaged, e.g. atrophy and necrosis
2. proliferative – in which cells grow in size or number, e.g. tumours
3. modified – in which the arrangement and relationships of tissues become deranged or infiltrated in some way, e.g. inflammation.

There are many diseases in which cellular pathology can be helpful but this short chapter concentrates on general pathology. The general aspects of diseases are covered because a thorough knowledge of general pathology is an essential prerequisite to understanding systematic pathology, in which individual pathologies are considered in a more comprehensive way usually arranged by organ system.

2.1   Inflammation

Inflammation is always included as a pathological subject and is present in most diseases, but it is strictly a normal response of the body. Whenever tissues are damaged inflammation occurs; however, the inflammation is not usually the disease itself but an attempt to limit and eliminate the disease and its cause. In histological sections the presence of inflammation is therefore often a good indicator of disease and the area of inflammation can be a good guide to the location and extent of the primary disease.

Inflammation can be the cause of disease in some instances, such as in autoimmune disease, but the vast majority of inflammation is just a normal and healthy response to injury or other disease. Without inflammation the human body would be vulnerable to many diseases and we would not survive. It is also extremely common and it would be unusual for someone not to have, and consider as minor, a mild localised inflammation, e.g. associated with minor cuts and abrasions or minor infections such as athlete's foot.

Inflammation can be divided into two types: **acute inflammation** (lasting a maximum of approximately 14 days) and **chronic inflammation** (which can last for months or years).

Box 2.1 // The naming of inflammation

The names for inflammation of specific tissues or organs are created by adding the suffix *-itis* to the histological name of the tissue or organ: so dermatitis, appendicitis and gingivitis are inflammation of skin, the appendix and the gums, respectively.

There is no distinction between chronic and acute inflammation except if the actual words are added to the description, e.g. acute appendicitis or chronic appendicitis.

In pathological terms the only difference between acute and chronic inflammation is the time of persistence, but for the general public the term acute is sometimes mistakenly confused with severity. It is possible to have mild acute inflammation e.g. a nettle sting and a severe chronic inflammation, e.g. syphilis or arthritis.

Acute inflammation

The effects of inflammation have long been recognised and the symptoms were first described in writing by Aulus Celsus in the first century AD. The four cardinal signs are **rubor** (redness), **calor** (heat), **tumor** (swelling) and **dolor** (pain), and they are mainly due to alterations in the blood vessels and blood flow. A further sign is loss of function (**functio laesa**) which is often present but does not always occur.

Most of the symptoms of inflammation cannot be seen directly in haematoxylin and eosin (H&E) stained sections. Pain, heat and loss of function are obviously not visible and swelling is often masked by the shrinkage occurring in histological preparations. The redness can be detectable indirectly as hyperaemia due to the altered blood flow pattern. Many of the features recognisable in tissue sections are related to these changes in blood flow (see *Table 2.1*). Acute inflammation in particular is characterised by changes in vascular size (so called vasodilation) and vascular permeability (see *Figs 2.1* and *2.2*).

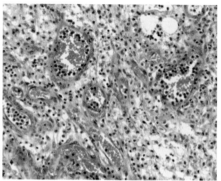

Figure 2.1
Acute inflammation. The blood vessels are enlarged and packed with blood cells with the white blood cells situated at the periphery of the vessels. The tissue is slightly swollen with oedema and has a higher content of WBCs than normal.

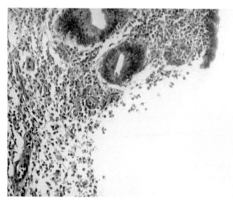

Figure 2.2
Inflamed appendix. A small remnant of the normal intestinal columnar epithelium can be seen at the top, but lower down the epithelium has been lost leaving an ulcerated surface. The underlying tissue shows increased cellularity and some changes in small blood vessels.

Vasodilation

Dilation mainly occurs in the smaller blood vessels (arterioles, venules and capillaries) as the increased luminal size of the blood vessels results in the slowing of the blood flow and a local increase in the volume of blood. It is this that contributes to the redness and the heat associated with inflammation. The changes in blood flow also increase the blood pressure inside the capillaries and this increased pressure then leads to an increased outflow of tissue fluid. The extra fluid passing through an intact capillary wall is called a transudate and it has a low protein content, similar to normal tissue fluid. Any increase in the fluid content of tissue is termed oedema and is a major cause of tissue swelling. Although tissue shrinks during histological processing, vasodilation is still recognisable in sections, though oedema due to transudation may be masked by tissue shrinkage.

Increased vascular permeability

The blood vessels become more permeable as the endothelial cells contract leaving gaps between the cells. These gaps are too small to be seen with the light microscope but can be visualised with the electron microscope. As a result of the gaps there is an increase in the permeability of blood vessels, especially venules, and they begin to leak protein as well as the normal tissue fluid of salts and water. The proteinaceous fluid is called exudate and again leads to oedema and swelling. Histologically, the areas of oedema can look pale because of the extra fluid in the swollen tissues.

The loss of fluid from blood by exudation and transudation leads to haemoconcentration ('sludging') of the blood inside the blood vessels. Sludging is where blood cells show clumping and appear more compacted and this is at least partly due to rouleaux formation. Haemoconcentration is quite evident in tissue sections.

The fluid in the oedematous region can take several forms:

- **Serous exudate** is a clear watery secretion with low protein content such as is seen in blisters.
- **Proteinaceous exudate** is rich in protein.
- **Fibrinous exudate** occurs if the protein includes fibrinogen which causes clotting of the fibrin due to activation by tissue thromboplastins, e.g. in lobar pneumonia.
- **Purulent exudate** (pus) occurs where the exudate becomes infected and contains bacteria and cells, particularly leucocytes. Purulent exudate is usually coloured and quite viscous, e.g. in boils.

Exudate is often beneficial and can have several functions. Serous exudate dilutes toxins, fibrinous exudate forms a mechanical barrier against bacterial spread, and proteinaceous exudate carries antibodies to the site.

Margination and emigration

White blood cells, particularly neutrophils and macrophages, are normally transported in the centre of the vessels and do not come into contact with the endothelial cells. In inflammation, where blood flow is slowed and fluid is being lost by exudation, the white blood cells come into contact with the vessel walls and are said to be **marginated**. The white cells can form almost complete layers in some cases, and so margination is also called **pavementing**.

The white cells can penetrate between the endothelial cells and 'emigrate' into the tissues towards the site of the injury. Both margination and emigration are visible microscopically.

In acute inflammation the major cell types which emigrate are the neutrophil polymorphonuclear leucocytes (commonly just referred to as 'neutrophils' or 'polymorphs'). They are relatively small cells, typically about 10–12 μm (micrometres) in diameter (about 50% larger than an erythrocyte). Because of their small size they are regarded as **microphages** to distinguish them from **macrophages** which are much larger. The cytoplasm contains fine granules which stain azure with the Romanowsky stains used for blood smears, hence the term neutrophilic because they are neither acidophilic nor basophilic (see *Section 7.1*). Neutrophils have lobed nuclei with the age and maturity of the neutrophils being indicated by the number of lobes (hence *polymorphonuclear* *neutrophils* or PMNs): recently formed immature neutrophils have only one or two lobes while older neutrophils may have up to 5 lobes. They are a short-lived cell type with a lifespan typically less than a week.

Other cell types such as macrophages, eosinophils and lymphocytes may also be increased in acute inflammation but are less common and much less prominent in the tissues. Eosinophils occur more often in acute allergic inflammations and are more prominent in the tissues.

Abscess formation

Some organisms, e.g. *Staphylococcus aureus*, can resist phagocytosis and this leads to the formation of infected exudate (purulent exudate or pus). These organisms may be referred to as pyogenic organisms and the infection is sometimes referred to as a **suppurative inflammation**.

Pus consists mainly of dead neutrophils which are undergoing autolysis, some bacteria and some exuded fluids.

If the pus is localised in a solid organ it can form an abscess. The tissue surrounding the inflammatory focus forms a collagenous *pyogenic membrane* around the pus which effectively walls off the infection and this can stop or slow down the spread of the bacteria. The abscess can still grow due to internal pressure and pus may track through the tissues producing a *fistula* (opening onto a surface) that eventually drains the abscess. If the pus tracks to an internal surface it can result in peritonitis or septicaemia. Lancing of abscesses results in an artificial drainage and reduces the pressure. Draining the abscess allows resolution and repair to occur.

Table 2.1   Histological changes visible in acute inflammation

Changes in the vascular system
- Dilation of blood vessels
- Engorgement of the blood vessels with RBCs (stasis and sludging)
- Margination and pavement of WBCs especially neutrophils
- Escape of fluid and proteins (possibly producing fibrin) through permeable capillaries

Presence of unusual cells and/or unusual numbers of cells in tissues
- Neutrophils – characteristic multilobed nucleus
- Eosinophils – bilobed nucleus, eosinophilic granules in cytoplasm
- Macrophages/histiocytes – large cells with abundant cytoplasm
- Giant cells – very large multinuclear cells formed from macrophages
- RBCs– if outside the blood vessels they indicate haemorrhage

Degenerative changes
- Signs of necrosis, opaque cytoplasm and loss of cellular detail. Pyknosis, karyorrhexis and karyolysis. Cell debris.

Reactive changes
- Active tissue repair. Cellular proliferation, mitosis, growing capillaries.

Acute inflammation can have several outcomes:

- *Resolution* where the tissue returns to normal with no scarring.
- *Healing* where the tissue returns to a healthy state but with scarring of the tissue.
- Progression to a different form of inflammatory response: chronic inflammation.

> **Box 2.2 // Virulence and chronic inflammation**
>
> The agents that cause chronic inflammation are usually less toxic/virulent than those causing acute inflammation. This may seem at odds with the common perception because syphilis, leprosy and TB are chronic inflammations but are regarded as severe diseases. However, a patient can suffer with these conditions for years or even decades before dying. With a highly virulent infectious organism such as *Corynebacterium diphtheriae* (the causative organism of laryngeal diphtheria) then either the disease organism is killed (by the immune system or drug treatment) or the infection kills the patient. No-one could live for several years with an active pathogenic diphtheria infection.

Chronic inflammation

Chronic inflammation is not simply a continuation of acute inflammation but is quite a distinct type of pathology. Chronic inflammation can arise from acute inflammation but it can also be a primary process without ever showing acute inflammation. Acute inflammation is short lived (less than about two weeks) but chronic inflammation can persist for many years (see *Table 2.2* for other differences).

The causes of chronic inflammation can be grouped into three main categories;

1. Persistent infections: TB, leprosy (both mycobacterial infections), syphilis, fungi, some viruses and metazoan parasites.
2. Non-degradable materials: silica (silicosis), beryllium (berylliosis), suture materials, talcum powder and other foreign bodies.
3. Autoimmune disease: rheumatoid arthritis, multiple sclerosis, type 1 diabetes.

Chronic inflammation differs from acute inflammation because the cardinal signs (pain, redness, heat and swelling) are usually absent with little sign of vascular changes (sludging, margination or oedema). The white cells in the inflammatory area are mainly the longer lived mononuclear cells (macrophages and lymphocytes) rather than the short lived neutrophils (see *Fig. 2.3*). The lymphocytes and the plasma cells (derived from the lymphocytes) indicate a strong immune response to an antigenic stimulus, often of a delayed hypersensitivity type response (type IV), e.g. ulcerative colitis, coeliac disease and Crohn's disease. Many of these are idiopathic with no single antigen clearly identified.

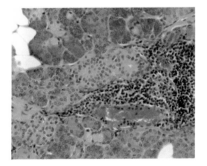

Figure 2.3
Chronic inflammation of a salivary gland. An area of lymphocytic infiltration is seen in the centre right of the image. There is less evidence of tissue oedema, but the blood vessels show signs of enlargement which is not characteristic of chronic inflammation.

The strongly phagocytic macrophages are often very active and can destroy the tissue – this damaging effect can be referred to as 'bystander' or 'collateral' damage. Although damaging to the healthy tissue the macrophages are often ineffective against the cause of the chronic inflammation (which therefore persists). The bacteria and fungi that cause chronic inflammation often have defences against macrophage activity and the non-living agents are often inert and non-degradable, e.g. silica. Collections of macrophages occur often with a rim of lymphocytes to form *granulomas* in some diseases (**granulomatous diseases**, e.g. leprosy).

The macrophages may have different forms in chronic inflammation. They can be described as **epithelioid cells** which are believed to be a type of macrophage but have some resemblance to epithelial cells (abundant eosinophilic cytoplasm, rich ER). These macrophages are not strongly phagocytic and are less mobile than normal macrophages. Their function may be to synthesize macrophage products (e.g. cytokines) and/or remove small molecules by engulfing them within small membrane-bound vesicles (**pinocytosis**). **Giant cells** (see *Fig. 2.4*) are large (>50 μm), multinucleated (>20 nuclei and perhaps up to 200 nuclei) and come in a variety of forms including foreign body, Langhans and Touton (*Fig. 2.5*). These can have characteristic forms in some pathologies, e.g. Crohn's disease and tuberculosis.

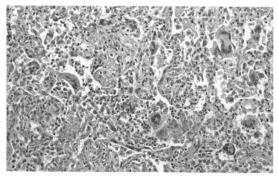

Figure 2.4
Giant cells. The tissue, taken from a case of measles pneumonia, shows significant numbers of multinucleated giant cells.

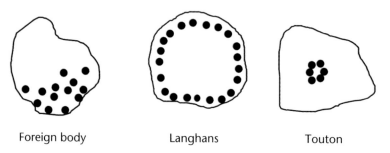

Foreign body   Langhans   Touton

Figure 2.5
Types of giant cell.

The destruction of tissue by the active macrophages evokes repair mechanisms and so the tissue shows signs of healing, including large numbers of active fibroblasts, new collagen formation and angiogenesis. This can lead to *fibrosis*. Fibrosis may itself be pathological, e.g. in lung with silica or in the pericardial sac with chronic pericarditis.

Eosinophils may occur as well as the usual leucocytes in chronic allergy states or in parasitic infections. The similarity between allergic responses and the response to parasites has led to the idea that the rise in allergies in western countries and the elimination of many parasitic infestations may be causally linked (this is known as the hygiene hypothesis). Necrosis, described later in this chapter, is also a common finding in chronic inflammation.

Table 2.2 Differences between acute and chronic inflammation

	Acute	Chronic
Duration	Short (days)	Long (weeks to years)
Cell types	Neutrophils, macrophages	Lymphocytes, plasma cells, macrophages, giant cells, epithelioid cells, fibroblasts
Vascular changes	Vasodilation, increased permeability	Angiogenesis (repair)
Cardinal signs and fluid exudation/oedema	Strong	Absent or weak
Necrosis	Usually absent (except in suppurative and necrotizing inflammation)	Usually present
Fibrosis	Usually absent	Usually present
Systemic symptoms	Fever, neutrophil leucocytosis	Low grade fever, weight loss, variable white cell changes, increased plasma immunoglobulin

2.2 Healing

Healing is a general term for recovery from injury. It can take three forms depending on the tissue and the extent of the injury.

1. **Recovery** – this involves no tissue loss. There is some minor damage, often with inflammation. The tissue returns to normal with no new tissue or scar being formed. Recovery is rapid and complete. This type of injury is common in minor superficial scratches.
2. **Regeneration** – this involves tissue loss. There is complete recovery with the original tissue being replaced by identical tissue. This is unusual in human tissues. It occurs with blood and bone marrow. A slight injury to epithelia may result in regeneration.
3. **Repair** – there is again tissue loss but the replacement tissue is significantly different. The replacement tissue is usually distinctive scar tissue, but it can also resemble the original tissue structure though at least some of the functions of the original tissue will be lost.

The ability to repair or regenerate depends on the type of tissue.

- Labile tissues which are constantly growing and replacing older tissues, such as epithelia, are capable of regeneration and so heal well.

- Stable tissues are those which are normally non-mitotic, such as many connective tissues and liver, but are able to become mitotic if there is damage or loss. Stable tissues will generally repair rather than regenerate resulting in altered tissue structure and with poorer healing.
- Permanent tissues are incapable of division even when damage occurs. Permanent tissues include heart muscle and neuronal tissues. Damage to permanent tissues results in replacement of the tissue with scar tissue that is very different. The scar tissue will fill any gaps or holes in the tissue but without restoring the other functions. So a major loss of heart muscle by myocardial infarction results in a connective tissue scar with the loss of physiological capacity and efficiency.

Skin wounds

These are amongst the commonest and clinically the most important type of repair. In other tissues and organs there will be similar processes but with some variations. With increasing tissue loss there will be larger amounts of replacement tissue needed and this results in greater scarring. Infection of wounds and the presence of foreign objects slow down repair and increase the chances of a large scar. Control of infection and debridement are therefore important in clinical practice to aid in effective healing.

Skin consists of two portions which are significantly different in their repair capabilities.

Epithelium

This is a labile (dividing) tissue and so repair can begin immediately. The cells at the side of the wound show an accelerated rate of division producing many new cells. The newly formed cells spread laterally across any gap by amoeboid movement. Only when the cells form a complete monolayer do they begin to mature in the normal way (contact inhibition of mitosis).

Provided the wound is relatively superficial, cells can also migrate from areas within the wound that contain epithelium, e.g. hair follicles, etc. This speeds up recovery even when the damaged area is large.

Box 2.3 // Skin grafting in severe burns

When damage is very deep and skin follicles have been completely lost (as occurs in 'full thickness burns'), the only source of cells is from the edge of the wound. The distances that cells must migrate in order to completely cover the surface become much greater and this slows recovery considerably. Skin grafts can be used to form new areas of seeding of cells and so imitate the widely scattered and closely packed hair follicles which act as natural areas of seeding. This speeds up regeneration and repair and reduces healing times considerably.

Connective tissues

These are stable tissues and take longer to activate in the various repair processes. During the first 24 hours the most obvious effects are the inflammatory changes and the emigration of macrophages then leads to the **demolition** phase where any dead tissue or clot is phagocytosed. Macrophages are relatively resistant to low oxygen levels so can work in the anaerobic environment of dead tissue. The phagocytes leave tunnels of cleared tissue into which the regenerating blood vessels can grow. The endothelial sprouts will fuse with other sprouts to form loops. The vessels are simply solid rods of cells at first and only later develop lumens and restore circulation.

Once a circulation is established, the more oxygen dependent fibroblasts can begin to grow into the damaged area and lay down collagen (see *Fig. 2.6*). The fibroblasts first produce matrix of type III collagen. This is weaker than 'normal' collagen but this is later replaced by stronger type I collagen over a period of weeks. The capillaries are initially quite fragile and bleed easily. This type of tissue is called granulation tissue. Although granulation tissue is mechanically weak it is very resistant to infection because of the presence of macrophages and other inflammatory cells. The fibrous tissue gradually replaces the blood vessels and the tissue becomes paler scar tissue. Eventually the scar contracts, forming a cicatrix.

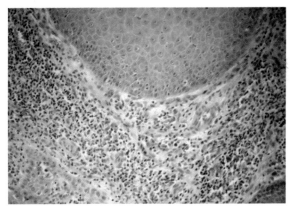

Figure 2.6
Tissue repair. The centre of the image shows an area with inflammation but also repair where there are fibroblasts beginning to lay down new collagen (granulation tissue).

Box 2.4 // Healing can cause problems

Although healing and repair are normally beneficial they can also produce problems.
- **Keloid scars** in which scarring is excessive. This is a genetic trait and it is more frequent in some racial groups.
- **Adhesions** in which the repair extends beyond the wound itself and scar tissue tethers organs together. This can be a major problem in organs which regularly move against adjacent organs such as the lungs and intestines. The tethering of the lung across the pleural cavity can be painful and dangerous.
- **Contractures and strictures** which are caused by the contraction of the scar. In muscle it results in permanent shortening of the muscle. If a ring of scar occurs around a hollow structure such as the gut then the lumen will be narrowed by the contraction and this may partially or completely obstruct the gut.
- **Fibrosis** where metabolic tissue is replaced with inert fibrous scar tissue. This is associated with chronic inflammation such as in the liver (e.g. following long term and excessive alcohol abuse). Gradually the healthy liver is replaced by isolated nodules of fibrosis resulting in cirrhosis and liver failure.

2.3 Tissue growth abnormalities

Organs may become relatively larger or smaller than normal. These changes may be associated with disease (pathological) or may be a normal response (physiological).

Atrophy

Atrophy is the shrinkage of an organ. Atrophy can occur for several different reasons and is often reversible if the atrophic stimulus or agent is removed. Atrophy may be pathological or physiological. Physiological atrophy is often called involution and involves the process of apoptosis (see later in this chapter), e.g. of the thymus gland at puberty.

Different forms of atrophy can be identified:

- **Disuse atrophy**: loss of muscle bulk when the muscles are immobilised (e.g. following fractures), loss of bone matrix in bedridden patients.
- **Neurogenic atrophy**: loss of muscle bulk following nerve damage, may be a subtype of disuse atrophy but may also involve a more direct effect involving neurotrophic chemicals.
- **Hormonal effects**: loss of trophic hormones (e.g. atrophy of the adrenal cortex following loss of pituitary ACTH) or excess hormone (e.g. skin thinning following prolonged use of topical steroids).
- **Ischaemic atrophy**: the organ adjusts to a size appropriate to the blood available.
- **Pressure atrophy**: when pressure is applied to specific areas they may diminish, especially if they cannot move to relieve the pressure. Probably due to diminished blood flow.
- **Brown atrophy**: prolonged disease may result in severe atrophy and the phagocytosed remains of the lost cells produces a high level of lipofuscin which makes the organ a deep brown colour.
- **Senile atrophy**.
- **Generalised atrophy**, e.g. in starvation.

Tissue overgrowth

The increase in size of an organ can be termed **hyperplasia** or **hypertrophy** and the distinction between the two terms is simple.

Hypertrophy involves an increase in cell size without any increase in cell numbers. This is the only form of enlargement possible in permanent tissues such as skeletal and heart muscle. The increase in size implies an increase in the cytoplasmic contents of the cell, suggesting an imbalance between breakdown and synthesis of proteins. An increase in the nuclear component is less common although polyploidy may develop in severely affected cells. Hypertrophy is common in muscle tissues (see *Fig. 2.7*) which have been worked harder than normal (e.g. heart muscle in stenosis or high BP, skeletal muscle in athletes, smooth muscle in the uterus in pregnancy, and smooth muscle in the gut in obstructive lesions). Hypertrophy can also occur in glandular tissues (due to extra work or hormonal stimulation). Hypertrophy is generally not connected with true tumour growth and is often, though not invariably, harmless. If the stress is removed then the tissue will return to normal size.

Hyperplasia involves an increase in cell number often without any increase in cell size and is usually only found in tissues which are capable of cell division. The increase in cell number involves an increase in mitosis or a decrease in cell loss (particularly apoptosis)

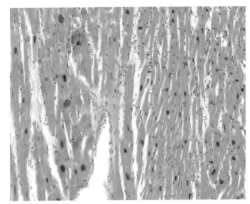

Figure 2.7
Cardiac muscle hypertrophy. The muscle fibres on the left of the image are larger and show hypertrophy compared to the smaller fibres on the right of the image.

or both (see **Fig. 2.8**). There is an increase in tissue production but it is not constrained to an area of repair and occurs in the absence of genuine tissue loss. Hyperplasia can be induced by increased work requirements such as in the pregnant uterus or a lactating mammary gland. The increased growth in these cases is often associated with an increased hormonal stimulation.

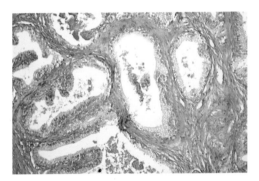

Figure 2.8
Prostate hyperplasia. This specimen is from a case of prostatic hyperplasia and shows increased cellularity but without increased cell size when compared to normal.

Hyperplasia can also be induced by chronic irritation of tissues. It has some association with true tumour growth because hyperplasia occasionally develops into true tumour growth, but this may simply be a reflection of having a larger population of cells at risk of conversion to tumour cells rather than having a distinct change in the cells themselves. The distinction between hyperplasia and neoplasia (tumours) is not always distinct, but in most cases a true hyperplasia will revert to normal if the cause of the hyperplastic change is removed. Some conditions are often considered to be hyperplastic changes but seem to be autonomous and will continue to grow in the absence of any obvious stimulation, e.g. psoriasis (epidermal hyperplasia) and desmoids (a fibromatosis with hyperplasia of myofibroblasts) and Dupuytren's contracture (also a fibromatosis).

Hyperplasia and hypertrophy may occur together.

> **Box 2.5 // Aplasia**
>
> Unlike the overgrowths of tissue (hypertrophy and hyperplasia) there is no distinction between loss of cell number, which implies cell death (particularly apoptosis), and loss of cytoplasmic bulk (which implies catabolism exceeding synthesis).
>
> The term aplasia is used in a different sense in pathology and means the lack of development of an organ during embryogenesis or childhood rather than shrinking of an organ.
>
> Aplastic anaemia, in which there is suppression of blood cell synthesis, is an exception to this pattern.

Metaplasia is an alteration in the type of tissue which is present. It occurs in epithelia, for example, stratified squamous epithelium (a skin-type epithelium) may replace columnar epithelium (columnar epithelium is found in many internal structures). This transformation is linked to irritation and may be a response to the irritation because skin-type epithelium is more resistant to irritation. Metaplasia may lead to true tumour production in some cases, though this is often only where metaplasia and dysplasia are both present.

Dysplasia is an abnormality of differentiation of the cells. It is seen in stratified squamous epithelium where the lower layers are dividing, the newly produced cells are then pushed upwards by further divisions in the lowest layer. Normally as the cells move upwards they mature, with the cells becoming flatter, the nucleus disappearing and keratin accumulating in the cytoplasm. In dysplasia, immature nucleated and even mitotic cells can be found in the upper layers of the epithelium. Dysplasia is an indication of an increased risk of malignancy. Dysplasia may disappear and tissues return to normal but its presence is significant and if detected it should be monitored to detect any early signs of true malignant change.

2.4 Tumours

These are difficult to define accurately in a simple description but consist of abnormal growths of tissue which exceed normal tissue limits and are not coordinated with normal tissue growth. Two types are usually distinguished:

- **benign tumours** (see *Figs 2.9* and *2.10*) are not usually life threatening nor a major health hazard but they can be disfiguring

> **Box 2.6 // Tumour size**
>
> The size of the tumour is not an indication of malignancy. Malignant tumours can be large but a large malignant tumour is less well tolerated by the body.
>
> The size of a benign tumour is not in itself a health problem and patients can survive well even with quite large tumours. In the case of ovarian cysts the tumours often become quite large before being detected and diagnosed. Ovarian cysts the size of a football are not that uncommon. The largest ovarian cyst ever removed weighed more than the patient.

- **malignant tumours** are often called cancers and are more aggressive (see *Figs 2.11* and *2.12*). Many (but not all) malignancies can lead to the death of the patient or to serious health problems if left untreated.

The differences between benign and malignant tumours in behaviour, growth and structure are summarised in *Table 2.3*.

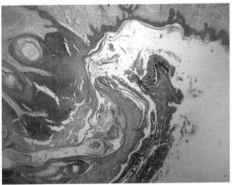

Figure 2.9
Warts are benign tumours caused by viruses. This shows an increase in the epithelial thickness and hyperkeratinisation. The warts are not invasive and show only the nuclear changes associated with the viral infection and not malignant transformation.

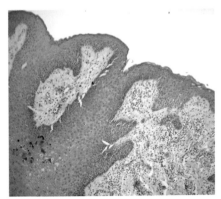

Figure 2.10
Benign pregnancy tumour or epulis. During pregnancy the hormonal changes can result in a swelling of the gingiva (gum around the teeth); this has several different names, but is a type of haemangioma usually associated with irritation.

Borderline tumours

Although the characteristics described in *Table 2.3* serve well in distinguishing between benign and malignant tumours, tumorigenesis is not an all or nothing phenomenon. Some tumours form intermediate or borderline lesions.

These intermediate tumours may remain borderline or may resolve and disappear, but a significant number progress to full malignancy, so early diagnosis and monitoring will improve the prognosis and clinical outcome.

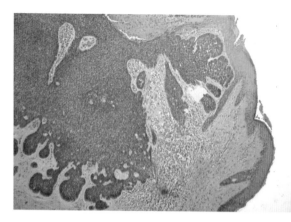

Figure 2.11
Basal cell carcinoma. This is a low grade skin malignancy associated with prolonged
exposure to sunlight. The bottom and right shows an area of normal skin epithelium which
is thinner and penetrates less into the connective tissues than the tumour. The upper part
shows the tumour which is spreading locally. It is not usually genuinely invasive and in this
case the edges of the tumour are clearly delineated and the basement membrane of the
epithelium is intact.

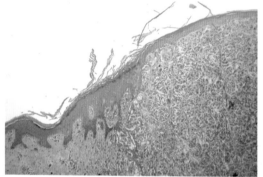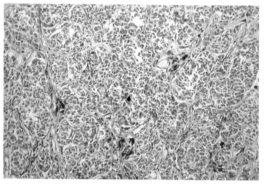

Figure 2.12
Malignant melanoma. The normal epithelial covering seen at the left hand side of the image on the left
becomes thinner over the tumour. The tumour is invasive and the abnormal cells in the connective tissue
show no clear border and are spreading into the adjacent connective tissue. Pockets of highly pigmented cells
can also be seen (right).

Tumour nomenclature

There are potentially at least as many different tumours as there are specialised cell types
in the body (i.e. more than 200). Some cell types can produce more than one type of
tumour and it is not too far-fetched to regard every tumour to be as individual as the
patient at a genetic level. This can make understanding nomenclature a serious challenge
(see **Box 2.7** for details). Fortunately for immunohistological detection purposes there
are only four main classes of tumour that relate to their tissue of origin: epithelium,
connective, muscle and nerve, plus primitive cells (germ cells, blast cells) and antibody
panels are constructed with this in mind.

Table 2.3 Differences between benign and malignant tumours

Characteristic	Benign	Malignant
Growth rate	Usually slower.	Usually more rapid, especially in the early stages.
	May spontaneously stop growing.	Never spontaneously stop growing.
Mitosis	Normal rate of mitosis for the tissue involved.	The rate of mitosis is significantly higher than normal.
	Normal mechanism of chromosome separation and distribution.	May be abnormal. Chromosomes may break. Chromosomes may be shared unequally between cells resulting in abnormal chromosome numbers, including chromosomes not separating and resulting in triploid (or more) cells.
Chromosome numbers	Normal (diploid).	Often abnormal (aneuploid, haploid or polypoid).
Structure	Relatively normal cell and tissue structure.	Abnormal tissue architecture and cell structure.
	Cells differentiate normally, most noticeable in stratified epithelium.	Cells are often poorly differentiated or undifferentiated (termed anaplasia or anaplastic).
	Equivalent cells have a uniform appearance.	Pleomorphic. Cells which are expected to look similar have very different size, structure, staining characteristics.
	The vascular system (blood vessels) has the normal arrangement.	The vascular system may be quite abnormal. Examples include vessels with abnormal walls which allow cells to pass from the tissues into the blood stream and, in extreme cases, blood sinuses with no retaining vessel wall. The growth of blood vessels within a tumour (angiogenesis) is a common feature.
Spread	Local growth in which cells add on to the tumour without breaking through tissue barriers such as basement membranes. The tumour remains discrete and not intermingled with surrounding tissues.	Growth is often **invasive**. Invasion involves cells moving away from the primary tumour site into surrounding tissues which often involves breaking through the normal tissue barriers.
	Never metastatic. Metastasis is the movement of cells to distant parts of the body.	Often metastatic. Cells migrate to distant organs. Not just adjacent tissues or organs but far distant from the original site which involves transport around the body by circulation systems such as blood or lymph. Tumours in distant sites are referred to as secondary tumours. A primary tumour of the breast may have cells travelling to the brain to produce secondary tumours within the brain tissue.
	May be encapsulated by having a distinct wall of connective tissue separating it from the adjacent normal tissues.	Never encapsulated.
Injury response	Normal repair. If the tissue is damaged, e.g. by abrasion producing a raw surface, then it will usually heal by the normal repair mechanism.	Poor repair. If damage occurs it may not heal, resulting in necrosis. Persistent ulcers are a frequent finding in tumours. Causes other than malignancy may result in persistent ulcers but they can be just as dangerous. So persistent ulceration is always a concern and needs investigation.
	Normal inflammatory response. Tumours will not be more prone to infection than normal tissues.	Poor inflammatory response and tumours may act as bridgeheads to allow pathogenic organisms to enter the body.

> **Box 2.7 // Naming of neoplasms**
>
> Tumours are named histogenetically (after their tissue of origin).
>
> *Differentiated cells*
> - Benign neoplasms are suffixed with -oma, e.g. fibroma, chondroma, lipoma, adenoma, papilloma.
> - Malignant neoplasms of mesodermal tissues (muscle and connective tissues) are suffixed with -sarcoma, e.g. fibrosarcoma, chondrosarcoma, liposarcoma.
> - Malignant neoplasms of epithelial origin are suffixed with -carcinoma, e.g. adenocarcinoma, basal cell carcinoma, squamous cell carcinoma.
>
> *Pluripotent cells (embryonic cells)*
> - Suffixed with –blastoma, e.g. retinoblastoma, nephroblastoma. These are typically malignant tumours of childhood.
>
> *Totipotent cells (germ cells)*
> - These are called teratomas and arise from the testis or ovary. Testicular tumours are commonly malignant and ovarian teratomas are usually benign.
>
> *Blood cells*
> - Leukaemias.
>
> *Eponymous tumours*
> - Some tumours are named after the person who originally described them, e.g. Wilms' tumour, Burkitt's lymphoma.
>
> However, some names are misleading and do not follow the usual rules:
> - Lymphoma and melanoma are actually malignant. To avoid confusion they should therefore be referred to as malignant melanoma and malignant lymphoma but this is not always the case.
> - Osteoblastoma and chondroblastoma are not tumours of pluripotent cells and are not malignant. The confusion arises because osteoblasts and chondroblasts are not pluripotent cells but well differentiated adult cells. So the name should be read as osteoblast-oma not osteo-blastoma.
> - Atheroma and granuloma are not true tumours. Atheroma is an abnormality of blood vessels with a thickened vessel wall but not a neoplastic thickening. Granulomas are associated with inflammation and can be found in infectious diseases, e.g. leprosy which is a granulomatous disease producing granulomas.

Genes and malignancy

Many viruses have been implicated in the onset of malignancy in humans and many other animals. Investigations into animal viruses identified individual genes that were causing the malignant transformation and these were called **oncogenes**. These genes were then identified in animals and humans and it seems that the viral genes were originally accidentally picked up from animal cells during viral replication. Many individual oncogenes have been identified and seem to cause the malignant change either by overexpression of the gene, e.g. the *abl* oncogenes which were originally identified in the mouse Abelson leukaemia virus, or by mutation of the gene, e.g. the *H-Ras* oncogene originally identified in Harvey rat sarcoma virus. A further set of tumour-related genes were found to cause malignancy when they were lost or inactivated allowing cells to replicate continuously; these are called **tumour suppressor genes**. These genes are now being targeted for the diagnosis, prognosis and treatment of malignancy. Identification of the abnormal gene activity by molecular biological methods, e.g. ISH, is becoming increasingly common in the histopathology laboratory (see **Chapter 16** for more details).

Cysts

Cysts are hollow fluid-filled structures, usually with an epithelial lining. Although cysts often cause swelling, not all cysts are true neoplasms. **True neoplastic cysts** involve an overgrowth of cells which secrete fluid, e.g. ovarian cysts which can contain a watery fluid (serous type) or a gelatinous fluid (mucinous type).

Three types of non-neoplastic cysts can be identified and vary in their causative mechanism.

- **Retention cysts** are caused by a blockage of a duct. They cause swelling because the secretion builds up behind the blockage and causes a fluid-filled lump but there is no overgrowth of tissue, e.g. sebaceous cyst.
- **Infective cysts** are caused by infective parasites, e.g. hydatid cyst.
- **Congenital cysts** are caused by embryological defects, e.g. branchial and thyroglossal cysts in which embryonic cells are left behind in tissues during development and later produce cysts.

Dangers of tumours

Benign tumours are called benign because they usually pose little risk to the patient. Some are less benign than others though; for example, tissue compression of benign tumours in hormonal cells may result in excessive hormone secretion. This occurs because the cells are well differentiated and will retain the ability to synthesize and secrete hormones. The dangers are indirect and arise from the excess hormone rather than from the tumour.

A second way in which benign tumours can cause health problems is by mechanical interference. This can occur in the brain where the benign tumour compresses the brain tissue which is retained in the cranium and has no way of avoiding the pressure from the growing tumour. An atrial myxoma that forms on a stalk within the upper chamber of the heart can eventually interfere with valve function and the passage of blood through the heart.

In addition, some benign tumours are not a risk to health but may be a source of embarrassment if they are visible externally.

Benign tumours are usually fairly straightforward to treat because they are discrete and well defined in size. They can be completely removed by surgery as they do not spread beyond the margins of the tumour. Surgery of small skin tumours for cosmetic purposes is often done as an excision biopsy, whilst larger tumours may need a more extensive resection.

Malignant tumours are more dangerous and more difficult to treat than benign tumours because they are invasive and metastatic.

Invasiveness

When growth becomes invasive the tumour edge will be less well defined than in benign tumours and complete removal is therefore more difficult. Even with a large resection that removes not only the tumour but also adjacent normal tissue, there is the risk that invasive growth has spread beyond the edge of the resection so that some tumour is left behind. In these cases it is important to check the whole of the cut surfaces of the tumour to minimise the risk of leaving malignant cells. An important role for histological examination is to determine if the resection margins are clear of tumour and so confirm complete removal of the tumour. The use of inking of surgically cut surfaces so that the true resection edge is recognisable after trimming and processing is used to help in this verification. If the edge of the resected tissue shows signs of spread beyond the edge of the removed tissue this may result in more surgery or other treatment being needed.

Metastasis

This is the process in which tumour cells become detached from the original tumour and travel in the blood or lymph to distant organs where they form secondary tumours. The ability to spread to several sites is a dangerous characteristic of malignant tumours and greatly hinders effective treatment.

The ability to metastasize varies from tumour to tumour. Some tumours form metastases early and often (e.g. malignant melanoma) whilst others rarely metastasize (e.g. basal cell carcinoma). In some cases the metastases become clinically important before the original tumour shows any symptoms.

If the tumour is metastatic then distant secondary tumours could be present in several organs throughout the body. When secondary tumours occur in many sites, e.g. in brain, bone, liver and lung, then surgery and localised radiotherapy become difficult or impossible. Whole body chemotherapy or radiotherapy or a combination of both becomes the treatment of choice. These treatments are often unpleasant for the patient with normal cells also being severely affected and are they are often less effective than complete surgical removal.

Tumours can reach secondary sites by being transported in the blood or lymphatic vessels. Lymphatic vessels have lymph nodes along their length which filter the lymph and remove bacteria and viruses to prevent the spread of disease. Unfortunately, the macrophages which can ingest bacteria are less capable of destroying intact malignant cells. The cells are still trapped but can continue dividing and growing (see *Fig 2.13*). They may eventually spread beyond the lymph node and so spread the malignancy into secondary tumours further downstream. It is therefore important to check lymph nodes in the area of the malignancy to see if metastatic malignant cells are present. The first lymph nodes draining the tumour are called sentinel lymph nodes and if these are negative for tumour cells it is then unlikely that the tumour has spread through the lymphatic system. An important role for cellular pathology is to determine the extent of metastatic spread from a primary tumour.

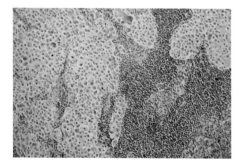

Figure 2.13
Secondary tumour in lymph node. This section of a lymph node shows the presence of paler staining tumour cells in a lymph node. The darker, intensely nucleated area at the bottom right contains large numbers of lymphocytes and is the more characteristic tissue found in nodes.

Many tumours have characteristic sites to which they metastasize more often and metastases at other sites are less common (**Table 2.4**).

Table 2.4   Sites of secondary tumours

Primary tumour	Common secondary sites
Breast	Lung
	Brain
Lung	Brain
	Adrenal gland
Prostate	Bone
Melanoma	Brain
	Lung
	Liver
	Ovaries

Some sites are easily colonised by a range of primary tumour cells (e.g. lung and brain) whilst others are rarely affected by any metastases (e.g. spleen and muscle).

In order to successfully metastasize the cells must be able to detach from the original tumour, enter a suitable transporting medium, survive in the circulation, migrate out of a capillary and finally grow in the new site. These are unusual properties and only a small fraction of malignant cells exhibit them all. Even in actively metastasizing tumours less than 1% of cells in blood can form secondary growths and finding malignant cells in the blood does not automatically mean that secondary deposits will be present.

Two different mechanisms have been proposed to account for the site specificity of metastases.

- The current hypothesis suggests that the site of secondaries is controlled by the pattern of blood or lymph flow in the body; such a hypothesis is a good explanation of lung and lymph node involvement. The route of drainage can be seen in tumours from the oesophagus: if the tumour is in the lower oesophagus then it is four times more likely to cause secondaries in the liver than a similar tumour in the upper oesophagus. This corresponds to the pattern of drainage with the blood from the upper oesophagus draining into the vena cava whilst the lower region drains into the hepatic portal vein.
- The 'seed and soil' hypothesis suggests that the site of secondaries is determined more by the specific growth requirements of the cells; this hypothesis accounts well for the secondaries in unusual sites such as bone or adrenal and the lack of secondaries in other organs.

Growth of tumours

Estimation of tumour growth rate

Tumour growth rate can be useful in prognosis and treatment as it indicates how rapidly the tumour is likely to spread. Growth rate can be difficult to measure accurately *in vivo*. In tissue samples different techniques can be used to measure different growth parameters but these do not always correlate well.

- **Mitotic count** – the number of cells in mitosis in a unit area (e.g. 10 high power fields). This is very simple to do but is only a crude measure because it takes little note of cell size, amount of connective tissue and extracellular matrix, etc. However, it is probably the most widely used estimate in clinical microscopy, particularly for malignant melanoma and leiomyosarcoma.
- **Mitotic rate** – the number of new cells produced per 1000 cells per hour. This is more difficult to measure than a mitotic count. The estimate of mitotic rate does not give doubling time because growth fraction is not estimated and not all new cells will remain viable and mitotic.
- **Mitotic index** – number of mitotic events per 100 cells examined. This is intermediate between mitotic count and mitotic rate – it is relatively easy to measure but the result is distorted by length of mitosis. Basal cell carcinomas can give high rates despite a low growth rate because of a prolonged metaphase.

Grading and staging of malignancy

These are attempts to quantify the degree of malignancy and give some indication of the likely clinical prognosis.

Grading is a histological indication of the degree of malignancy based on the morphology of the cells and particularly on how well differentiated the cells appear. Degree of histological invasion and vascularization of the tumour and structural features may also be used. Grades run from I (the least malignant) through to IV. Often the degree of malignancy is given as a description (e.g. well-differentiated adenocarcinoma with no evidence of invasion of lymphatics) rather than a numerical grade because the criteria of grading are somewhat vague and vary from organ to organ. More specific organ-related grading systems have been devised for different types of cancer.

Staging involves an estimate of the degree of spread of a tumour and is based on size and metastatic spread. Different systems are used, often varying with site, e.g. Dukes' for colorectal tumours and FIGO for ovarian malignancy. More generalised systems exist with specific criteria, e.g. the TNM (tumour-node-metastases) system:

T = tumour size and invasion (T1–T4)
N = node involvement (N0–N3)
M = metastatic spread (M0–M1)

The system is still organ specific and has different criteria for different sites. The criteria for tumour size (T) are the most variable in definitions for different tumour types.

Histological recognition of malignancy

Malignant change is mainly diagnosed on the basis of nuclear abnormalities present in the cells. The cytoplasmic changes are frequently less useful in recognising malignant change, but once malignancy is diagnosed the appearance of the cytoplasm can be important in determining which type of malignancy is present. The use of immunocytochemistry is often critical in identifying markers in the cytoplasm of cells that indicate the exact type of tumour present.

The most striking feature of malignancy is the loss of uniformity. The range of cellular morphology (called **pleomorphism**) is much greater in malignancy than in normal tissue.

There is no single criterion of malignancy and a diagnosis depends on several features being present. The more features that are present then the greater the likelihood that malignant change has occurred; if only one or two features are present then malignancy is less likely. Some changes are more significant than others, e.g. a disturbed chromatin

---

Box 2.7 // Clonal origin of tumours

Tumours have two different origins:

Polyclonal origin
Several cells within a tissue could be altered at the same time to produce a tumour.

Monoclonal origin
Only one cell becomes altered and this gives rise to all the cells in the tumour. The monoclonal theory has long been considered to be the case in most malignancies and this is supported by evidence:

- Single parental source of the heterochromatic X chromosome in tumours in women (the heterochromatic X chromosome is usually a random mix of paternal and maternal X chromosomes).
- Lymphoid tumours show a single antibody type and are monoclonal.
- Chromosomal translocations, which are associated with the malignant transformation, are the same in all cells in each individual tumour, but different translocations are found in tumours from different individuals.

A polyclonal origin may be more common in benign tumours where large numbers of cells share the same genetic make-up and are exposed to the same environmental influences. There is also a possibility that the widely held view of the monoclonal origin of malignancies is not entirely correct, the evidence may be interpreted in other ways so the monoclonal origin is being reassessed, especially in the field of pluripotent stem cell mutation.

---

pattern is much more significant than simple nuclear enlargement. The following histological characteristics are all significant: nuclear changes, cytoplasmic changes and changes in tissue grouping.

Nuclear changes compared to equivalent normal cells

- **Variation in nuclear size**: enlargement is most common but small malignant nuclei do occur.
- **Increased nucleo-cytoplasmic (N/C) ratio**: although this is related to altered nuclear and/or cytoplasmic size it is much easier to spot the altered ratio as an approximate percentage than to actually measure nuclear and cytoplasmic size in absolute units. An N/C greater than 60% is indicative of inflammatory change.
- **Increased chromatin content**: this may not always be apparent. For example, if the nucleus is greatly enlarged then the chromatin is more spread out and so the nucleus appears paler, suggesting that it contains less chromatin.
- **Irregularity of nuclear shape**: the nuclear outline is often irregular and less well defined.
- **Disturbed chromatin distribution**: this is almost always present and is very significant.
- **Prominent nucleoli**: these are not always present, but if present they are often irregular in shape in malignant cells and often in increased numbers.
- **Altered staining**: can be hyperchromatic (darker staining), hypochromatic (lighter staining) or polychromatic (altered colours).
- **Multinucleation**: this is a significant finding but needs to be considered carefully because some cells are normally multinucleate e.g. giant cells.
- **Cell division**: often more frequent than in normal tissues but as explained earlier this can be difficult to measure. There may also be abnormal or irregular mitosis, though most divisions in tumours appear relatively normal.

Cytoplasmic changes

- **Variation in size** – larger with increased production in secretory cells.
- **Variation in shape** – changed morphology in metaplasia.
- **Vacuolation** – usually indicates cellular damage and/or apoptosis.
- **Abnormal staining** – often indicates an increase in synthesis of either normal or unusual products, e.g. mucin.
- **Loss of identifying features** – lack of differentiation, anaplasia.
- **Altered antigens** – malignancy is often associated with altered gene activity and this can lead to changes in protein production. Detection of altered surface antigens may help not only in the identification of tumours but also in their monitoring and treatment.

Changes in tissue grouping

- **Inflammation**: often present but not significant by itself.
- **Abnormal maturation**: cells appear immature in places where mature cells would be expected, e.g. upper layers of a stratified squamous epithelium.
- **Invasion**: cells breaking through normal tissue barriers such as basement membranes. This is a highly significant finding.
- **Metastasis**: the presence of tumour cells in other tissues such as lymph nodes is highly significant.

Because malignant tumours are extremely varied, their treatment needs to be similarly flexible. Many tumours which, in the past, were diagnosed purely on the appearances of the cells can now be subdivided by immunotechniques into several subtypes. The subtypes may respond differently to different therapeutic strategies. By better selection of the most effective treatment for the individual subtype, the prognosis for the patient has been improved, and sometimes the improvement has been dramatic.

Even within a single tumour there may be cell subtypes with different susceptibilities to chemotherapeutic drugs. A single drug may kill a large proportion of the malignant cells resulting in a decrease in tumour size and improvement of the patient's health. However, some tumour cells may be resistant to the initial drug and will continue to grow leading to a recurrence of the tumour. Using a sequence of different chemotherapy drugs is more effective. The first drug will kill many cells, resulting in some remission, then switching to a different drug will destroy many of the remaining malignant cells, and further switching to a third or even a fourth drug can result in all (or almost all) of the cancer cells being destroyed. This can give a longer lasting remission or even complete cure.

Laboratory recognition of these subtypes and their susceptibility to specific chemotherapeutic agents can be important in tailoring the therapy to a patient's cancer to give the best possible prognosis. The use of genome analysis of an individual tumour and the response of cells extracted from the tumour to cytotoxic drug regimes is being investigated as a feasible approach to tailored therapy.

2.5 Necrosis

Necrosis is the death of cells. There is a general misconception that cells 'just die', whereas in reality necrosis is quite complicated. In the early stages of necrosis the cells are reacting to oppose the deadly effects of whatever agent or process is damaging and killing them. The cellular metabolism continues even after the cells are damaged beyond repair.

There are normally two phases of necrosis:

> **Box 2.8 // 'Pins and needles'**
>
> The early stages of ischaemic deprivation (lack of blood flow leading to anoxia) are extremely common. Simply lying on an arm can result in temporary loss of circulation. The early stages represent the cells' responses rather than a genuine pathological change and these changes are easily reversed by removing the blockage to circulation. There is then the 'pins and needles' feeling as the blood enters the tissue and carries away the waste products. Following the restoration of normal circulation the normal sensation returns to the nerves.
>
> The ability of tissues to resist anoxia varies. The glycogen content of cells is a critical factor. Most skin, muscle and connective tissues have a good cellular content of glycogen and they can withstand several minutes of anoxia as the glucose from glycogen keeps glycolysis working to produce ATP. The cells become irreparably damaged only when the pH falls and inhibits enzyme activity. Nerve cells have only small stores of glucose so glycolysis stops as soon as the internal supplies are exhausted and nervous tissue suffers irreparable damage in 1–2 minutes.
>
> So, it is relatively safe to put a tourniquet on an arm or a leg for a few minutes to stop bleeding, but it is NOT safe to put a tourniquet around the neck.

- **Intrinsic phase** – in which the cell structure is destroyed by the cell's own activities. This involves swelling and lysis due to ionic imbalance and cellular degradation by the release of lysosomal enzymes into the cytoplasm.
- **Extrinsic phase** – in which macrophages and fibroblasts invade and replace the degrading tissue with fibrous scar tissue; this is a form of secondary healing.

In most cases both phases occur, but in some circumstances only one may occur, e.g. if the tissue is killed by excessive heat then all the enzymes will be inactivated by the heat and the tissue will be degraded solely by extrinsic mechanisms. In tuberculosis the bacterium (*Mycobacterium tuberculosis*) not only kills cells but it interferes with the normal activity of the macrophage, so any debris produced during the intrinsic phase will not be removed. Resolution is delayed and the nodule (known as a tubercle) persists for a long time and often becomes calcified and permanent.

Nuclear changes in necrosis

Three types of nuclear change can be identified in necrosis:

- **Pyknosis** – the nucleus becomes small, homogeneous and densely staining. This change is reversible and also occurs naturally in some cells (e.g. normoblasts/erythroblasts). It involves a compaction of the chromatin and is possibly caused by a fall in pH.
- **Karyorrhexis** – the nucleus breaks up into fragments which may be large when the nuclear membrane is disrupted and the chromatin becomes compacted, but the internal structure of the chromosome is not disrupted. This will give large fragments of complete chromosomes. If the chromosomes themselves also break down then very small fragments will be produced. This is sometimes referred to as 'powdering' of the nucleus. Karyorrhexis is an irreversible change and does not occur in normal cells.
- **Karyolysis** – the nuclear material dissolves completely and is lost during processing leaving a blank area in the cell. This is irreversible. Karyolysis often occurs after karyorrhexis or pyknosis.

Ischaemic changes in necrosis

A common cause of necrosis is ischaemia. The loss of blood circulation results in anoxia in cells because of the anaerobic environment. Ischaemic necrosis is one of the commonest causes of death and is the major factor in many heart attacks and strokes (see *Fig. 2.14*). The changes below relate particularly to ischaemic necrosis.

Figure 2.14
Myocardial infarct. The specimen is from a patient who died from a myocardial infarct. The tissue in the centre shows necrosis due to obstruction of the blood flow to the myocardium. Nuclei have been lost and the cells are becoming paler as they become necrotic. The surrounding tissues are showing signs of inflammation and repair but the patient died before the repair could become effective.

- Loss of oxygen causes a decrease in ATP synthesis by the aerobic mitochondria and the resulting fall in the ATP/ADP ratio stimulates an increase in anaerobic ATP production by glycolysis.
- Glycolysis is a less efficient way of generating ATP and produces lactate and CO_2 as by-products. The waste from glycolysis cannot be removed because the tissue is ischaemic, so the concentration of both lactic acid and CO_2 increases. Lactic acid and CO_2 are both acidic and so the cellular pH falls. The drop in pH causes nuclear clumping and pyknosis may occur.
- The membrane of the cell suffers from tension due to swelling of the cell and the cell contours are altered by the swelling. Eventually, the cell pH falls so low that all enzymes are inhibited and glycolysis stops. The loss of glycolysis results in virtually no ATP being available for metabolic activity and so the majority of the cell's metabolism stops.
- **Once the cell moves to the next stage it is beyond recovery.** Removal of the metabolic block or other noxious influence will not result in recovery.
- The permeability of the mitochondrial inner membrane increases and mitochondria swell ('high amplitude swelling'). The cell membranes become unstable and break down. The rupture of the membrane gives a massive increase in cell permeability so cell proteins can leak into body fluids – the leaking of cellular protein can be detected in blood samples and is used to show that infarction has occurred and the amount of cellular proteins indicates the size of the infarct.
- With the disintegration of the membranes, cell junctions will separate and tissue structure begins to break down.

- Only at this late stage do the lysosomes swell and burst releasing their degradative enzymes. Autolytic changes can then occur. The cell proteins become denatured and the cell becomes strongly eosinophilic.
- Eventually the membrane systems become completely disrupted, the ribosomes detach from the endoplasmic reticulum and breaks in the cell membrane become visible by electron microscopy.
- A dense homogeneous eosinophilic material (protein and lipid) accumulates in the cytoplasm as enzymes break down the cell contents. Individual cell outlines are lost but the general structure of the tissue remains recognisable. Karyorrhexis and karyolysis occur so nuclei become indistinct or there is a visible 'hole' in the cytoplasm indicating the original position of the nucleus. Tissue repair now occurs and the second stage of extrinsic necrosis begins with the production of collagen that follows phagocytosis and the removal of cellular debris. Any tissue debris from the cellular degradation is removed by phagocytes and replaced by newly formed collagen in a similar way to secondary healing.

The above scheme forms a good basis for necrosis from any cause, not just ischaemia. Toxic compounds may interfere with respiration (e.g. cyanide) or with membrane integrity (e.g. carbon tetrachloride) or other vital functions and produce similar results. The rate at which the changes occur may be altered but the general sequence of changes is similar.

Several different forms of necrosis can be recognised.

- **Coagulative necrosis** – the cell retains its outline, often for several days. The protein is denatured and staining is strongly eosinophilic. The internal structure of the cell is lost and the cell appears homogeneous. The nuclei are disrupted (karyorrhexis and/ or karyolysis). The denatured protein is more resistant to lysosomal enzymes hence its retention for several days.
- **Caseous necrosis** – the tissue is denatured as in coagulative necrosis, but cell outlines are lost. Fibrin is more prominent and the protein mass becomes an unrecognisable 'cheesy' mass. These necrotic remains often become calcified. This form of necrosis occurs in tuberculous nodules.
- **Liquefactive necrosis** – this involves complete dissolving of the cell's contents by lysosomal enzymes, leaving a fluid. It occurs in the brain, e.g. following ischaemia. The remnants of the dead neurones are removed by microglial cells. This is a slow process and it takes a long time to remove the large amounts of dead tissue. During the removal the brain tissue is spongy with large fluid-filled spaces. It may also occur in pus formation (suppurative inflammation) by the action of proteolytic enzymes released by neutrophils (heterolysis).
- **Fibrinoid necrosis** – occurs in connective tissue and involves loss of normal structure and its replacement by homogeneous eosinophilic necrotic material, which resembles fibrin microscopically, but consists of a mixture of proteins and their breakdown products including immunoglobulins, albumin, collagen and fibrin.
- **Fatty necrosis** – this consists of breakdown products of lipid which have become complexed with calcium ions to form soaps and appears chalky white. It occurs in the pancreas (due to lipases) or in areas with high fat content such as the breast and subcutaneous fat (may occur following trauma).
- **Gangrene** – this is necrosis with infection and putrefaction of the tissues by anaerobic bacteria such as *Clostridium perfringens*. The tissue becomes blackened as the iron in haemoglobin is converted to iron sulphide by the bacteria and the anaerobic conditions.

2.6   Apoptosis

Apoptosis is also known as **programmed cell death** and is a mechanism for eliminating unwanted cells, or cells that have been damaged, by inducing them to commit a form of cell suicide. The cells initiate a precise suicide sequence which results in a quick and clean cell death without causing any inflammation. See *Table 2.5* for a comparison of apoptosis with necrosis.

Apoptosis can be induced in cells by some cytokines (e.g. tumour necrosis factor), some drugs (e.g. vinca alkaloids) and some physical agents (e.g. ionising radiation). During apoptosis the cells activate the specific proteolytic enzymes known as CASPases (Cytosolic Aspartate Specific cysteine Proteases) which cleave intracellular proteins (particularly cytoskeletal fibres). The cells also activate non-lysosomal endonucleases which split DNA between the nucleosomes. Nucleosomes are the basic unit of chromosome organisation and consist of a DNA strand wrapped around a core of histone protein. Nucleosomes are spaced quite regularly with a repeat length of about 200 base pairs along the DNA strand. DNA cleaved by these endonucleases produces fragments with exact multiples of the 200 base pair repeat (200, 400, 600..., etc.) and this gives a characteristic laddering when the DNA is subject to electrophoresis. The multiple fragments formed during apoptosis can be detected by the TUNEL assay (terminal deoxynucleotide dUTP nick end labelling) which uses the incorporation of labelled dUTP (see **Chapter 16** and **Fig. 16.5**) to detect the large number of nicks in the DNA of apoptotic cells.

The nucleus condenses, often forming a crescent-shaped structure, and the cell fragments and forms apoptotic bodies. The cell apoptotic bodies have sugar and lipid changes in the membrane which are associated with increased phagocytosis.

The apoptotic bodies are rapidly ingested, not only by phagocytes but also by many otherwise non-phagocytic cells. The whole process is rapid, taking a similar time (1–3 h) to mitosis.

Table 2.5   Comparison of apoptosis and necrosis

	Apoptosis	Necrosis
Cells affected	Isolated cells	Groups of cells or tracts of tissues
Morphology	Chromatin margination	Pyknosis, karyorrhexis and karyolysis
	Cell fragments into membrane bound structures	Coagulation of cell contents
	Shrinkage of cells and nucleus	Swelling of cell
	Cell organelles, including lysosomes, remain intact	Cell outline often recognisable but cell contents disrupted
		Lysosomes rupture and break down cell contents
Inflammation	Absent	Present
Phagocytosis	Rapid: many cell types	Slower: only ingestion macrophages as part of inflammatory process
Occurrence	Physiological and pathological	Always pathological
Mechanism	Active	Passive
	ATP used but ATP/ADP ratio maintained	Degradation by enzyme catabolism – ATP low
	RNA and protein synthesis	pH and ionic balance disturbed

Apoptosis is a normal process and has many different uses in the body. It is found in a wide variety of animals and possibly in plants, indicating its important role in normal cell physiology. Apoptosis is a useful mechanism in a variety of situations because it helps to maintain a steady cell number by eliminating unwanted excess of cells. Apoptosis is essential in embryonic development where it sculpts the body by removing unwanted areas of tissue leaving behind the structures in their final form. Apoptosis helps to prevent autoimmune diseases by elimination of self-reactive lymphocytes in the thymus.

As well as these normal physiological uses, the apoptotic mechanism is also important in some disease states, for example, in pathological atrophy the loss of cells is often due to apoptosis. In many malignancies large numbers of cells can be observed to undergo apoptosis and this limits the growth rate of the tumour. Many treatments for malignant tumours, including radiotherapy, seem to induce apoptotic death of the cells.

2.7 Ageing and senescence

Old age is a time when tissues lose their effectiveness and degenerate. Approximately 30 brain cells die per minute during adult life so the brain gradually becomes less effective. By the age of 90 up to 50% of nephrons have been lost which compromises kidney function.

In the 1950s Sir Peter Medawar distinguished between **ageing** and **senescence**. Ageing is simply how old the person has become; by itself this has no connotations of ill health. Medawar identified senescence as "that change of the bodily faculties, sensibilities and energies which renders the individual progressively more likely to die from accidental causes". In senescence most tissues become less effective and less active.

Every system and organ of the body changes with increasing age and this change is usually associated with increasing impairment of their function (see **Table 2.6**). Most people do not die of old age per se but from complications from the loss of function associated with senescence.

Old age and senescence are associated with increasing disease. Screening of the aged has shown that most older people are medically ill. In one survey medically ill patients had, on admission to hospital, an average of six significant pathologies per patient

Box 2.9 // What is normal?

The term normal has several different meanings.

It is used to indicate that a person or organ is healthy and free from disease. For example "I was ill but I'm back to normal", or a normal healthy person.

It is used in a different way to mean the average of the population, i.e. it is *usual* and is commonly given as a range of values (e.g. an average or 'normal' person will have a heart rate between 60 and 80 beats per minute). It is considered normal (i.e. usual) to become senescent in old age but this is not the same as healthy.

It is also used to describe the usual findings for an individual, so a highly trained ('conditioned') athlete may have a resting heart rate of 40 beats per minute, which is normal for the athlete but would be considered bradycardic in most people.

To get around these differences in meaning, it is a common practice in medicine not to report things as normal but to use phrases such 'within normal limits' for such measurements as sodium concentrations in blood. Or to use the abbreviation NAD (variously translated as No Appreciable Disease, No Abnormality Diagnosed or No Abnormality Detected). This is more of a description of the results of an investigation rather than a description of actual disease-free status.

Table 2.6 Effects of senescence and ageing on selected organs

Organ	Tissue changes in senescence	Diseases associated with advancing age
Skin	Atrophy, dermal elastosis, dryness	Keratosis, carcinoma
Heart	Atrophy, amyloidosis, accumulation of lipofuscin (brown atrophy), valve calcification	Ischaemic heart disease, aortic valve stenosis
Arteries	Calcification of the tunica media, elastic fibre loss	Atherosclerosis, hypertension
Respiratory system	Emphysema, decreased ciliary activity	Carcinoma
Brain	Cortical atrophy, accumulation of lipofuscin, amyloidosis	Alzheimer's disease, Parkinson's disease, stroke
Bone	Osteoporosis	Fractures, vertebral collapse
Joints	Degeneration of cartilage	Osteoarthritis
Kidneys	Basement membrane thickening (glomerulus), amyloidosis, loss of nephrons	Glomerular sclerosis, renal insufficiency
Prostate	Hyperplasia	Carcinoma
Testis	Atrophy	
Ovary	Atrophy	Carcinoma
Uterus	Atrophy, polyps	Prolapse, carcinoma
Breast	Atrophy, fibrosis	Carcinoma
Pancreas	Atrophy, fibrosis	Diabetes, carcinoma
Eye	Presbyopia	Cataract
Immune system	Impaired immunity	Infections, autoimmune disease

and even in the general 'healthy' population of over 65s there was an average of three significant pathologies per person.

Senescence and death are considered to be normal (see **Box 2.9**) results of ageing, but the cause of senescence and death is still disputed and research in this area is intensive. It remains to be seen if the research will result in a treatment to abolish senescence and death.

Theories regarding the causes and mechanism of senescence and death

One group of explanations is that it is 'wear and tear' during life that results in the deterioration of function. Criticism of these wear and tear concepts include the fact that some animals live far longer than others (mice live less than 3 years but humans regularly live to be over 100 years old), yet the wear and tear rates will not differ by the same factor. Another argument against simple wear and tear mechanisms are the inherited progeria syndromes in which the affected person becomes prematurely senescent and suffers the same degenerative processes at the age of 7 years that we would normally associate with being 70 years old.

An alternative concept is that senescence and death are programmed into cells. This was suggested by the finding that normal cells in tissue culture could only divide a limited number of times (the so called Hayflick limit, named after the discoverer of the effect). An important finding in support of the idea of 'programmed senescence' is that the ends of chromosomes have repeated sequences of 'telomeres' which become shortened with each mitosis and when they reach a critical length the cell can no longer divide and becomes senescent.

Suggested further reading

Finlayson, C. and Newell, B. (2009) *Pathology at a Glance.* John Wiley and Son.

Lakhani, S., Dilly, S.A. and Finlayson, C.J. (2009) *Basic Pathology*, 4th edn. CRC Press.

Lowe, S. and Scott, I. (2009) *Core Pathology*, 3rd edn. Mosby Elsevier.

Underwood, J. and Cross, S. (2009) *General and Systematic Pathology.* Churchill Livingstone.

Young, B., Stewart, W. and O'Dowd, B. (2009) *Wheater's Basic Pathology.* Elsevier Health Sciences.

Suggested websites

http://library.med.utah.edu/WebPath/GENERAL.html

http://www.cumc.columbia.edu/dept/curric-pathology/pathology/pathology/pathoatlas/introduction.html

http://www.pathologyoutlines.com

http://homepages.ihug.co.nz/~luong/Files/BHB%203/General%20Pathology%20Notes.pdf

Self-assessment questions

1. Explain the difficulty of defining 'health' and indicate a suitable description.
2. What are the four major signs of inflammation? Describe the processes in tissue that produce these symptoms.
3. Describe the changes in inflamed tissues that can be seen in histological preparations.
4. How does chronic inflammation differ from acute inflammation?
5. With regard to healing, distinguish between repair, regeneration and recovery.
6. Explain the meaning of atrophy and name four causes of atrophy.
7. Distinguish between hypertrophy, hyperplasia, metaplasia and dystrophy.
8. Outline the histological differences between benign and malignant tumours.
9. Outline the growth and spread characteristics that make malignant tumours more dangerous and difficult to treat than benign tumours.
10. Explain and differentiate between staging and grading as indications of tumour prognosis.
11. Explain the meaning of the term 'cyst' and briefly outline the causes of cysts.
12. Necrosis and apoptosis are both forms of cell death. How do they differ?
13. Give a brief description of gangrene.

Learning objectives

After studying this chapter you should confidently be able to:

- **Describe why tissue removed from the body will change**
  Tissues can degenerate by three mechanisms. Putrefaction is destruction by micro-organisms, autolysis is degeneration by lysosomal enzymes and abnormal metabolism occurs in isolated tissues.

- **Describe how fixatives stop post-mortem changes**
  Fixation is the attempt to stop tissue degenerating and preserve the tissue in a lifelike condition. Fixatives are usually enzyme poisons.

- **List the properties of a perfect fixative**
  Fixatives penetrate tissues and kill cells quickly and evenly, prevent autolysis and putrefaction, do not alter the tissue adversely and prepare the tissue for sectioning and staining. They must be economical, safe and easy to use.

- **Discuss how different fixatives have special properties and uses in histology**
  All fixatives have drawbacks and advantages. Some are good for general structure, some are better for cytological detail and some for histochemistry. The fixative should be chosen to match the investigation.

- **Describe how simple fixatives can be used in combination to improve preservation**
  Since all fixatives have drawbacks, there is no single perfect fixative. Using two or more fixatives in combination can sometimes improve preservation. Fixative solutions must also be mixed with salts and buffers to control osmotic pressure and pH.

3.1 Tissue degeneration

Human cells and tissues in the body can only remain alive because they are connected to the blood stream, which provides them with the oxygen and nutrients that they need and which also removes the toxic waste products of metabolism. If the blood circulation stops, or if the tissues are removed from the body during an operation, then the cells will be cut off from this essential life-support system. Once cells are deprived of oxygen for long enough, they will die. In histological investigations we would like to see the cells looking exactly the same as they were in the living body so the tissue must be stabilized as quickly as possible. This is achieved by appropriate fixation.

Some effects of tissue degeneration can be seen in *Fig. 3.1.*

Types of tissue change

The way in which tissues change can vary depending on the organ involved but we can identify three types of change that occur in all tissues:

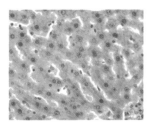 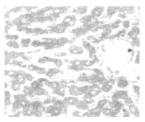

Figure 3.1
The left-hand photograph shows a normal fixed piece of liver. The right-hand photograph shows a piece of liver left unfixed for 48 hours before being processed but otherwise treated identically. The loss of structure and stainability is obvious.

1. The cells do not die instantly when the blood circulation stops but remain alive by drawing on their energy supplies and respiring anaerobically (see **Box 2.8**). This is a useful response since cells often get temporarily cut off from circulating blood; for example, lying for a long time on your arm in bed restricts the blood flow. When you move and allow the blood to recirculate into the area it results in 'pins and needles' due to the accumulation of waste products including lactic acid and carbon dioxide, both of which are acidic. If the cell was left without a blood supply for many minutes, the cells would die; this is the cause of bedsores. The affected cells would then be different to the state they were in before the blood supply ceased. They would have exhausted their store of glycogen and would be more acidic than normal tissues due to the lactic acid and carbon dioxide. Both of these can result in visible changes in the cell.

2. When the cells die, they begin to break down. The most important part of this breakdown occurs when the lysosomes release their enzymes into the cell cytoplasm. These enzymes include proteases, which destroy proteins, nucleases, which destroy nucleic acids, and lipases, which destroy lipids. This will eventually result in the complete disintegration of the cell and its contents. This process of self-digestion is called **autolysis**.

3. If bacteria or fungal spores alight on tissues, particularly dead tissues, they can begin to grow and this will result in microbial spoilage or **putrefaction**, which will drastically alter the structure of the tissues.

Putrefaction is not the most important change in histology

Collectively these changes are sometimes referred to as post-mortem changes. It is putrefaction that tends to be identified by most people as the most important form of change in pieces of tissue, but in histological terms this is a slow and therefore less important form of change than autolysis and abnormal metabolism (see **Box 3.1** for extra comments). Timely, appropriate and rapid fixation are therefore of the utmost importance, as are the efficiency and accuracy of the sampling procedure.

3.2 Fixation

Fixation is an attempt to preserve the tissues in a lifelike condition (see **Box 3.2** for extra comments). Fixation must stop abnormal enzyme activity and metabolism, in other words kill the cells, and it must stop autolysis by inactivating the lysosomal enzymes.

Box 3.1 // Putrefaction and rate of degeneration

Putrefaction is important in dealing with meat in everyday life since bacterial contamination can lead to food poisoning. Fortunately for butchers, the process is very slow so that, provided the meat is kept cool, it will not spoil by bacterial growth. Most tissues removed from the body are quite sterile and need to be contaminated by spores from the air before bacterial growth can even start. Autolysis is more rapid but is not a disadvantage for culinary meat; indeed, it is an advantage since it helps to tenderize the meat and improve the flavour. The traditional 'hanging' of meat for several days makes use of autolysis to improve its eating qualities. The different types and rates of change allow pathologists to estimate how long a body has been dead and this can be important in forensic pathology.

Box 3.2 // Preservation of food

The process of fixation is not new. Preservation of meat for eating is an old technology. In food preservation, the processes can be slowed by removing water. In the absence of water, enzymes cannot act, metabolism is stopped and bacterial growth is greatly retarded. Water can be removed by simple drying in air, or by using salt to extract the water. Both will work with human tissue as well as with meat for food. The process is called mummification after the Egyptian mummies, which were dried using salts and preserved using oils and resins. However, the tissue becomes desiccated and shrunken, making mummification a poor method of preservation for histology. Although simple drying of tissues is not a good method by modern standards, some tissue structures can still be seen and identified in mummies over 5000 years old.

Most fixatives are therefore enzyme poisons. Destroying enzymes will also stop bacterial and fungal growth and so prevent putrefactive changes.

Fixation is usually achieved by immersing the tissues in a chemical solution; hence, it is referred to as chemical fixation, but it can also be achieved by heat. Bacteriologists have traditionally fixed their bacterial smears by passing them through the flame of a Bunsen burner. High temperatures destroy enzyme activity and so act as a form of fixation. This is not common in histology since it is difficult to control, but it has been used for rapid fixation or in some histochemical investigations where the use of chemical solutions would interfere with the chemical reactions. Microwave ovens have made this form of fixation easier to apply and control and are probably the best way to heat-fix tissues. Microwave fixation is becoming increasingly popular for some applications. Microwave ovens are also being used to speed up fixation by chemical fixatives, but care is needed as heating formaldehyde in a microwave produces a lot of dangerous vapour. Glyoxal-based fixatives do not produce vapours and are much safer.

There are many chemicals that act as enzyme poisons but only a few are useful as histological fixatives. No fixative can perfectly preserve any tissue exactly the way it was in life, but some fixatives can give very good histological results for certain restricted purposes.

A perfect fixative needs several properties

Before considering some commonly used fixatives, it is useful to identify the properties that we would like to see in a perfect fixative.

A perfect fixative should:

- Penetrate tissues quickly and evenly; this makes all parts of the tissue look the same. If penetration is uneven, then the edges of the tissue will look quite different to the centre of the specimen.
- Kill cells quickly and evenly; the killing stops abnormal metabolism.
- Prevent autolysis.
- Prevent putrefaction.
- Not add any extraneous material to the tissue.
- Not swell or shrink the tissue.
- Prepare the tissue for later treatments such as staining and not prevent any later investigation that might be needed.
- Prevent desiccation and drying of tissue, which would cause shrinkage and distortion.
- Be safe to use (non-toxic, non-flammable).
- Be reasonably priced.
- Be convenient to use (shelf life, storage, etc.)
- Be easy and safe to dispose of.

The first eight properties are related to how the fixative affects the tissues, whilst the last four are related to their practical use in the laboratory.

No reagent has all of these properties and fixation is always a compromise and is always less than perfect. In particular, the ability to kill cells quickly and yet be safe to use are conflicting requirements. Many laboratory reagents used in histology and cytology are chosen to react efficiently and selectively with human tissues and this is especially true of fixatives.

3.3 Fixatives

Commonly used fixatives

The following are a few of the common fixatives that are used for preserving tissues. Photographs of tissues treated with different fixatives can be seen in *Fig. 3.2*.

Formaldehyde (HCHO)

This is a pungent toxic gas that is soluble in water. It is usually sold as a 37–40% solution of formaldehyde in water when it is called **formalin** (in the UK, formalin is a general term, but in some countries it is a trade mark). It is the most widely used fixative for routine light microscopy sections, when it is used as a 4% solution of formaldehyde. Note that 4% formaldehyde is made by diluting the formalin solution (i.e. 40% formaldehyde) using one part formalin to nine parts water (a 1:10 dilution) and it is often called 10% formalin. The terms formalin and formaldehyde are both still in common use but the concentrations are quite different. Thus, 10% formalin is the same strength as 4% formaldehyde. The use of formaldehyde concentrations rather than formalin concentrations is the more accurate and preferred method.

Formaldehyde is a very reactive chemical and undergoes several changes in storage. Formalin solutions are therefore usually impure and contain a number of contaminants, some added by the manufacturer and others formed during storage.

In aqueous solution, formaldehyde reacts with water forming methylene glycol:

$$HCHO + H_2O \rightarrow CH_2(OH)_2$$

Formic acid (methanoic acid) can be formed by oxidation of formaldehyde by the oxygen in air:

$$2HCHO + O_2 \rightarrow 2HCOOH$$

and also by the Canizzaro reaction:

$$2HCHO + H_2O \rightarrow HCOOH + CH_3OH$$

Methanol is also formed in this reaction. Methanol is deliberately added to formalin solutions by the manufacturers, partly to inhibit the Canizzaro reaction by driving the dissociation reaction in reverse and also to prevent polymerization.

Finally formaldehyde polymerizes to form 'paraformaldehyde':

$$2nHCHO \leftrightarrow HO(CH_2O)_nH + (n - 1)H_2O$$

This occurs especially in the cold conditions found in most chemical stores.

The paraformaldehyde does not harm the fixative except by reducing the formaldehyde concentration. Provided this is adjusted to take account of the loss, the final solution will still fix tissues very well. If the polymer formed is a short chain of less than eight formaldehyde molecules, then the paraformaldehyde remains soluble and will not be noticed, but if the chains are longer than eight carbons, it precipitates as a white powder. The paraformaldehyde will depolymerize if warmed or diluted. Depolymerization is more rapid if the solution is around neutral pH.

Thus, the commercial formalin solutions are really quite impure. The most important contaminant is the methanoic acid, which will affect the pH and the fixing properties. Acid formalin will, for example, produce a brown deposit around degenerating blood cells. This is an artefact and is therefore a nuisance. To prevent this occurring, only neutral formaldehyde solutions should be used and most laboratories normally use buffered formaldehyde as their routine fixative.

If greater purity is needed for critical applications, such as electron microscopy, then it is better not to use commercial formalin solutions. Instead, formaldehyde can be generated by depolymerizing paraformaldehyde. The paraformaldehyde is unadulterated and not contaminated in the same way as the commercial formalin solutions and is more reliable and purer.

Fixation properties of formaldehyde. Formaldehyde in solution is almost entirely in the hydrated form of methylene glycol and it is the methylene glycol that acts as the fixative. The glycol adds on to proteins, forming single and double additions.

$$CH_2(OH)_2 + R\text{-}NH_2 \leftrightarrow R\text{-}NHCH_2OH + H_2O$$

$$R\text{-}NHCH_2OH + R_2\text{-}NH_2 \rightarrow R\text{-}NHCH_2HNR_2 + H_2O$$

Single additions are easily reversed by washing in water for a few hours. Double reactions are more stable and produce methylene bridges.

These reactions occur slowly over a period of days, so formaldehyde is a slow-acting fixative and full fixation takes about a week, although adequate fixation occurs in 12–24 h.

As well as its main fixative properties, formaldehyde has several other advantages as a routine fixative. The effect of formaldehyde on tissues is to cause them to swell slightly; this means the organs remain soft and pliable. This soft fixation is good for dissection and for trimming of tissues into blocks for processing. Many other fixatives harden the tissue making trimming and dissection difficult or impossible. The tissue does, however, harden during processing, so the final paraffin block is very hard.

Formaldehyde is a very tolerant and forgiving fixative. Tissues cannot be overfixed and can be stored in formalin for many years. Many other fixatives can overfix the tissue and formaldehyde is often used to store tissues originally fixed in such fixatives.

Most fixatives alter the colour of tissues, often turning them a dull greyish colour. Formaldehyde is the only fixative that allows natural colour to be restored by soaking in 70% ethanol and so it forms the basis of all museum fixatives.

Formaldehyde is not perfect; it has its drawbacks. Cytoplasmic staining is duller after simple formaldehyde fixation than it is after many other fixatives. Formaldehyde is also toxic and causes formalin dermatitis and is a suspected carcinogenic agent (see **Box 3.3** for more details).

Glutaraldehyde

This is a bifunctional aldehyde that reacts more quickly than formaldehyde (it binds to 90% of NH_2 groups in 2 hours; formaldehyde binds to only 70% of such groups in 7 days). It polymerizes very easily and glutaraldehyde solutions should be depolymerized before use. It is difficult to remove the glutaraldehyde from tissues and this makes histochemistry difficult; in particular, many aldehydes remain active since each molecule has two aldehyde groups and often only one is actually bound to the tissue so the second group remains active. These extraneous aldehyde groups prevent glutaraldehyde being a useful fixative when the periodic acid–Schiff reaction and other aldehyde-detecting reagents are being used. Glutaraldehyde inactivates all enzymes, thus totally preventing enzyme histochemistry. It also binds and masks many antigens, making immunotechniques difficult or even impossible.

Glutaraldehyde fixation gives excellent morphology but staining with acidic dyes is depressed. Like formaldehyde it is very dangerous if inhaled or in contact with the skin. Glutaraldehyde is widely used as a fixative in electron microscopy.

Mercuric chloride

This is white crystalline solid soluble to about 7% in water. It is highly toxic as a cumulative poison, so small doses over a long period can gradually build up to a toxic dose.

Mercuric chloride fixes tissues by attaching to SH groups in proteins and it cross-links them forming a coarse coagulum of protein. Coagulating fixatives such as mercuric chloride leave relatively large fluid-filled spaces within the protein and allow rapid penetration of reagents. Staining is not depressed and the cytoplasm stains brightly with acid dyes but it

---

Box 3.3 // Formalin toxicity

Formaldehyde is a very dangerous chemical and should be handled carefully. Contact with the skin can result in an allergic reaction due to the formalin attacking skin proteins and modifying their structure sufficiently to make the body identify them as foreign agents. The immune system will then start attacking these proteins and this results in severe disease. It is not uncommon for laboratory workers in histology having to avoid formalin (and xylene) completely. In sensitized individuals the immediate immune response can be as severe as anaphylactic shock.

Always use protective gloves and clothing when dealing with formaldehyde.

Formaldehyde is also a respiratory poison and must only be used in a well-ventilated area. The ventilation must carry the fumes **away** from the face, so downward or backward ventilating systems are needed. Upward ventilation can actually pull the vapour towards the mouth and nose.

> **Box 3.4 // Problems with the use of mercury salts**
>
> Mercuric chloride was previously called 'corrosive sublimate'. The 'corrosive' refers to the fact that it attacks virtually all metals (including gold), so care is needed when using mercuric chloride as it will corrode jewellery (including wedding rings), scalpels, metal lids and forceps. It will corrode metal piping, so prolonged disposal down laboratory sinks has been known to rot metal waste pipes and result in the flooding of laboratories.
>
> Mercury and its compounds should not in any case be disposed of down the sink as it is toxic to sewage bacteria and will contaminate the effluent from sewage works.

also leaves the tissue excessively hard and shrinks the tissue considerably. Historically it has been an important fixative but its cost, waste disposal and toxicity have reduced its popularity (see **Box 3.4**).

Mercuric chloride leaves a black precipitate (mercury pigment) in tissues, which interferes with the appearance and needs to be removed. The pigment is easily removed with iodine, which converts it to mercuric iodide, followed by sodium thiosulphate (often referred to as 'hypo'), which removes the mercuric iodide.

It has been suggested that zinc salts may be useful as a replacement for mercuric chloride, but their mode of action is uncertain. The fixative B5, used especially for haematopoietic tissues, has a zinc formulation but it is not yet widely used.

Osmium tetroxide

This is very expensive (up to £334,000 per kg), so is usually restricted to fixing very small pieces of tissue. It is a powerful oxidizing agent and adds on to double bonds and in so doing it blackens them. Since most lipids contain double bonds it fixes lipids, and is the only fixative that really does so. For this reason it is widely used to fix lipid membranes for electron microscopy.

Fixation in osmium tetroxide makes proteins lose their acidophilia. It is volatile and can react with the cornea resulting in blindness. It must be handled only in fume cupboards.

Osmium tetroxide is widely used in electron microscopy as a fixative and stain for membranes, but is rarely used as a light microscopy fixative.

Acetic acid (ethanoic acid)

Although the correct chemical name is ethanoic acid, this is rarely heard in histology laboratories and the older, less systematic name of acetic acid is used. Unlike all of the preceding fixatives, acetic acid is a non-additive fixative and does not bind to any chemical groups in the tissues. Its main effect is to denature nucleic acids as a result of the pH change. It does not fix proteins and causes them to swell (especially collagen) and it effectively destroys mitochondrial staining.

Acetic acid is widely used to improve nuclear staining since it preserves DNA very well without blocking any dye-binding sites. The poor cytoplasmic staining from the loss of protein is an advantage in some types of work, e.g. chromosome studies where poor cytoplasmic preservation leaves the chromosomes more easily observed against a clear background. If a good cytoplasmic fixation is achieved, the cytoplasm will stain strongly and this will obscure the fine chromosome structure.

Acetic acid has often been included in fixative mixtures to combat the shrinkage of other fixatives (see **Box 3.5** for extra comments).

> **Box 3.5 // Acetic acid and alcohol have a long history of use**
>
> The use of alcohol and acetic acid has a long history in food preservation. Peaches in brandy and pickled onions are both highly regarded delicacies by many people.
>
> Alcohol is also one of the oldest embalming techniques. Following his death at the battle of Trafalgar, Nelson's body was placed in a barrel of brandy to preserve it during the journey back to Britain.

Alcohols (methanol, ethanol, propanol, butanol, etc.)

These are also non-additive fixatives and denature proteins and nucleic acids by removing their bound water and replacing it with the alcohol. This alters the shape of molecules (tertiary structure) without altering the reactive groups. Thus, alcohols preserve the chemical reactivity of many cellular materials and are popular in histochemical investigations.

The tissues do shrink considerably and may become overhard and brittle if long alcohol fixation is used. Alcohol is therefore less commonly used for fixing blocks of tissue, but is common for fixing smears or fresh sections (e.g. cryostat sections; see *Chapter 6*) where it is less of a problem if the cells become brittle as no further sectioning is needed.

Alcoholic fixation is probably more widespread than is acknowledged because a poorly fixed specimen is often processed unknowingly. The alcohol used in processing then acts as the fixative for any unfixed parts of the block.

Other fixatives

Many other chemicals have fixative properties, and fixatives in reasonably common use include picric acid, potassium dichromate and chromic acid. They all have some useful characteristics but their properties will not be dealt with in detail here.

Photographs of tissue treated with different fixatives can be seen in *Fig. 3.2*.

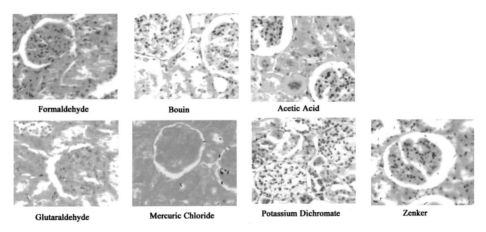

Formaldehyde	Bouin	Acetic Acid

Glutaraldehyde	Mercuric Chloride	Potassium Dichromate	Zenker

Figure 3.2
Photographs of pieces of kidney fixed in different fixatives. All the tissues were taken from the same kidney and processed identically. The effects of altering fixation are clearly seen in the staining reactions. Shrinkage effects can be gauged by observing the space around the glomeruli.

Non-formalin aldehydes and fixatives

Moves have been made over the past few years to improve the safety of the laboratory working environment by formulating fixatives that do not contain formalin. This has been accelerated by the need to glean the maximum amount of information from paraffin wax embedded tissues by the use of antibody and molecular techniques; this can only be achieved if the fixative used does not degrade or mask the target sites, i.e. antigens and gene base sequences.

These new formulations may contain glyoxal, the simplest di-aldehyde (CHO.CHO) or substitute formalin with chloroform and/or simple alcohols. Glyoxal has the benefit of being far less hazardous to use when compared with formalin, whilst retaining many benefits.

3.4 Practical fixative solutions

Although the simple fixatives are the major component of most practical fixative solutions, they are rarely used as pure solutions and are usually made up with other compounds. These other components are often very important to the final result but are not directly involved in the fixation. They are used to control the environment around and within the tissues whilst fixation occurs.

The various types of additive can be grouped into three classes:

1. **To control tonicity (osmotic pressure).** Living cells are osmotic systems and will shrink in hypertonic solutions (high salt concentration) and swell and even burst in hypotonic solutions (pure water or low salt concentration). Blood plasma and cell cytoplasm have a tonicity of about 340 mOsm and this is said to be isotonic. The final tonicity of many practical fixatives is made hypertonic at around 400–450 mOsm. This is to ensure that, even when the fixative is removed by binding to tissues during the fixation process, the residual solution is still isotonic and will not cause cellular lysis due to osmotic imbalance. If the fluid became hypotonic before fixation was achieved, the cells might absorb water by osmosis resulting in swelling and eventual lysis.
2. **To control pH.** The use of buffers is preferred. Many buffers can be used including phosphate, Tris and cacodylates. The choice of buffer is controlled by other components. Phosphate buffers will precipitate if calcium or magnesium ions are added, whilst Tris and barbiturate buffers will react with aldehyde fixatives.
3. **Other additives for specific reasons.** For example, calcium is sometimes added to help preserve phospholipids, ammonium bromide is said to enhance the fixation of nervous tissue (particularly neuroglia) and detergents have been used to remove membrane lipids and aid penetration of the fixative and other reagents such as lectins and antibodies. Alcohol improves the retention of simple carbohydrates, e.g. glucose.

Physical conditions play a major role and the temperature, volume and duration of fixation can also be critical.

Short fixation times are common in histochemistry, especially when fixing frozen sections (2 min at 4°C, for example). Prolonged fixation in many reagents will result in overfixation and the tissue may become unusable. Even formaldehyde, which is often used for long-term storage of specimens, slowly alters the tissue, e.g. the ability of the tissue to stain with acid dyes becomes diminished.

The volume of fixative should be very large compared with the volume of tissue (20 times is a safe margin) since the fixative gets depleted or diluted by the tissue. This is often overlooked with very large specimens, which may be crammed into too small a container that is then topped up with fixative. A large ovarian cyst (whose total volume may be a litre or more) may arrive in the laboratory with less than 100 ml of fixative poured over it (see *Box 3.6*).

If the specimen is large and penetration is likely to be poor, the block should be cut into pieces small enough for diffusion of the fixative to reach all parts in a reasonable time. Alternatively the fixative can be perfused into the blood vessels and pumped through the vascular channels so that it reaches all parts of the tissue very quickly, usually with the specimen suspended in a bucket of fixative, e.g. fixation of brain with perfusion through the basilar artery. The adverse effects of those components that cause rapid hardening and prevent efficient fixation are effectively by-passed as the central areas are able to react at the same time as the periphery.

Box 3.6 // Fixation by non-histologists can be done badly

Unfortunately fixation is often left to people who are not histologists. These include operating theatre staff who all too often mistakenly believe that the most important form of tissue breakdown is bacterial decay and so put tissues into antiseptics or disinfectants. These will certainly kill bacteria, but they do little else to preserve the tissues. Whenever possible, either do the fixation yourself or make sure that the person who does it understands what is needed; otherwise poor fixation and poor preservation will be the result.

I have personally received specimens from operating theatres in antiseptic solutions such as Savlon with dreadful results and even once in Domestos solution, which certainly killed the germs but did little to enhance the tissue preservation. I also received one specimen, from a general dental practitioner, in gin. This gave adequate preservation but I would not recommend it as a general fixative.

Compound fixatives

Since every simple fixative has deficiencies, many people have tried to improve fixation by using more than one fixative in a solution. The intention is to use the good properties of one fixative to counteract the bad properties of the other components.

For example, Zenker's fixative is based on mercuric chloride but has acetic acid and potassium dichromate added to modify the results.

Compound fixatives do have problems. Mixing fixatives with different penetrating powers may result in very uneven fixation with only the outer part of the tissue being fixed in the full mixture. The slowest penetrating fixative may lag so far behind the other fixatives that the inner parts are fixed only by the faster-diffusing components.

The use of several component fixatives means that it becomes difficult to predict all of the chemical reactions occurring during fixation. Thus, compound fixatives are not usually the best fixatives for critical histochemistry.

For general diagnostic histopathology, many of the compound fixatives have fallen out of favour and simple formaldehyde fixation has become standard. However, there are variations between countries with some compound fixatives still being popular. In France Bouin's fluid is more popular, possibly because it was developed by a Frenchman

---

> ### Box 3.7 // Preservation in history and art
>
> Fixation is usually associated with science, but Damien Hirst achieved notoriety by using animals in formaldehyde as examples of modern art. He exhibited a sheep in formaldehyde titled 'Away from the Flock' at the Serpentine Gallery in 1994, which was reputedly sold for £25,000. The mummified remains of the pharaohs have been icons of history for generations and the embalmed body of Lenin is an important tourist attraction in Moscow. Thus, preservation is also important outside the laboratory.
>
> More recently, Gunther von Hagens has become something of a celebrity by exhibiting whole body anatomical preparations that have been preserved by the process of plastination. This involves fixation with formalin, extraction of water with cold acetone and finally infiltration with a liquid polymer that sets into a rigid plastic with heat or UV.

(Pol Bouin). Also, many commercial firms now offer their own fixatives and do not always declare either the components or their concentration.

Sequential fixation

This is a similar concept to the use of compound fixatives with the benefits of one fixative being added to those of another, but the fixatives are used sequentially rather than as a single solution. The primary fixative is almost always an aldehyde but the secondary fixative varies, with post-chroming (potassium dichromate), post-subliming (mercuric chloride) and post-osmicating (osmium tetroxide) all having their place.

Their use is designed to improve the clarity and intensity of subsequent staining methods, e.g. mercury salts with trichrome stains.

The importance of correct fixation

When choosing a fixative it is important to get it right. Properly fixed tissues are easier to prepare and the final results are better. Poor or inappropriate fixation makes later processing, sectioning and staining more difficult and may make certain investigations impossible. This is especially true when working with very fatty tissues, e.g. breast, lipomas, brain. Thus, before fixing tissues, and indeed before starting to remove them from the body, the final purpose of the sections must be considered and the fixative chosen to match the requirements of the technique. Trying to match a staining technique to an inappropriately fixed tissue always gives inferior results (see *Table 3.1*).

Reversal of fixation

The chemical effects and tissue changes caused by most of the coagulant fixatives are to a large extent irreversible. This should not be surprising since their mode of action is through the denaturation of protein.

Samples fixed by the cross-linking addition reaction characteristic of formalin can be recovered to a certain extent by washing for 18–24 hours in running tap water. Investigations on samples fixed with ^{14}C-labelled formaldehyde have shown that radioactivity continues to be washed out after 48 hours in running tap water. This property is useful in permitting frozen sections to be taken from fixed material once sufficient formalin has been washed out. One application could be the demonstration of lipids in storage disease or thecoma specimens that need to be shown for confirmation that processing had removed them. If there is reserve fixed material then frozen sectioning followed by a fat stain would be a valid approach.

Table 3.1 Conflicts between fixatives and staining methods

Fixative	Stain adversely affected	Recommended alternative fixative
Neutral buffered formalin	Copper and iron (Mallory stain)	Alcoholic fixative
	Masson trichrome (connective tissues)	Bouin
Bouin	Schultz (cholesterol)	Neutral buffered formalin
	Feulgen (DNA)	Alcohol
	Mallory phosphotungstic haematoxylin	Zenker
	Fibrin (Weigert)	Bouin
	Haemoglobin (Dunn–Thompson)	Neutral buffered formalin
	Pancreatic $\alpha, \beta \& \delta$ cells (Trichrome–PAS)	Neutral buffered formalin
Zenker	Schultz (cholesterol)	Neutral buffered formalin
	Fats/lipids (oil red O, Sudan black)	Neutral buffered formalin
	Haemoglobin (Dunn–Thompson)	Neutral buffered formalin
	Pancreatic $\alpha, \beta \& \delta$ cells (Trichrome–PAS)	Neutral buffered formalin
All aqueous fixatives	Glycogen (PAS, Best's carmine)	Carnoy, alcoholic formalin
Alcohol	Melanin (DOPA oxidase)	Fresh frozen or neutral buffered formalin
Glutaraldehyde	Mucoproteins (PAS)	Neutral buffered formalin
	Silver stains	

Suggested further reading

Bancroft, J., Layton, C. and Suvarna, S. (2012) Chapter 4: Fixation. In: *Bancroft's Theory and Practice of Histological Technique*, 7th edn. Elsevier Health Science.

Bignold, L.P. (2002) Hypothesis for the influence of fixatives on the chromatin patterns of interphase nuclei. *British Journal of Biomedical Science*, **59**: 105–113.

Boon, M.E. and Kok, L. (1988) *Microwave Cookbook of Pathology: The Art of Microscopic Visualisation*, 2nd edn. Coulomb Press.

Hopwood, D. (2002) Fixation and fixatives, In: *Theory and Practice of Histological Techniques* (eds Bancroft, J.D. and Gamble, M.), 5th edn. Churchill Livingstone.

Kiernan, J.A. (2008) *Histological and Histochemical Methods: theory and practice*, 4th edn. Scion Publishing.

Suggested website

http://library.med.utah.edu/WebPath/webpath.html

Self-assessment questions

1. In what ways will tissue removed from the body change with time?
2. How can post-mortem changes be prevented?

3. List the properties of a good fixative.
4. Dettol will kill bacteria including those that cause putrefaction. Will Dettol be a good fixative?
5. Schiff's reagent is used to detect naturally occurring aldehydes in tissues. It can be used easily on formaldehyde-fixed tissues but not on glutaraldehyde-fixed tissues. Why is there this difference?
6. Why should all fixatives be handled and treated carefully?

04 Processing and microtomy

> ## Learning objectives
>
> *After studying this chapter you should confidently be able to:*
>
> - **Describe how specimens are described and trimmed**
>   Specimens need to be accurately recorded to ensure that results are reported correctly and can be matched to the correct patient. Tissues need to be trimmed, oriented and labelled to ensure accurate diagnosis.
>
> - **Describe why tissue needs support during sectioning**
>   Tissues are soft and will flex and not cut cleanly. Tissues are most commonly embedded in paraffin wax for ease of producing thin sections.
>
> - **Describe how tissues can be impregnated with wax using intermediate reagents**
>   Processing tissues involves dehydration to remove water, clearing in an organic solvent to remove the alcohol and finally impregnation in molten wax.
>
> - **Name some common and useful dehydrating and clearing agents**
>   Ethanol is commonly used for dehydration but acetone, methanol and propylene oxide are alternatives. Clearing agents include xylene, chloroform and petrol hydrocarbons.
>
> - **Explain the advantages and use of automated tissue processing**
>   Automated processing machines offer greater speed, consistency and safety.
>
> - **Describe the preparation of mineralized tissues for sectioning**
>   The mineral content can be removed by treatment with acids and the tissue can then be processed as usual. Alternatively, undecalcified sections can be prepared, for example by grinding the hard tissue to give a thin slice.
>
> - **Describe the uses of different types of microtomes and knives**
>   Microtomes differ in the size and hardness of blocks which can be easily cut. Knives can be solid and need regular sharpening or they can be disposable and be discarded when they become blunt.

4.1 Examination, sampling and trimming of tissue

Specimens need to be prepared before processing can begin. It is essential that the specimen containers are first checked for details of the patient and cross-referenced to the form and clinical details to prevent any errors. If there is a lack of necessary detail, or if there is a discrepancy, then it needs to be resolved before the specimen is processed. Once checked, the tissue is then taken for 'trimming' (sometimes called 'cut-up' or 'grossing') and sampling.

Details of the specimen are recorded, such as size, shape and any significant features such as sutures, foreign bodies, tattoos, etc. There is also increasing use of digital cameras to record the specimen. These details may be important if any identification problems arise; it is very unusual but occasionally specimens can be mislabelled or confused with another specimen. In these cases the description can be invaluable to identify the sample. A full record is also essential if there is a forensic case where a 'chain of evidence' is required.

The specimen is then examined by a competent person. Historically, this was a pathologist but with increasing workloads, increased demands on pathologists' time and pressure on budgets, some of the workload is being delegated to an advanced practioner biomedical scientist.

Specimens can be graded into different categories depending on their complexity (*Table 4.1*).

Table 4.1   Specimen grading categories

Category of specimen	Example	Sampling requirement
A	Small biopsy samples; contraceptive vasectomy specimens to confirm they contain vas deferens	Tissue is transferred intact into cassette (see *Fig. 4.1*) for processing; no dissection or subsampling is needed.
B	Appendix	Tissue needs to be cut into a small number (1–3) of pieces or subsamples but these may be processed together to produce a single block. The sampling is routine and simple.
C	Cervix	Sampling requires dissection using a standard protocol which determines how many samples and from which locations. May require more than one cassette and block.
D	Lengths of colon removed for malignancy	Sample requires interpretation to determine representative samples, samples of more suspicious areas and possibly associated sentinel lymph nodes. May require multiple cassettes and blocks.
E	Large complex resection specimens of malignant tumours	Require extensive sampling including resection margins to determine if the tumour has been completely excised. May need large numbers of separately identified samples and blocks.

Which categories may be dissected by an advanced practitioner would depend on their experience, training and local agreements. The examination, reporting and sampling are usually specified within a laboratory by a Standard Operating Procedure (SOP) based on the recommended minimal requirements of the Royal College of Pathologists.

With large resections of malignant tumours it is expected that the surgical margins are examined to see if they are tumour free. After specimens have been cut from the resected tissue it becomes difficult to determine the surgical margin from the laboratory created margin. In these cases the surgical margins can be marked by 'inking'. This may be done by the surgeon in the operating theatre or in the laboratory during gross examination. Inking can be done with Indian ink, or dyes such as alcian blue, or coloured gelatin. The colours can be helpful in indicating the orientation of the specimen. With very small specimens such as biopsy cores, or very pale specimens, it may be necessary to dye the whole block to facilitate handling and sectioning after processing.

Figure 4.1
Tissue processing cassettes. These consist of perforated plasic or metal containers with snap on lids. They come in differing sizes to fit the different sizes of tissue to be processed.

Sample size and orientation

The physical size of a block is an important factor to be considered. If the specimen is small then it may not need to be cut and can be processed whole. However, some small samples may need to be cut to expose the lesion. This can happen with small excisional skin biopsies where, in addition to the lesion, there is removal of surrounding normal skin. Transecting the specimen has the advantage of helping orientation of the block during blocking out ('cut surface down' in the mould) because skin needs to be sectioned as a vertical slice through all the skin layers rather than a horizontal section parallel to the skin surface. It also exposes the lesion so that it shows a cross-section through the lesion and adjacent normal skin without needing to examine several levels during sectioning.

Larger specimens may need to be trimmed to keep the processed tissue to a reasonable thickness as overthick tissue samples slow down penetration of reagents and can result in prolonged processing times or poor processing. In some cases very large cross-sectional areas are needed, called mega blocks, e.g. whole sections of intestine or prostate which allow examination of the entire cross-section of the original sample.

Any materials which will interfere with normal sectioning should be removed before processing. This can include metallic objects such as surgical clamps, stents, radiotherapy implants ('seeds' or 'needles'), staples or even bullets. Calcified tissues should be noted because decalcification may be needed.

A specialized form of block production is the use of microarrays in which a large number of small cores punched out of different specimens are combined into a single block. Sectioning of this block produces a single section with perhaps 100 different specimens, so a single stain/technique gives 100 results. These blocks can be used in research to identify genes, mutations or enzyme changes in cancer. It is more usual to take cores from paraffin wax-embedded material.

Safety is a major consideration and avoidance of breathing formaldehyde vapour is essential. This may entail the use of perforated metal benches and a down draft of air to draw fumes away from the face while examining unprocessed material.

4.2 Paraffin wax impregnation

Tissues need support to allow sections to be cut

Fixation hardens tissues but even fixed human tissues are generally quite soft and need extra support to allow thin sections to be cut. Sections are usually cut between 3 and 6 μm thick and the easiest way to prepare sections of this thickness is to use a hard embedding medium to support the cells and to hold the tissue firmly. **Processing** is the method by which tissues are converted into a block for **microtomy.**

Paraffin wax impregnation is the usual method of support

The commonest embedding medium in routine histology is paraffin wax. Although called a wax, it is actually composed of long-chain paraffin hydrocarbons and is similar to the material used to make candles. Paraffin wax is molten above 60°C but is solid at room temperature and has a hardness similar to the tissues once they have been impregnated. However, the specimen cannot simply be immersed in the wax as wax and water are immiscible and putting the tissue in molten wax will result only in the tissue being coated in wax (see *Box 4.1*). The water in the tissue is instead replaced by wax so that the support runs completely through the tissues and supports them internally as well as externally (wax impregnation). The replacement of the tissue water with wax involves several steps and is generally called processing. One or more intermediate processing reagents are needed to remove the water, replace it with a reagent miscible with wax and then finally impregnate with the wax. Formalin fixation followed by paraffin wax embedding (FFPE) is a routine procedure used to prepare histological samples for sectioning.

> **Box 4.1 // Support needs to be deep into the tissue**
>
> Originally the support for tissues was much less intimate and did not penetrate into the tissue but simply surrounded it. Botanists still often support stems and leaves during cutting by wrapping them in elder pith to make a more rigid structure. This is less easy to do with animal tissues that are much more irregular in shape and are less intrinsically rigid than plant tissues. Histologists therefore used wax. The tissues were dipped in molten wax that moulded itself to the irregular shape of the tissue and helped support it. This was still not totally successful and it was not until the wax was permeated into the tissue as well as around it that effective sectioning was possible.

Processing tissues to wax with a single reagent

Although there are some reagents that are miscible with water and wax, they are not popular in histology. For example, dioxane (diethylene dioxide) can be used as a single bath and will mix completely with water and paraffin wax. The tissue can simply be taken from the fixative (which is usually a water-based solution) and placed in dioxane. Once all of the water has been removed, the tissue can be transferred into molten wax, which then replaces the dioxane. Unfortunately, dioxane has problems, In particular, it is quite toxic, it is relatively expensive and the final results are poorer than using other methods. For these reasons, it has never become a common or popular method despite its relative simplicity.

Isopropanol, which can act as a dehydrating agent as well as being miscible with some waxes, is being used more frequently, especially in automated processing machines in order to avoid the use of xylene. Isopropanol is most effective if used at much higher temperatures than room temperature, and so it is mostly used in totally enclosed tissue processing machines where high temperatures can be used safely.

Replacement of tissue water with wax using two intermediate steps

Instead of using a single reagent, double-stage processing is the norm. The first stage removes water (**dehydration**) and the second stage replaces the dehydrating agent with a reagent that is miscible with wax (termed **clearing**), because the tissue becomes translucent and fat becomes transparent. The tissues are then impregnated with wax, placed in a mould and the wax is solidified. Sections can be cut from the wax block and then stained. Calcified tissues such as bone are extremely brittle in their natural state but can be decalcified to leave only the soft tissue matrix of the bone (osteoid), which can then be processed normally. *Fig. 4.2* shows how the steps are arranged. Each of these steps will be considered in more detail but it is important to grasp the general scheme of processing tissues.

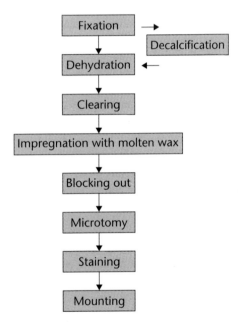

Figure 4.2
Flow of the specimen during processing. The fixed specimen is either transferred to dehydrating agents directly or, if there is any mineralized bone present, it is decalcified first before being dehydrated. The dehydration, clearing and impregnation with wax are often done by a machine and can be considered as a single stage called **processing**

Choice of dehydration reagent

The first step in processing is dehydration of the tissue. In choosing a reagent for dehydration, there are various properties that need to be considered. These are cost, safety, speed of action and its effect on tissues. The speed of action is controlled by the viscosity of the solvent. Dehydration, like most of the processing steps, is basically a diffusion process with the solvent diffusing into the tissue and the water diffusing out. The rate of diffusion into a relatively compact structure such as tissue is more dependent on the viscosity of the liquid than on simple molecular weight. Water has a viscosity of 1 cP; higher numbers mean more viscous (thicker) fluids, which will penetrate more slowly.

Safety considerations include both toxicity and flammability. Most of the solvents used in dehydration will burn but their flash points and fire risks vary. The properties of some dehydrating agents are listed in *Table 4.2.*

As with many steps in histological processing, the more rapid the removal of water, the harsher the effect on the tissues. This applies equally to the inherent speed of the reagent and to the graded series of strengths of the reagent. Very gradual removal with a slow acting reagent may damage the tissues less than rapid removal with a harsher reagent. It is always a trade-off between speed and quality.

Table 4.2   Properties of dehydrating agents

Reagent	Viscosity	Toxicity	Fire hazard	Comments
Ethanol (IMS, 74OP)	1.2	?	+	Commonest reagent (see **Box 4.3**)
Methanol	0.6	++	+	Faster and harsher
Propanol	2.5	+	+	Slower
Acetone	0.4	++	++	Faster and harsh
Cellosolve (2-ethoxy ethanol)		+	+	Slower and gentler but destroys some elements
1-Epoxy propane (propylene oxide)		++ (carcinogen)	+++	Used in electron microscopy; inhibits some stains

Ethanol

Ethanol has long been used for removing water and is so ubiquitous in laboratories that if a method uses the term alcohol it will mean ethanol rather than any other alcohol. It is a fairly gentle reagent compared with many of the other agents that can be used. Its almost universal use is probably due to its low cost and low toxicity with little tissue damage. Ethanol will remove some materials from the tissue, but usually these same materials are removed by all dehydrating and clearing agents so this is not a problem unique to ethanol. Ethanol can be regarded as almost non-toxic in use (see **Box 4.2**) since the amounts liable to be ingested or inhaled, provided reasonable care is taken, are well below the levels known to cause damage.

During the process of dehydration it is usual to remove the water in a series of steps by using several baths of alcohol of gradually increasing strength (a 'graded series'). If the specimen is placed directly into concentrated alcohol, the tissue shrinks more and may become excessively hard. The actual concentrations of alcohol used in the graded

> **Box 4.2 // Safety of ethanol**
>
> Alcohol is unusual amongst laboratory chemicals in also being a foodstuff. It is included in the diets of many people. Alcohol is not seriously damaging to the skin and is the basis of many perfumes and aftershave lotions. Alcohol will slightly dry the skin by removing some of the natural oils and is also extremely painful if applied to cuts.

series vary from laboratory to laboratory. Some processing schedules start with relatively weak mixtures (30% alcohol) and increase the strength very slowly, whilst others begin at 70% and move quickly to even higher concentrations. Most laboratories get good results, which suggests that any differences in the methods are small. It is usual to treat delicate tissues such as embryos with a more gradual removal as they are more easily damaged. Very tough tissues, however, may also benefit from a slower and gentler processing to prevent them becoming even tougher. Tough tissues can also be softened by adding other reagents to the dehydrating agents. The addition of phenol can have a beneficial effect on very tough materials.

Methanol

Methanol is important not only as a dehydrating agent in its own right but also because it is added to ethanol to give industrial methylated spirit (IMS), which is not drinkable (see *Box 4.3*) and so has a lower cost and is less strictly controlled than pure ethanol. Methanol is somewhat harsher than ethanol but is also faster and is sometimes used for rapid processing.

Propyl alcohol comes in two isomeric forms, propan-1-ol and propan-2-ol. Many of their properties are similar (e.g. viscosity), but their boiling point and flash point are significantly different. In histological techniques the commonest reagent form is propan-2-ol, also known as isopropanol. It is increasingly used in 'xylene free processing' in automatic machines.

Higher alcohols such as butanol can be used as dehydrating agents but have no great advantages and are only occasionally used for specialized techniques where ethanol and methanol are unsuitable.

Box 4.3 // Ethanol and excise regulations

Ethanol is cheap to produce but has the problem that the Revenue and Customs people take a great deal of interest in, and duty from, anything alcoholic. Although laboratories can get the alcohol with reduced duty, they must keep accurate records to show that the alcohol is not being used illicitly. This explains the use of IMS instead of pure ethanol. Sometimes alcohol is referred to as 74OP spirit. This again is due to HM Revenue and Customs who used to test alcoholic spirits to 'prove' them. Originally, a spirit defined as 100° proof was the lowest strength of alcohol that, if poured on to gunpowder, would allow it to ignite. This was the original 'proof' of the spirit's strength. Anything stronger could be diluted and still pass the test so it was overproof (OP) spirit. Pure alcohol is equivalent to 174° proof, hence the term 74OP.

Acetone

Acetone, like methanol, is occasionally used as it will dehydrate a tissue more quickly than other reagents, but it does shrink and harden the tissue much more than ethanol. Acetone is also used more in electron microscopy (EM) than simple ethanol, as it is more readily miscible with the plastic resins used in EM. In EM, the hardening of tissue is less of a problem, as the resin gives stronger support than wax and shrinkage is usually less as the fixatives used harden the tissues more than the usual fixatives for wax processing. Acetone will dissolve more materials from the tissues than ethanol and is more of a danger as it has a low flash point and is more toxic than ethanol.

Propylene oxide

The use of this is almost entirely restricted to EM and even there some workers prefer to avoid it as it has a very high toxicity compared with other dehydrating agents. The miscibility of propylene oxide with resins is its only real advantage over other agents.

Cellosolve

Cellosolve has the reputation of being gentler than other reagents. Tissues may be left in cellosolve for long periods without becoming excessively hard and it has been recommended for embryos. Generally, it is not popular because of its cost. It readily absorbs water from the air so careful storage is essential.

Ensuring complete dehydration

Complete dehydration is essential, as the clearing reagents and waxes are usually intolerant of even small amounts of water. To ensure complete removal of all of the water it is important to have at least two baths of 100% alcohol (three or more are often used). The first 100% alcohol bath rapidly becomes contaminated with small amounts of water carried over from the previous lower concentration (e.g. 95% alcohol) bath. The second bath effectively removes all of the water. It is still essential to replace the reagents regularly as there will always be a gradual transfer of water from one bath to the next. It is possible to use anhydrous copper sulphate to test for water in the alcohol baths. When the alcohol is dry the copper sulphate remains white but when water gets into the alcohol it will turn blue and the reagent needs to be changed. To save on reagent costs, it is common simply to move these reagents one step lower down the chain when they become slightly contaminated. Thus, only the final 100% alcohol is replaced by fresh reagent. The old final bath becomes the penultimate bath and so on.

Safety is very important with dehydrating agents as they are used in such large quantities. Alcohol has a relatively low toxicity, whilst propylene oxide is very toxic. For this reason many people are turning away from propylene oxide, even for EM. Fire risks are also important since most laboratories now use electrically driven machines and these may produce sparks. I know of at least two laboratories that have caught fire and burnt down from fires that have started in the processing rooms.

Clearing

This is the next step and involves the removal of the dehydrating agent, usually alcohol, and its replacement with a solvent miscible with wax. It was originally named thus because many of the reagents had a similar refractive index to the tissues and so cleared them optically (as xylene does). Optical clearing means that the tissues become transparent once they have been completely permeated by the reagent. Although most modern reagents do not optically clear the tissues, they are still usually referred to as clearing agents (perhaps because they 'clear' the alcohol out of the tissues). Other names have been suggested (e.g. antemedia or dealcoholizing reagents) but the term clearing is still the most common. For hand processing of tissues, there is still an advantage in optical clearing, as it allows the person processing the tissue to see how clearing is progressing; when the tissue is completely transparent, clearing is complete.

Clearing agents

Many organic solvents can act as clearing agents. Many different agents have been used and they all have their own advantages and problems (see *Table 4.3*). Several properties are crucial in evaluating the usefulness of reagents:

Table 4.3 **Properties of clearing agents**

Clearing agent	Boiling point (°C)	Refractive index	Toxicity	Flammability
Toluene	110.6	1.5	++	++
Xylene	138	1,5	+++	++
Chloroform	61.5	1.45	++	–
Cedarwood oil	Variable	1.5	0	Weak
Petroleum hydrocarbons	157	1.3–1.5	+	++
1,1,1-Trichloroethane	75	1.43	+	–

1. Speed. Two factors are included: firstly, the speed of removal of alcohol from the tissues, and secondly, the speed of evaporation of the clearing agent from the wax bath. The speed of removal from wax is related to the boiling point of the clearing agent.
2. Harshness of their action and how much shrinkage or hardening they cause.
3. Flammability.
4. Toxicity.
5. Cost.

Whereas alcohol is universally used as a routine dehydrating agent, there is still variation in which clearing agent is used routinely in different laboratories. Clearing agents are generally much more toxic than alcohol and, being volatile, they need to be treated with caution (see **Box 4.4**). It is generally toxicity that forces the change from one solvent to another rather than poor section quality. For many years, benzene was the clearing agent of choice in many laboratories, but the increased risk of leukaemia in people exposed to the vapour made it less popular for use and today it is rarely used.

Box 4.4 // Disposal of solvents

As well as being dangerous to use, clearing agents are also dangerous to dispose of. Many are combustible and can be burnt, but this requires very high temperatures and is not practicable for most laboratories. Passing the waste on to a specialist disposal firm is often the best alternative but does carry a high cost. This cost of disposal should be added to the purchase cost when considering which agent to use.

An alternative for large laboratories is recycling of the waste material. Specialist closed stills are available that will redistill solvents including alcohol and most clearing agents. The capital costs can be high, as they need a special room and careful monitoring to ensure safety. Providing the throughput is high enough, they can repay the costs in a couple of years. Recycling can be dangerous if picric acid is used, as it is explosive.

Not even very small amounts of clearing agents should be allowed to get into the normal waste disposal system, as they will not be effectively rinsed away and diluted. They are not miscible with water and will remain in the traps of sinks. Xylene tends to float, so a layer of xylene will remain floating in the trap and the fumes will be released. Constant exposure to small amounts of solvents will degrade the rubber and plastic components of the waste system and produce leaks.

Toluene (methyl benzene) was used as a replacement for benzene but again suffers from the problem of toxicity. It is similar in many respects to xylene but not as popular.

Xylene. Many people still use xylene, which has a lower toxicity than benzene but is still quite toxic. Xylene is used less in preparing blocks, as its harsh action leaves tissues more brittle and harder than many of the other clearing agents. However, it is still very popular for clearing sections after staining where the hardness and brittleness are less of a disadvantage.

Chlorinated solvents have also been used. Carbon tetrachloride (tetrachloromethane) was quite popular at one time, as it is almost non-flammable (it was used in fire extinguishers), but it has severe hepatotoxicity problems and is a cause of liver cancer. Chloroform (trichloromethane) has a similar toxicity but is still used by some laboratories. 1,1,1-Trichloroethane (used as a dry-cleaning fluid) has a lower toxicity; however, its use is now restricted as it adversely affects the ozone layer and is a substance of abuse.

Chlorinated hydrocarbons are usually denser and heavier than other clearing agents. This has two consequences. Firstly, and most obviously, they are heavier to lift and secondly, many tissues will float in them. This is worth remembering if a cassette holding tissues falls open in the processor, as you will need to look in the top of the reagent to retrieve it not at the bottom as you might expect.

Petrol hydrocarbons/aliphatic hydrocarbons are now being promoted as a safer alternative to xylene. They appear to be suitable for tissue processing but have problems as clearing agents before coverslipping, as they are less miscible with common mounting media.

Cedarwood oil. This is a slow-acting clearing agent, which has very little hardening effect on tissues. It is a very traditional clearing agent and was highly rated in the past, but is hardly ever used today because of its cost and slow speed of action. It has a distinctive odour that never seems to disappear completely from the wax, so the blocks have a faint but pleasant smell. Although it will burn, it does not readily catch fire and is not a real fire risk.

Many other reagents have been tried as clearing agents including simple petrol (inflammable but cheap), cooking oil (cheap and non-toxic) and essential oils such as lavender oil or sandalwood oil (very expensive but the laboratory smells lovely!), citrus oils (limonene reagents) from orange and lemon skin (they have a low toxicity but a very pungent smell).

Choice of clearing reagent

No one reagent has so far emerged as a clear winner. Each laboratory has different priorities. For some laboratories, price is more important than getting the very best quality, whereas others want perfection regardless of cost. Some laboratories use such small amounts of clearing agent that toxicity and flammability can be controlled and so this is less of a worry. Other laboratories use solvents on an industrial scale and need to consider safety in a different way.

The dehydrating agent must be removed completely from the tissue so three or more baths of clearing agent are needed to ensure complete replacement. Reagents need to be changed regularly as mentioned above for the alcohols. Again, the final bath can be reused as an earlier bath to save replacing all of the reagents at once.

Impregnation with molten wax

After clearing is complete, the tissues are passed through two to four baths of fresh molten wax. The wax is held at no more than 2–3°C above its melting point, rather than at much higher temperatures, as heat is very damaging to tissues.

Different waxes have different melting points. Usually a wax with a melting point in the range of 50–60°C is used. As a general rule, the higher the melting point of the wax, the harder the wax is at room temperature. Ideally, the hardness of the wax and the hardness of the tissue should be the same so that the texture of the wax and the block are even and when a section is cut, the knife moves smoothly through the block and does not 'jump' on encountering a different texture. In most laboratories a compromise is used, with only one hardness of wax being used for most specimens. The slight differences in texture between one tissue and the next are usually small enough to be ignored.

There are also specialist waxes that contain additives (such as plasticizers) or microcrystalline waxes to aid in the cutting. These are available ready prepared and are usually known by their commercial names. They often have excellent properties and have replaced not only simple paraffin wax but also some of the specialized embedding techniques (see **Chapter 5**).

Removal of clearing agents

The clearing agent needs to be completely removed from the final wax bath. If some persists in the block, it will later evaporate from the surface of the block. The block may initially cut quite well but after a few days or weeks the tissue will shrink, leaving the surface concave and below the surface of the wax.

The clearing agent is not just washed out of the tissue by the wax bath, it is also removed by evaporation. The wax bath is often at a high enough temperature to cause significant amounts of the clearing agent to 'boil off'. This evaporation is often the most sensitive way to tell when the final wax bath needs changing. If the final wax bath smells of clearing agent, then it needs changing.

Evaporation of the clearing agent and removal of air bubbles can be helped by using reduced pressure in the final wax bath(s). This is called '**vacuum embedding**', although a true vacuum is not needed, simply a significantly reduced pressure (about 50–75% of atmospheric pressure will suffice).

Blocking out of tissues

Once all of the clearing agent has been removed, the tissue is blocked out in a mould (see **Box 4.5**) containing fresh wax. It is oriented so that the correct surface will be cut during microtomy and the block is then cooled quickly. Rapid cooling prevents larger wax crystals forming and makes the block more homogeneous, improving the cutting quality of the block.

Once the tissue is embedded in the wax, it remains stable, so the next stages can be performed slowly, allowing careful preparation of fine sections. The block can also be stored for many years without deterioration. For unusual or important specimens, the blocks can be stored indefinitely and often prove a useful source of retrospective material for research. The use of archival tissue is subject to The Human Tissue Act 2004 (see **Chapter 20**).

Box 4.5 // Moulds for blocking out

Blocking out originally used L-shaped brass blocks (Leuckhart's angles), which could be moved to make a variable-sized mould. The wax rapidly cools and solidifies so a fluid tight seal is not needed.

After cooling, the blocks were trimmed with a scalpel or small knife and then attached to a fibre or wood block by melting the lower surface and pressing it on to the surface. It was easy to get the labels mixed up with so many separate steps.

Next came disposable moulds and the integrated cassette and mould system, which is now commonly used. The blocks are directly mounted on the cassette they were processed in and the cassette fits into a special holder in the microtome. There is less scope for a mix up, the blocks are dealt with more rapidly and throughput is greater.

4.3 Hand versus automatic processing

It is possible to process the tissues manually or using an automated schedule on an automatic tissue processor.

Hand processing

Tissues can be processed by hand using a series of containers with the various reagents in sequence. This requires somebody to transfer the tissues. It is tedious and time-consuming but very flexible (each tissue can be treated differently). It is generally slower since there is not constant agitation of the reagent. If left without agitation, the reagent diffusing out of the specimen will form a static stagnant layer around the tissue and slow down the processing.

Hand processing is cheap to implement if staff efficiency and labour costs are not significant, and is still used by laboratories with a very small throughput of blocks. Even large laboratories may occasionally use hand processing for unusual specimens where setting up an automatic processor for a single specimen would be impractical.

Automated processing

Automatic processing uses a machine to carry out all the changes and includes some method of programming so that the time in each reagent can be controlled. Times in the baths can be varied from a few minutes up to several hours. The machines are intended to run unsupervised so processing can be done overnight. The start can also be delayed so that the machine can run over a weekend without the specimens being left too long in the final reagent. The machines usually have facilities for thermostatically controlled wax baths and many can apply a partial vacuum to one or more steps. This allows the machine to carry out the complete processing schedule, taking the specimens from fixative right through to wax impregnation. Schedules can be changed quickly and easily and may be computer/microprocessor controlled. This makes processing very easy. Specimens are placed in the machine and can then be left unattended (see *Box 4.6*). Some schedules are arranged so that a specimen placed in the machine on an afternoon will have completed the processing schedule by the following morning and will be ready for blocking out into moulds when staff arrive. Large and/or fatty samples, e.g. breast or brain, can be run on

> **Box 4.6 // Fire risks of processing**
>
> The use of flammable liquids such as alcohols and hydrocarbons in processing makes it a fire risk. This is especially true once the system becomes automated. Automatic processors are electric and have electrically heated wax baths. A spark from an electric motor or thermostat can easily result in a major fire. To prevent this, most modern machines are completely enclosed and the vapours are prevented from escaping by always using a partial vacuum so that the air flow is into the chamber. The reagents are pumped into and out of this evacuated chamber and stored in sealed containers when not in use. Older machines, which used relatively open baths of reagents, were always a fire risk. Some laboratories were destroyed by fire, which also meant the loss of irreplaceable diagnostic tissue samples.

extended schedules (e.g. 48 hours) while small urgent biopsy cores can be processed on urgent 3 hour schedules. Paraffin processing often looks complex when the full schedule is written down, but is extremely easy with a machine to do the tedious work.

Many specimens (up to hundreds in large machines) are processed simultaneously. Each specimen is placed in a labelled cassette to keep it separate from all of the others and a batch of cassettes is placed in the machine. The actual structure and mechanism of the machines differ. In some, the reagents are in large beakers and the specimens are held in a basket, which is then moved from one beaker to the next. In other machines, there is only one bath into which the specimens are placed and the reagents are pumped into and out of the chamber. This allows a sealed system and a vacuum can be applied to the bath. Both types have similar applications; only the cost and number of specimens is different.

Machine processing has become much faster with advances in technology. Machines using different reagents, either isopropanol or specially blended reagents, and with the application of heat mean that small pieces of tissue can be processed rapidly. The total processing times can be reduced, from the traditional overnight 14 hour cycle, by a factor of ten to little more than an hour. Most of the improvement is achieved in the dehydrating and clearing stages of the processing. This increase in speed has been further aided in some machines either by using microwaves to heat the reagent or ultrasonic vibrations to agitate the baths more effectively in the closed container. The reagents are continuously microwaved, with an energy intensity that is only a fraction (10%) of the energy intensity of a domestic microwave oven, and this raises the temperature to about 50°C. Although machines greatly reduce the exposure of staff to reagent fumes, many of the manufacturers have optimized their systems to use xylene-free processing to further reduce the risk.

Automatic processing is generally faster, more reliable and can be done overnight. It is less flexible since all the specimens in one batch get the same times regardless of size, etc. Some newer machines are incorporating microwave treatment which speeds up the process, allowing blocks to be fully processed in a few hours. If the laboratory is large, then it may be able to justify several automatic processors set up with a variety of schedules so individual blocks can get more specialized treatment.

4.4 Decalcification

Tissues such as bone and teeth are strengthened and hardened by the addition of the mineral hydroxyapatite, a calcium phosphate salt, with smaller amounts of magnesium, potassium, chlorides and carbonates. The mineral content of the tissue is about 70% in bone, about 97% in dentine and over 99% in tooth enamel. Hydroxyapatite is hard and makes routine sectioning difficult. The tissue can be softened by removing the mineral; this process is usually called decalcification, although it also removes other materials in addition to calcium.

Decalcification with acid solutions

Hydroxyapatite is soluble at low pH and can be removed by treating the tissues with acid:

$$Ca_{10}(PO_4)_6(OH)_2 = 10Ca^{2+} + 6PO_4^{3-} + 2OH^-$$

The mineral is usually in equilibrium with the body fluids, but if the pH is lowered, then the excess hydrogen ions combine with the hydroxyl ions to give water. This removal of the hydroxyl ions drives the reaction to the right, giving a net result of:

$$Ca_{10}(PO_4)_6(OH)_2 + 20H^+ \rightarrow 10Ca^{2+} + 6H_3PO_4 + 2H_2O$$

The calcium ions will form salts with whatever acid is being used (e.g. calcium chloride with hydrochloric acid).

Box 4.7 // Decalcification of teeth

The very high mineral content of enamel makes it almost impossible to decalcify teeth and still retain the enamel layer. The enamel has almost no organic content to hold it together, so once the mineral disappears the enamel layer becomes very fragile and falls apart. Enamel and dentine can also be decalcified *in situ* in the mouth by the acids produced by bacteria. Teeth are also attacked by highly acidic drinks such as lemon juice or cola. Cola drinks contain phosphoric acid and may have a pH of around 4. This is quite low enough to be used as a decalcifying solution.

Soft tissue damage caused by acid solutions

Mineral acids such as hydrochloric acid and nitric acid are generally faster in action than organic acids but cause more **maceration** of the tissues. Maceration involves swelling of tissues, loss of staining ability and generally poorer morphology. Maceration is the result of the hydrogen ions causing hydrolysis of proteins and nucleic acids and is worse if the tissue is inadequately fixed. Although thorough fixation minimizes the effects of maceration, it cannot totally prevent some deterioration. Maceration is also increased by leaving tissues in acid for longer than necessary or by using higher temperatures to speed up decalcification. Decalcification is always a compromise between speed and quality of morphology.

Acid decalcifying solutions

Different acid decalcifying solutions give a choice between quality and speed. Acids that are used in practice include:

Nitric acid (e.g. Perenyi's fluid). This is very fast acting but causes severe maceration if used for more than a few hours. It is only suitable for very urgent cases. Old stocks of nitric acid often turn a yellowish colour and this exaggerates the disruption and also stains the tissue. The addition of urea is sometimes recommended to help prevent the unwanted disruption and discoloration.

Hydrochloric acid has been used and is often a part of commercial or proprietary decalcifying solutions. Although it works well, it does cause maceration, although not as much as nitric acid. It should not be mixed with formaldehyde as it can produce a carcinogen.

Formic acid is probably the most common decalcifying acid and is used at a concentration of between 5 and 30%. It is significantly slower than the mineral acids but causes less damage.

Acetic acid and peracetic acid can also be used but are even slower than formic acid. However, they do not adversely affect nuclear staining.

Some acids are unsuitable as decalcifying agents. Sulphuric acid, for example, forms insoluble calcium sulphate, which is extremely difficult to remove.

Chelating agents as non-acidic decalcifying solutions

An alternative to using acids is to use a chelating agent. These decalcify by continuously removing the small amount of free ionic calcium that is always present around the bone mineral. This removal acts in the same way as the removal of hydroxide ions drives the equilibrium to the right in the equation given above. However, chelating agents have a much slower action. The main chelating agent used is EDTA (ethylenediaminetetraacetic acid). Since this reaction can occur at neutral pH, there is no damage to fixed tissue and preservation is excellent, although decalcification is very slow (it may take months to completely decalcify larger pieces of bone or teeth).

Acceleration of decalcification

Many attempts have been made to speed up decalcification, but some methods also accelerate maceration. The most effective action is regular replacement of the decalcifying solution. This replenishes the acidity and speeds up decalcification. Simple agitation to prevent the layer of fluid around the specimen becoming saturated with calcium is also very effective. Constant agitation is not required; an occasional shake seems to work well, although any gentle automatic agitation will be beneficial.

Heating the solution certainly speeds up the reaction but also speeds up the maceration reactions, so gives no real benefit. The use of electrical currents (electrolysis) and ultrasonic treatment seem to speed up decalcification, but this seems to be mainly due to their heating effects rather than any more subtle effect. If the solutions are kept cool, then there is no real increase in speed.

By using an organic acid in a buffered state, it is possible to keep the hydrogen ion concentration more uniform. As the hydroxide ions neutralize the hydrogen ions in an unbuffered solution, the rate of decalcification falls as the pH rises. In a buffer, the hydrogen ions are replaced at the same rate as they are removed, so decalcification is more evenly paced.

The inclusion of an ion-exchange resin in the solution has been used to try and give the same replenishing effect. This seems to be rather more tedious to use than simply buffering the solution since the resin needs to be regenerated after use. Ion-exchange resins can only be used with organic acids and they prevent the use of the chemical end-point test (see below).

Buffering is advantageous, and one of the best decalcifying agents is a mixture of formic acid and trisodium citrate. The citrate acts as a buffer salt and also as a chelating agent. The pH is quite moderate and causes minimal maceration, but it is still quite a slow decalcifying solution.

Decalcification in fixatives

Acids must of course be avoided if the bone is to be examined as an undecalcified specimen (see *Box 4.8*). Sources of acid in processing include acetic and peracetic acids, picric acid and unbuffered formaldehyde solutions (these may contain formic acid). Acetic acid- and picric acid-containing fixatives will decalcify very small fragments of bone during fixation. This can be useful for bone marrow specimens where there may be a few tiny fragments of bone present but it is not worthwhile doing a full decalcification. Fixation in an acetic acid-containing fixative softens the tiny fragments allowing the tissue to be cut without delaying the specimen.

Box 4.8 // Undecalcified sections

Undecalcified sections are needed when the actual process of bone mineralization is being investigated. This occurs in bone diseases such as osteoporosis. There is no theoretical difference between undecalcified bone sections and soft tissue sectioning, but the practical problems and difficulties are much greater with the rigid and brittle mineralized tissues. They can also be prepared by the totally different method of ground sections. In this technique, rather than removing a section from a block by microtomy, the thin section is prepared by removing all of the unwanted block by grinding it away. One side of the specimen is ground flat and then attached to a slide. The exposed side of the block is then ground away until the required thinness is achieved. This is quite wasteful of tissue as only one section is prepared from each block. Machines to grind sections automatically are available as this is the standard way of making geological slides from rock samples. In fact, geologists consider bone to be very soft rather than hard. I suppose it is if you compare it with granite.

Accurate determination of the end point of decalcification

The 'end point' of decalcification is the moment when the last calcium deposit is removed leaving only the soft tissues. The end point needs to be determined accurately so that the specimen is removed from the destructive acid bath as soon as possible. The end point can be determined by several methods.

X-rays

This is probably the most accurate method but is inconvenient for most laboratories. Tissues are placed on top of an X-ray film (held in a light-tight cassette) and exposed to sufficient X-rays to make the film very dark but not quite completely black. Any calcium deposits show as white areas on the developed X-ray film, whilst soft tissues appear as a faint grey background. This can be expensive as it requires X-ray films. It can be used to show how decalcification is proceeding, as well as indicating when it is complete.

Chemical testing

A sample of the decalcifying fluid is taken and neutralized with hydroxide solution. Adding ammonium oxalate to this neutralized solution will show whether calcium is present by producing a precipitate, which appears slightly hazy if viewed against a dark background. If calcium is detected, this means that decalcification is still proceeding and the decalcifying fluid should be changed and retested a few hours later. Once there is no longer a faint precipitate, it can be assumed that decalcification is complete.

Manual testing

This includes testing by bending or poking with a needle (NOT recommended, but see **Box 4.9**). Although it is possible to detect calcified tissue by its rigidity, this method can cause damage to the tissue, especially if it is repeated many times.

Box 4.9 // Using a spare piece of bone to test for the end point

A slightly better method of manual testing by poking is to set up a decalcification using a test piece of bone the same size as the specimen and decalcify it in parallel with the specimen. This piece of bone can be bent, needled and even cut to see if it is completely decalcified. When the test sample is decalcified the other piece is assumed to be complete. This method is not particularly accurate as the density of the bones may not be identical, but it is useful as a guide.

Washing

Following decalcification the block should be washed free of acid in either a sodium sulphate solution or alcohol (simply washing in water causes further swelling) and the tissue can then be processed as for soft tissue.

Surface decalcification

If a wax-embedded block is found to have small calcified fragments at the time of cutting then the surface of the block can be decalcified. Surface decalcification is achieved with a tissue soaked in 1% HCl held on the surface of the block.

4.5 Microtomy

The wax blocks can be cut using a microtome. This is a finely engineered machine capable of slicing sections regularly at 5 μm or less. There are many types of microtome from small hand-held ones to extremely large ones that can fill a small room. They all have certain characteristics in common:

1. A mechanism for holding the block firmly. The block is usually attached to a holder or chuck and this is clamped in a jaw mechanism.
2. A knife holder.
3. A mechanism to move the knife across the surface of the block to cut the section.
4. A feed mechanism to advance the block by a predetermined amount so that each subsequent section can be cut at the required thickness.

The engineering ingenuity that has gone into achieving these characteristics is quite remarkable and most microtomists have their own favourite type of microtome.

Some microtomes are better for particular purposes. The factors that make one microtome better than another, or at least better for one purpose, depend mainly on the specimen and what is required from it. Considerations include:

1. How big is the block? The bigger the block, the larger the microtome needed, but large microtomes can be inconvenient for small specimens.

2.   How hard is the block? Very hard blocks of bone or tooth need a specialized microtome that is engineered for improved strength and rigidity, but they are usually somewhat tedious to use for soft or small specimens.

3.   Are serial sections needed? Serial sections are where several adjacent sections are cut in order to follow structures, e.g. nerves, through the tissue or where exactly adjacent sections are used to apply more than one staining technique to the one specimen. A microtome that can produce ribbons is best for such work.

4.   Cost.

Microtome types

Hand microtome

These are very simple with a small flat circular table around a central hollow pillar. The specimen is placed in the centre of the pillar and can be inched gradually up above the table. (It should be 'micrometred' up but that doesn't sound right in English.) By slicing across, using the table as a guide plate, thin sections can be cut. These are not suitable for serious histology and are mainly used for relatively thick sections of botanical specimens.

Rocking microtome

These are quite small microtomes with two rocking arms. One arm gives the cutting action, whilst the other acts as a feed mechanism. These small microtomes were once common but have largely disappeared from laboratory use. They are still used as 'student microtomes' as they are cheap and robust. They are limited to cutting small blocks and soft tissue since the actual cutting action is done by a spring. They were also very popular in cryostats (see **Chapter 6**) as they have no sliding parts (the only movement is a rocking movement), so they had no problems with lubrication at the low temperatures used in a cryostat (even thin oils become sticky at –20°C).

Rotary microtomes

These are more robust microtomes with a rotating handle. They are very popular in laboratories and come in a variety of sizes. They can produce nice ribbons of sections and are quite reliable if well maintained, but can be prone to wear if badly maintained, when they then become quite erratic in use. Once you are familiar with them, they can be quick and easy to use. They are not quite as good as a sliding microtome for very large or very hard blocks. Motorized versions are available for producing high quality thin sections of resin-embedded material, e.g. bone marrow trephine cores.

Sliding microtomes

These are usually much heavier and larger machines than the other types of microtome. The cutting action is by sliding a heavy 'sledge' under a knife. They can cut very large and hard blocks, but are heavier to use as a consequence. They are quite popular and in some laboratories are used to cut all specimens, even small soft ones.

Microtome knives

The knife used in a microtome is equally important. It must be very sharp and the edge must be free from blemishes. Knives can be solid and need sharpening or they can be disposable. Solid knives can be sharpened by hand, or alternatively by machines.

Sharpening machines now use a soft metal plate into which a diamond paste is rubbed. The plate becomes studded with very fine diamond particles and the knife is sharpened

> **Box 4.10 // Microtomes**
>
> There is nothing that brings out the argumentative side of histologists more than suggesting that the microtome or knife they are using is inferior. The relative benefits and shortcomings of these machines can be discussed at great length.
>
> In reality, the only criterion is whether it produces a good section. Most microtomes in the hands of a good microtomist will cut good sections from most blocks as long as the tissue sample has been adequately fixed, processed and embedded. Only with the very hard or very large blocks does the choice of microtome become crucial.
>
> Microtomy has improved greatly over the years and sections are now routinely cut much thinner than a few years ago. The advent of the disposable knife has been a major factor in this improvement.
>
> There are a few unusual types of microtome in some specialist laboratories. The Tetrander and Jung K microtomes were produced for very large and very hard blocks, respectively, and would not look out of place in a heavy machine shop, as they are pretty substantial instruments.

on a thin layer of diamond, which is regularly renewed; thus, the plate itself does not wear away. This requires the use of several plates, however, as each plate represents one grade of abrasive and you cannot mix coarse and fine abrasives on the same plate.

Disposable knives have become almost universal in routine histology. They involve less capital cost (for example, no knife-sharpening equipment is needed) and a fresh knife is always available. The time spent not re-sharpening blunt or damaged knives is also a cost saving. One of the horrors for a microtomist is when all of his knives are damaged at the same time and he is stuck until they are resharpened. Disposable knives possibly cost more in day-to-day running expenses, but the ease of use and the fact that it does not matter if the knife gets damaged usually outweigh this in a busy laboratory where urgency is paramount. A knife can be expected to cut about 20 blocks depending on the hardness of the specimen and the number of sections needed from each block.

The blades themselves are thin and flexible but are held in blade holders which support the blades leaving the cutting edge free. The base of the holder resembles the non-disposable knife base so that disposable and non-disposable blades can be used in the same microtome with no adaptation required.

Much of the technology to produce high quality disposable blades has been developed for razor blades and then adapted for the histological market. The cutting edges of blades are often coated to reduce friction making sectioning smoother and more consistent. Other coatings such as platinum sputtering can increase hardness and prolong the working life of a blade or make it possible to cut harder materials. Different grades of blade are available to give the best results for different specimens and requirements. Some are harder for harder tissues, some allow even thinner sectioning, some have a longer edge for bigger blocks and some are better for cryostats.

Disposable knives are equally effective when cutting FFPE or frozen material in a cryostat with section thicknesses approaching 2–3 μm achievable on well-processed material such as lymph node.

For very hard specimens, special knives may be required. These can be made of extra-hard materials, usually tungsten, and are not disposable. Also, very large specimens are not suitably catered for by the manufacturers of disposable knives, so non-disposable knives are needed for very large and very hard blocks. This is also true for ultra-size blocks in plastic cassettes which are best sectioned using a sledge microtome.

Glass knives can be used to produce sections from resin-embedded material, e.g. the Ralph knife is used to produce 1–2 µm sections of glycol methacrylate embedded samples. Commonly these would be highly cellular, e.g. lymph node or bone marrow trephine, where good nuclear detail is needed from a crowded field of view.

Cutting the section

Cutting is not automatic; it needs skill and sometimes can be difficult. It is a practical skill and is best learnt by actually doing it. Although full automation of sectioning is not yet available, many microtomes can be motorized to make sectioning smoother than by hand and thus minimizing 'chatter'.

Problems are usually caused by:

1. Poor impregnation. This means that the tissue is inadequately supported and will crumble or fold instead of cutting nicely. Poor support can also result from using wax of the wrong temperature (usually too soft) or if the room is too hot causing the wax to soften. Cooling the block with a block of ice may help.
2. A blunt knife. This is usually quite easy to spot in practice.
3. The wrong knife angle. This is probably one of the few things that can be tried quickly during cutting but once the angle is right it will not usually need changing between blocks.
4. Slackness in the microtome. This results in alternate thick and thin sections or in chattering, depending on how much slackness or unwanted movement there is.

N.B. Sections will always vary slightly in thickness depending on the speed of cutting. A faster transit across the knife edge will increase the compression of the block and the relaxing wax may then produce a thicker section on the next pass.

There are many tricks and dodges, which are learnt with practice (e.g. see **Box 4.11**) and experience, and no amount of 'book learning' will ever replace simply sitting at a microtome and cutting.

Sections can be cut at different thickness. Thick sections are useful for low-power microscopy, for tracing long structures such as nerves as they travel through tissues and for some histochemical techniques. Thin sections are needed for fine intracellular detail and high-power microscopy (especially in photomicrography). For routine wax sections, a thickness of 3–6 µm is usual.

Box 4.11 // Use of Sellotape in cutting

To help cut difficult specimens, it is possible to give extra support by covering the surface of the block with a material such as Sellotape or specialized films that polymerize. The Sellotape is pressed firmly on to the face of the block and as the section is cut, the Sellotape and section are lifted off. The sections are then attached to slides using an adhesive but without being floated on to a waterbath. The Sellotape is usually removed using a solvent to dissolve the glue before the section is stained. This method is used for cutting undecalcified bone sections.

A solution of nitrocellulose in alcohol can also be used by painting it on to the block face. This leaves a thin layer of nitrocellulose on the block. This gives less support than Sellotape but is easier to remove.

Floating and mounting

All microtomes have a chopping rather than a cutting/slicing action and this has a tendency to compress the section in the direction of cutting. Often this will result in wrinkles or folds in the section. To get rid of the wrinkles, the section is floated on to a bath of warm water. This does not melt the wax but softens it sufficiently for the tissues to recover their original shape and size. Gentle prodding with a soft brush can also help to flatten the section. For stubborn wrinkles, it is sometimes useful to float the section on to a few drops of alcohol on a slide first and then float it on to the warm water. The turbulence produced as the alcohol mixes with the warm water may sometimes gently ease out the wrinkles.

Creases in large curved structures, e.g. blood vessels and heart valves, can be eased by cutting out excess wax from the greater and lesser curve using a scalpel before the section is floated onto the waterbath. Sections can then be lifted on to a clean slide. The slide is held in the waterbath vertically and the section lifted on to it by withdrawing the slide still held vertically. This works much better than trying to slip the slide horizontally under the section and then trap it on the slide. Withdrawing the slide vertically also traps less water and is less likely to trap air bubbles under the section. It should be noted that there is a smooth (shiny) side and a dull side to any section. The shiny side is the surface that has contacted the knife edge during sectioning and, being smoother, should always be treated as the 'base' of the section as it will adhere more effectively to the surface of the slide.

If a ribbon of sections has been cut, then either individual sections or a short length of the ribbon containing several sections can be cut from the ribbon either before floating out or on the surface of the waterbath and then picked up as for single sections. The slide should then be labelled. Some slides have ground glass areas that will accept a soft pencil writing on them but plain slides should be labelled by scratching with a diamond-tipped stylus. Ordinary felt-tip pens, paper labels and most forms of ink will not survive the staining and mounting process and the labelling will be lost.

Once picked up on to a slide, the sections are drained well and then dried on to the slide to achieve adhesion. This can be done at any temperature from 37°C to temperatures high enough to melt the wax. Lower temperatures cause less damage to the tissue. If the wax melts, then the tissue may disintegrate completely. However, at 37°C many hours may be needed for complete drying. If drying is incomplete, the sections will detach during staining. Drying can be done in an oven or on a hotplate. The oven method is somewhat slower than using the same temperature with a hotplate, but there is less contamination from dust settling on the slide if it is dried vertically in an oven rather than horizontally on a hotplate on an open bench.

Drying for a few hours or overnight in an oven is preferred for slides intended for immunostaining as is the use of adhesives (see below).

If there are problems with sections adhering or if the staining method is particularly harsh and liable to remove the section, adhesion can be improved by coating the slides. Special slides coated with silane (3-aminopropyltriethoxysilane) are available; these have a surface charge, which helps the section cling to the slide. It is possible to coat ordinary slides with 2% silane in acetone and then drain and dry them, but this reagent is quite dangerous.

It is also possible to use chemical adhesives in the form of protein, such as egg albumen, serum, gelatine or poly-L-lysine, which are used to coat the slides prior to picking up the sections. Although these do help adhesion, they may stain with some techniques, causing background problems. One commonly used approach is to coat pre-cleaned slides by processing 50 at a time in a glass dish containing a 1% albumen solution and then set the albumen with a surface charge by dipping in a chrome alum solution.

Suggested further reading

Bancroft, J., Layton, C. and Suvarna, S. (2012) Chapter 6: Tissue processing and Chapter 7: Microtomy. In: *Bancroft's Theory and Practice of Histological Technique*, 7th edn. Elsevier Health Sciences.

Culling, C.F.A. (1975) *Handbook of Histopathological and Histochemical Techniques*. London: Butterworths.

Suggested website

http://library.med.utah.edu/WebPath/webpath.html

Self-assessment questions

1. Why are tissues commonly embedded in wax for section cutting?
2. Name, in order, the steps needed to prepare a paraffin wax block.
3. What is a 'graded series' of alcohol and what is its purpose? Name three alternatives to alcohol.
4. How did clearing agents get their name?
5. Name four clearing agents and compare their advantages and problems.
6. What is meant by vacuum embedding?
7. How can mineralized bone be softened for sectioning?
8. Outline how undecalcified bone sections can be prepared.
9. A piece of bone marrow is fixed in a mixture of methanol and acetic acid. Sections of the bone fragments show no sign of calcium. Does this indicate that the patient had osteomalacia (calcium deficiency)?
10. Outline the advantages of automated processing of tissues.

Other embedding techniques

Learning objectives

After studying this chapter you should confidently be able to:

- **Name an alternative to paraffin wax embedding and outline the advantages of the medium**
  Plastics (resins) give firmer support and allow thinner sections to be cut.

- **Briefly describe the preparation of plastic (resin)-embedded sections in light and electron microscopy**
  Processing tissues involves dehydration to remove water but clearing is not always needed. Tissues are impregnated in a fluid plastic monomer, which is then hardened by polymerization. The polymer is not always easy to remove and often remains there in the final section. Sections are commonly cut using glass or diamond knives.

- **Compare the use of methacrylate and epoxy resins**
  Methacrylate resins can either be removed from tissues or are permeable to dyes and so stain similarly to wax sections. Methacrylates are therefore popular for light microscopy where very thin sections are needed or for very hard tissues. They are not stable in the electron microscope beam and are not used for electron microscopy. Epoxy resins are more stable and are the routine embedding materials for electron microscopy. They are more difficult to stain, so are less popular than methacrylates for light microscopy.

- **Discuss alternative embedding materials**
  Ester, polyester and water-soluble waxes have been developed but their cost and difficulty in use have restricted their popularity.

Although paraffin wax embedding is the commonest method of supporting tissues for sectioning, there is a variety of alternative embedding media that can be used. Some of these are of mainly historical interest, whilst others are more recent and are now used routinely. Even the 'historical' methods have a surprising durability, with most of them still being used for specialized purposes and they are still useful to overcome some modern research problems. They range from the use of very hard plastics and resins to produce very thin sections, through to wax-like materials with some quite different properties to paraffin wax, to the soft and pliable celloidin sections.

5.1 Plastic or resin embedding

Plastics are very widely used as embedding media in histology. They are the main embedding method used in electron microscopy and have many applications in light microscopy. The terms plastic and resin tend to be used somewhat loosely in histology. To a chemist not all resins are plastics and not all plastics are resins but for most histologists the terms resin embedding and plastic embedding are simply alternatives.

Materials such as epoxy and methacrylate plastic are usually very much harder than waxes and so give much better support to the tissues. This is better for hard materials such as undecalcified bone and allows much thinner sections to be prepared from soft tissues. Sections may be thinner than 90–100 nm (about one-fiftieth of the thickness of paraffin sections). This thinness is essential for electron microscopy and can be very useful for light microscopy as there is less confusion from overlapping cells and organelles. One slight problem is the loss of staining contrast in the sections due to their thinness. This means that the results of staining must be made more intense and so many staining techniques must be changed for plastic sections. The traditional haematoxylin and eosin is often replaced with toluidine blue to stain the nucleus blue and give a stronger cytoplasmic stain.

Plastics are quite unlike waxes in one major property. Waxes are converted from a solid to a liquid form by melting and will solidify again on cooling. Thus, tissues are impregnated in molten wax at about 60°C and converted to a solid block simply by cooling. The block can be gently remelted and reblocked without damaging the tissue. Plastics, however, are solidified by chemically converting them from a liquid monomer to a solid polymer by polymerization. The efficiency and strength of the chemical cross-linking achieved during this curing process cannot be reversed which makes re-blocking extremely difficult. The best option is reprocessing after the complete removal of polymerized resin which is at best tedious and time consuming, and at worst impossible.

The monomers of the plastics are often miscible with one or more dehydrating agents, particularly acetone and propylene oxide, so that clearing agents such as xylene are not often needed. Some monomers are even miscible with water and the tissue can be impregnated with the monomer directly from aqueous solutions without any need to dehydrate. The miscibility of the monomer and the solubility of the polymer are, however, totally different and once converted to a polymer they are often insoluble. Thus, for example, glycol methacrylate monomer is miscible with water but once polymerized to the solid polymer, the plastic is not only no longer soluble in water but is also insoluble in most other common solvents.

Methacrylates

These are related to Perspex and were the first plastic materials to be used for tissue embedding. They were introduced into histology during the early 1950s and have been used in various forms ever since (see **Box 5.1**). The early methacrylates were the butyl and methyl methacrylates and later glycol methacrylate was used (2-hydroxyethyl methacrylate), but since the 1980s, aromatic polyhydroxy dimethacrylate resins have been available.

$$
\begin{array}{lll}
CH_3 & R = CH_3 & \text{Methyl methacrylate} \\
| & & \\
CH_2{=}C & R = (CH_2)_3CH_3 & \text{Butyl methacrylate} \\
| & & \\
COOR & R = (CH_2)_2OH & \text{Glycol methacrylate}
\end{array}
$$

The tissue requires dehydration but not clearing since the monomers are miscible with alcohol and acetone. The dehydrating agent must be completely removed by several changes of monomer before attempting to polymerize the block. Acetone is more of a problem than alcohol for interfering with polymerization and so alcohol is recommended.

Methacrylates polymerize by chain elongation using an active radical. Adding a catalyst such as benzoyl peroxide produces active radicals. This molecule is highly unstable and is

> **Box 5.1 // Plastic embedding for museum specimens**
>
> Methacrylates can also be used for museum specimens since they can be produced in quite large blocks and are crystal-clear. The specimen is dehydrated and impregnated as usual, although with large specimens the impregnation time is extended. Polymerization must be slow. The block is then trimmed and polished. These embedded specimens are good for teaching as they are less easily damaged than the conventional glass jars filled with fluid. They can get scratched or slightly chipped but do not shatter if dropped. The occasional polish to remove scratches is all that is needed.

explosive but can be obtained as a more stable paste mixed with dibutyl phthalate. When the molecule is heated, it can split to provide two stable active radicals (see **Fig. 5.1**).

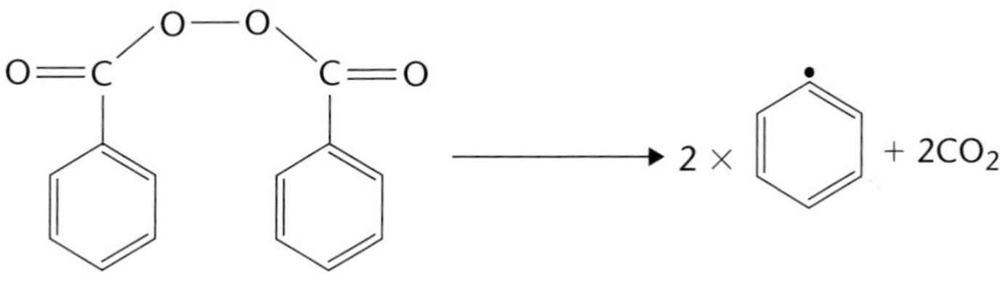

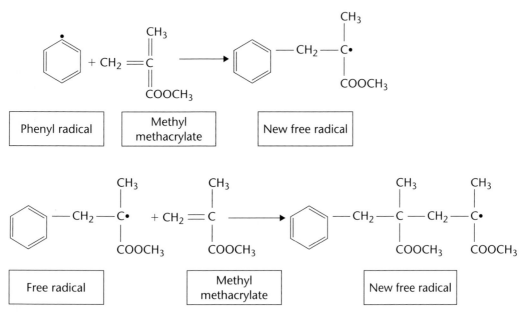

Figure 5.1
Splitting of benzoyl peroxide to give two phenyl radicals

The active radical then adds on to a monomer forming a new larger active radical, which in turn adds on to a monomer and hence the chain elongates (see **Fig. 5.2**). The chain elongation will continue until either there are no monomers to react or the free radical is inactivated by a free radical scavenger such as oxygen or acetone.

Figure 5.2
Polymerization of methacrylates

Polymerization of the resin is better if oxygen is removed by bubbling nitrogen gas through the liquid monomer just before use. If the block is not polymerized in a sealed mould, then the upper part of the block may not polymerize fully due to the inhibitory action of atmospheric oxygen. Some extra monomer should be used so that this upper soft layer can be removed without harming the embedded tissue.

Rapid high-temperature curing will result in more free radicals and the block will suffer, since the polymerized chains will be shorter. A slower polymerization will give a better result. This is difficult since the reaction needs heat to start it off and once started it is exothermic and will get hotter and hotter. Thus, it is a delicate problem to heat the block enough to start the polymerization and then cool it to remove the heat and slow down the polymerization.

Problems with methyl and butyl methacrylates

Methacrylates have significant problems in use. First, they are often sold in an impure form with quinol materials added to help to stabilize them. These can be removed by extracting the acrylic monomer with sodium hydroxide. The quinol dissolves in the sodium hydroxide solution, which can be discarded. The resin is then washed free of the hydroxide and finally dried with calcium chloride.

Chemically linking the monomer is a strongly exothermic reaction and will get very hot during polymerization. To minimize damage to the tissues, it is better to do some of the initial polymerization before impregnation. Thus, pre-polymerized monomer is made by starting the polymerization and then cooling the partially polymerized material to 0°C. An alternative is to use some low molecular weight beads dissolved in the monomer. If the heat is not controlled, then the heat may get so intense that the monomer may vaporize, causing cavities in the block.

Glycol methacrylate

Glycol methacrylate (2-hydroxyethyl methacrylate) is a useful addition to the range of resins since the monomer is hydrophilic and complete removal of water is not essential before impregnating with the monomer. The tissue is often only dehydrated to 90% alcohol before being impregnated with the glycol methacrylate monomer. Once completely impregnated with monomer, which can take two or three changes, it can then be polymerized. Some forms of glycol methacrylate are contaminated with methacrylic acid, which can take up stains, and this can cause the problem of a heavy background colour. Low-acidity monomers are preferable. Although the monomer is water-soluble, the cured resin is insoluble in most reagents and is not easily removed.

Aromatic polyhydroxy methacrylates

Aromatic polyhydroxy methacrylates have been available since the 1980s and have the hydrophilic characteristics of glycol methacrylate but are more stable in the electron microscope beam. These resins can also be cured by exposure to UV light and can be cured at very low temperature, allowing some enzyme histochemistry or immunocytochemistry to be performed after sectioning.

Cutting and staining for methacrylates

Acrylic resins can be cut with ordinary steel or disposable knives but better results are usually obtained if they are sectioned using glass knives (see **Chapter 19**) for thin sections of soft tissue or tungsten knives for sections of hard, undecalcified tissues. Sections can be stained as free-floating sections without removing the plastic or they can be attached to slides using an adhesive and the methacrylate then softened or removed.

Methyl and butyl methacrylate can be extracted from sections using xylene and so they can be stained using a wide variety of techniques similar to those used for paraffin sections. These methacrylates are no longer popular for electron microscopy because the resins disintegrate slowly in the electron beam, but they are still very useful for light microscopy and especially when working with hard tissues such as undecalcified bone and teeth. They are excellent for light microscopy as the resin is easily permeated by aqueous dyes and in the case of methyl methacrylate can even be removed by soaking in xylene. (The epoxy resins, which are preferred for electron microscopy, are more difficult to stain and impossible to remove.)

Glycol methacrylate is more difficult to remove and tissues are always stained with the resin still in place. The glycol resin, however, remains hydrophilic and allows penetration of many aqueous dyes, so staining is still fairly easy. It is this flexibility of staining with dyes that has meant that acrylic resins have retained their popularity for light microscopy, whilst electron microscopists have discarded them.

Epoxy resins

These are commercially produced resins and their exact chemical nature is often a commercial secret. They include Araldite, Epon and Spurr resins. They are generally much more viscous than the acrylic resins but have the advantage of causing very little shrinkage and do not produce damaging heat during polymerization.

The epoxy resins polymerize by a different mechanism involving the cross-linking of two different components and are more likely to produce cross-links between the extending polymer chains. Instead of using just a monomer and a small amount of catalyst, they need two major components that interact to form the polymer (see *Fig. 5.3*). Other materials can be added as accelerators and plasticizers such as dibutyl phthalate.

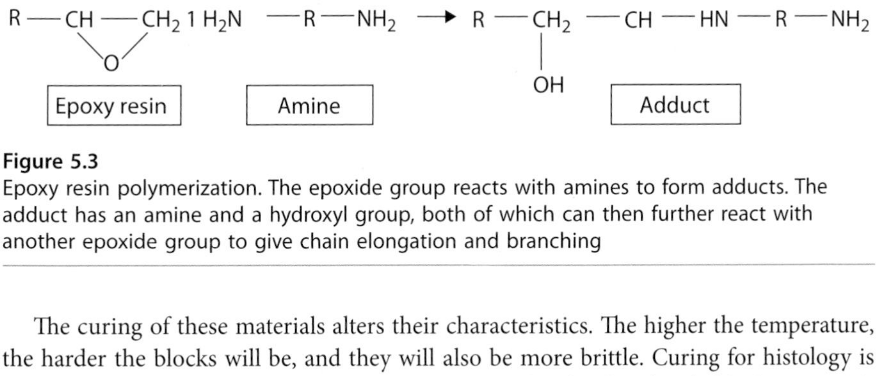

Figure 5.3
Epoxy resin polymerization. The epoxide group reacts with amines to form adducts. The adduct has an amine and a hydroxyl group, both of which can then further react with another epoxide group to give chain elongation and branching

The curing of these materials alters their characteristics. The higher the temperature, the harder the blocks will be, and they will also be more brittle. Curing for histology is usually done at 60°C for 12–24 h, which gives a tougher but less-brittle block. At 120°C, the block will be cured in 1 h but will be hard and brittle.

Since the components of the resin can bond to amine groups, there is the possibility of the plastic bonding directly on to the proteins and nucleic acids in the tissues. The tissue becomes permanently attached to the resin and the degree of cross-linking means that, especially for small blocks, the whole block is essentially one large extended molecule including the tissues. The result of this is that staining is inhibited and antigens are blocked, so staining with dyes and immunohistochemical techniques are poor. It is almost impossible to remove the resin without destroying the tissue in the process. Some improvement can be obtained by breaking down some of the cross-links or etching the block with sodium hydroxide in alcohol. Staining is still reduced compared with other embedding techniques.

Epoxy resins can include some very toxic materials and should be treated with care. They are capable of linking on to the tissues in human skin and acting antigenically as a hapten group. This slightly alters the nature of the proteins in skin so that the immune system will treat them as foreign proteins and produce an immune reaction. This results in an allergic hypersensitivity to the resin materials. They should only be used inside a fume cupboard and gloves are needed when the monomers are being handled.

Cutting and staining for epoxy resins

As with the methacrylates, it is better to use tungsten or glass knives to cut these materials, although it is possible with steel knives. Sections are usually dried on to slides using an adhesive and stained as for paraffin wax sections. As mentioned above, a pretreatment with sodium hydroxide may help the dyes to penetrate a little further, but most staining will only occur at the surface rather than throughout the section.

Because they essentially form a single large molecule, epoxy resins are less liable to disintegrate in the electron beam. This means that they will remain intact for longer and will not contaminate the vacuum column inside the electron microscope.

5.2 Alternative waxes

These are materials that look and feel like wax but have quite different properties. They have lower melting points than standard paraffin wax but are still strong enough to act as a tissue support. These waxes often need less processing and can accept tissues from dehydrating agents or even water. The result often is fewer artefacts and this can be a big advantage, but they have never become widely used for routine work because of the cost involved and even in research they are no longer used because the use of plastic embedding has made them redundant.

Water-soluble waxes

These are not really waxes at all but polyethylene glycols. They usually have a lower melting point than paraffin wax but are still similar in hardness to the common paraffin waxes used in histology. They are completely soluble in water, so tissues can be transferred into them without dehydration. The blocks are easy to cut and the tissue shrinks less since there are no dehydration and clearing steps and the impregnation is at a lower temperature. The main reason that they are not used very often is that after cutting the sections are very difficult to handle.

The sections cannot be conveniently floated on to a water bath for the tissue to relax and flatten after the rough action of being 'chopped' on a microtome. Tissues impregnated with water-soluble wax are difficult to store as either blocks or sections as they are hygroscopic and will gradually dissolve away in the moisture present in the air. For these reasons, they were never popular and are very rarely used today.

Ester and polyester wax

These were introduced by Steedman and consist of ethylene glycol esters. Polyester wax is sometimes referred to as Steedman's wax. They are based on glycol distearates.

> **Box 5.2 // Preservation using water-soluble waxes**
>
> Water-soluble waxes are also used for impregnating many other things to protect and support them. They have the advantage of being easily removed if this becomes necessary and can be applied from either water or from alcoholic solutions. These waxes are being used to protect the 16th century warship *The Mary Rose*, which sank in the Solent in 1545 and was raised from the seabed in 1982. Because the timbers were waterlogged and liable to disintegrate if they were allowed to dry, they have been kept moist and then gradually impregnated with polyethylene glycols, which coat and support the wood.
>
> A different medium, hydroxypropyl cellulose, was found to be the best way to solve a similar problem with books. The leather bindings of many old books have been restored in the famous old Long Room Library of Trinity College, Dublin, using this method. They did try using methacrylates to do the same supporting role for the drying and splintering covers but they tended to cause discoloration.

These waxes are extremely hard, despite their low melting point, and ester wax will even allow sectioning of small pieces of undecalcified bone. These waxes are rarely used in animal histology but are more favoured by plant scientists when sectioning tough, lignified cell walls.

Celloidin and low-viscosity nitrocellulose

This material is a derivative of cellulose (see *Box 5.3*) and although completely insoluble in water, it is soluble in a mixture of ethanol and diethyl ether. It is used in a different way to both wax and plastics by being used as a solution and then hardened by evaporating the solvent. This is a slow process and is very sensitive to the rate of evaporation.

The tissue is dehydrated in ethanol, treated with 1:1 ethanol and diethyl ether and then gradually infiltrated with a solution of celloidin in the same mixture. The tissue is then blocked out in a large mould and the solvent is evaporated slowly. The hardened block is then stored in alcohol. Storage in air results in the block becoming brittle as the remainder of the solvent evaporates.

Celloidin is good for tissues with widely differing textures. I have found it useful for whole eyes where there is a tough outer coat around a soft inner material (the humours) and it is the only method I have found that will reliably keep the retina attached.

> **Box 5.3 // Celluloid**
>
> Celloidin or collodion is cellulose dinitrate and is a close relative of the explosive gun cotton. Although it is not explosive itself, it is highly flammable. It was widely used as a plastic when it was usually mixed with camphor to form celluloid. It was used to make the backing material for the early cinematic film. It is not stable when completely dry, so many of these old 'nitrate' films are now falling apart. The solution was also used as a wound dressing by painting an alcoholic solution over wounds to give a clear protective covering.
>
> The decline in its commercial use has raised its price substantially and it is now a specialist chemical. It can be used for covering sections that are lifting, protecting microincinerated sections and coating crumbling blocks to enable sectioning. Although a valuable material for many tasks in histology it is unfortunately becoming too costly for routine use.

Thus, the support of tissues can be achieved in a number of ways. Each method will have its own particular advantages and problems. The choice of technique is often critical in obtaining good sections, so it is important to know that alternatives are available. In routine histology, they are not used as extensively as they might be because paraffin wax impregnation is readily available and used every day. Many routine histologists will often 'put up with' a paraffin block because it would be too much hassle to set up a processing schedule for one block.

Research applications may demand the use of double embedding where alternative embedding agents are used first and then followed up with paraffin wax processing and embedding.

Suggested further reading

Bancroft, J., Layton, C. and Suvarna, S. (2012) Chapter 8: Plastic embedding
– light and electron microscopy. In: *Bancroft's Theory and practice of
Histological Technique*, 7th edn. Elsevier Health Sciences.
Culling, C.F.A. (1975) *Handbook of Histopathological and Histochemical
Techniques*. London: Butterworths.
Hand, N.M. (2002) Plastic embedding media and techniques. In *Theory and
Practice of Histological Techniques*, 5th edn (eds J.D. Bancroft and M.
Gamble). Edinburgh: Churchill Livingstone.

Self-assessment questions

1. What are the advantages of embedding in plastics?
2. Why are epoxy resins preferred for electron microscopy?
3. Why do epoxy resin sections not stain as well as wax or methacrylate sections?
4. Water-soluble waxes (polyethylene glycols) support tissues well during cutting but are difficult to flatten. Why?

06 Cryotechniques

> ## Learning objectives
>
> *After studying this chapter you should confidently be able to:*
>
> - **Outline the principles involved in the preparation of frozen sections**
>   The water in tissues is frozen to form ice, which acts as a support for section cutting.
>
> - **Compare different methods of freezing tissues**
>   Tissues must be frozen rapidly to avoid ice-crystal damage. Fast freezing can be achieved with liquid or solid CO_2 or liquid nitrogen. Slow freezing is only useful for macroscopic work and can be done using a deep-freeze cabinet.
>
> - **State the advantages of frozen sections compared with other embedding techniques**
>   Frozen sections are quicker to produce and allow rapid diagnosis. Frozen sections also retain more materials in the tissue and are good for histochemistry and some antibody techniques.
>
> - **Outline the process of freeze-drying**
>   Freeze-drying involves rapid freezing followed by sublimation of the ice at low temperature and reduced pressure to remove water with minimal damage, loss of material or diffusion of cell components.

It has been emphasized in the previous chapters that the main methods for preparing tissues needs a rigid support for the tissue in order to cut thin sections, particularly from soft human tissues such as liver or kidney. The previous methods involved removing the water from the tissues by treatment with dehydrating agents and replacing it with wax or other hard materials. However, there is an alternative to this chemical replacement of the water in the tissues with a firm wax and that is to physically convert the water itself into a solid (ice) by lowering the temperature.

6.1 Cryotomy

Cryotomy is the term used to describe the use of ice as a support for section cutting. Freezing is quicker than replacement (see *Box 6.1*) and does not involve treating the tissues with chemicals such as alcohol, which may remove important materials (such as lipids) from the tissues or may alter the chemical properties of tissue components (such as enzymes). Such techniques are usually called **frozen section techniques** and are an important way of preparing tissues for rapid diagnosis and for histochemistry.

There are several methods of cutting frozen sections, but the most important difference is whether the microtome is entirely enclosed in a deep-freeze cabinet (usually at −15 to −30°C) or on a laboratory bench at room temperature with only the block being kept cold. The first technique is the most popular and the instrument, which consists of a deep-freeze cabinet and the enclosed microtome, is called a **cryostat**.

Box 6.1 // Freezing techniques allow rapid diagnosis

The main use in hospitals for frozen sections is for the rapid diagnosis of tumours. Tumours can be benign (relatively innocuous) or malignant (the cancers) and the treatment of the two types can be very different. Benign tumours are removed with very little excess normal tissue but malignant tumours require a more extensive removal including local lymph nodes. Mohs surgery to remove basal cell carcinoma from the face relies heavily on frozen sections to support precise surgery.

Clinical diagnosis is not always possible and a biopsy sample may be needed to confirm the diagnosis. With modern anaesthetics it is safer to keep a patient anaesthetized for an extra quarter of an hour than to sew them back up, send them back to the ward and give them a second operation 1 or 2 days later. The speed of the cryostat technique allows tissue to be frozen quickly, sectioned and stained ready for diagnosis within 15 min, so the surgeon can decide the extent of the removal once a microscopic diagnosis is made. This limits the patient to one trip to the operating theatre instead of two with a worrying wait whilst the tissue is processed.

Freezing and tissue damage

Freezing of tissues for frozen sections can damage the tissues and, although frozen sections involve less treatment of the tissue, artefacts are still produced and these depend on how the tissue is frozen.

Slow freezing and mechanical damage

The rate of freezing alters the size of the ice crystals. At slow freezing rates, the ice crystals will grow quite large and the crystals themselves expand as they freeze. This expansion results in mechanical damage to the tissue. Ice crystals may occur intracellularly and if they grow sufficiently large they can cause noticeable crevices in the tissue (ice-crystal artefact).

Water at atmospheric pressure is converted to ice at any temperature below 0°C, but it is in a dynamic state and will constantly be changing shape and interacting with adjacent ice crystals. Ice-crystal damage gets worse as blocks are stored, as the ice remodels and changes its shape and size. Only when the temperature of the ice drops very low, about −130°C for pure water, does it become stable and not recrystallize. The point at which recrystallization in tissues, which are filled with a salt solution, is inhibited is not known but it is probably somewhere below −90°C. Storage of blocks must be done in liquid nitrogen at −196°C to minimize this damage and cells should preferably be treated with a cryoprotectant solution such as 30% glycerol to reduce freezing damage.

Rapid freezing gives better tissue structure

Rapid freezing results in many smaller ice crystals that are less likely to cause visible alteration to the tissues. It follows that the higher the magnification that is going to be used on the tissue the smaller the ice crystals must be to avoid visible damage (see **Box 6.2**). Rapid freezing is needed for high magnifications. For low magnifications or macroscopic work (i.e. using the unaided human eye), the rate of freezing can be slower since the ice-crystal damage will not be seen. If the tissue is fixed before freezing, it will be able to resist damage more easily, so slower freezing of fixed tissue is less damaging than similar freezing of fresh tissue.

> ### Box 6.2 // Fast freezing kills cells
>
> Although fast freezing is best for tissue structure, it is actually much more dangerous to living tissues that are not protected. Slow freezing causes less tissue damage than very rapid freezing. In slow freezing, most of the ice crystals form outside the cells and are made of pure ice. This increases the osmotic pressure of the fluid, as the same amount of salt is now in a smaller volume. The cells lose fluid to the extracellular spaces but there is no necrosis. Fast freezing freezes the cell cytoplasm as well as the extracellular tissues and this is intensely necrotizing.
>
> The freezing effect is not the cause of the necrosis seen in cold injuries such as frostbite and trench foot. These injuries are caused by the capillaries in the affected area constricting and reducing oxygenation of the tissues. Ischaemic damage results and this can cause the loss of fingers and toes.

Slow freezing and osmotic damage

The fluids in tissues are not pure water but a solution of various salts, so when freezing occurs it is the effects on a salt solution that need to be considered, not simply the effects on water. Slow freezing of a salt solution results in the production of ice crystals that are pure water. It is only when the temperature gets below –21°C that the salt solution will freeze as a whole. Since water is being removed from the cellular fluid, the remaining solution becomes more concentrated. Water will be drawn out from the cells and they will shrink as a result.

Depression of freezing point

Salts dissolved in water lower the freezing point of the solution (depression of freezing point) and it requires a temperature of about –10°C before tissue fluid will freeze. The liquid left behind as the water crystallizes as pure ice may eventually become so concentrated that even at very low temperatures it may not freeze. The unfrozen areas will therefore remain liquid and the tissues in these regions will be unsupported. The result is that these areas will not section well and the sections may have distinctive holes where the pools of unfrozen tissue occurred. The tissue appears to have holes in it and this can make it look like a waffle; this artefact has been called a **waffle artefact** (see *Fig. 6.1*). The production of this artefact is dependent on the rate of cooling. If the rate of freezing is high enough, then the tissue freezes as one intact block without separating into water and salt solutions.

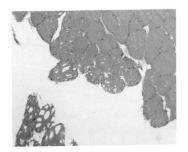

Figure 6.1
Frozen section of muscle showing a freezing ('waffle') artefact. The tissue at the bottom left shows holes where the tissue has failed to freeze properly and has been lost from the section.

Rate of cooling and heat conduction

The rate at which tissues freeze depends on conduction from the warmer tissues to the coolant. The rate of conduction is dependent on the temperature difference between the coolant and the tissue, so the lower the temperature of the coolant, the faster the tissue freezes. Thus, even though the tissue may need to be cut at –20°C, it is usual to cool it with a cooling system that will be at a much lower temperature and then allow the block to warm up to the cutting temperature. This means the tissue reaches –20°C much quicker and so there is less deterioration.

Conduction depends on the area in contact with the coolant

The rate of freezing is also dependent on the area in contact with the coolant. With a solid coolant packed around the microtome chuck or a Peltier module, the only contact is with the base of the block. Cooling rates will spread through the tissues from the base and this can be seen as a white line moving up the block. Cooling will be slightly slower at the top of the block since the heat has to transfer through the block as well as through the chuck. Anything that slows down the transfer of heat will similarly slow down the rate of freezing. Microtome chucks should therefore be made of metal and putting thick layers of insulating materials such as cork on top of the chuck should be avoided.

With liquid coolants, the whole block can be immersed in the coolant and this increases the surface area for heat transfer and can speed up cooling. The disadvantage is that the fluid may itself damage the block. Liquid coolants can, of course, also be used simply to cool the chuck without totally immersing the specimen.

Box 6.3 // Freezing ruptures cell membranes

The effects of freezing can be seen in frozen strawberries that retain their flavour but lose their texture. The firmness of strawberries is due to turgor pressure. This is a higher pressure inside the cell pressing against the cellulose cell wall like an inner tube and tyre. When frozen, the ice crystals rupture the cell membrane. The cell deflates like a punctured tyre and the strawberry loses its firm texture. Chemically it is unchanged but its physical integrity is lost. Freezing human tissue to keep it alive is not easy and requires replacement of the normal tissue fluid with cryopreservative and controlled freezing. The people who have paid thousands of pounds to have their bodies cryogenically stored are unlikely to be able to be restored, even with advancing technology, since all of their cells are likely to be damaged and repairing punctures in every cell in the body is totally impractical.

6.2 Methods of cooling tissues for sectioning

Various techniques have been employed to reduce the temperature of the tissues below freezing.

Slow freezing methods

Slow freezing is taken here to mean a freezing technique in which it takes several minutes to freeze the tissue into a solid block and will produce larger crystals that may produce damage that is visible with high magnifications.

Freezing tissues in a deep-freeze cabinet

Simply putting the tissue into a deep-freeze cabinet at –20 to –30°C will eventually freeze the tissues. This is usually done histologically by putting the tissue into the chamber of a cryostat that is normally held at –20°C. Some cryostats have an extra-cold shelf or chamber that may be as low as –40°C that freezes the tissue more rapidly but this is still relatively slow. The use of a deep-freeze cabinet or a cryostat is a cheap and easy method of freezing tissues since it involves no special reagents or extra equipment. However, it is very slow and may take hours for anything other than very small blocks. This very slow freezing causes quite a lot of damage and is only suitable for macroscopic work such as whole-body autoradiography. Even quite low magnifications will reveal some damage from the frozen ice crystals which gives a lace-like appearance to otherwise solid tissues.

Freezing tissues with a thermoelectric module

Peltier modules or thermoelectric modules are electrically activated heat pumps. When an electric current is passed through them, they absorb heat on one side and emit it on the other surface. By stacking several thermoelectric modules together, efficient heat removal can be achieved. The rate at which heat is extracted is controlled by the strength of the electric current. Increasing the current means a faster rate of freezing. Since the rate of heat pumping depends on the current, it means that the temperature of the block can easily be controlled by altering the current. Reversal of the current reverses the direction of the heat flow and is a convenient way to melt the lower surface of the block and allow it to be removed. These are not very popular as simple cooling devices but are used in some cryostats and other specialized instruments.

Rapid freezing methods

Rapid freezing techniques are now the most popular and the choice of method is often a matter of availability rather than strictly scientific choice. Most of these methods will give excellent results if carried out carefully.

Freezing using a fast-evaporating liquid

This method uses commercial spray cans containing fast-evaporating liquids (e.g. chlorofluorocarbons). As a liquid evaporates it will absorb latent heat of evaporation. This property applies to all liquids, but the faster they evaporate the faster they absorb heat. The liquids chosen for these sprays evaporate very quickly and therefore absorb heat very quickly. They tend to be expensive to use on a regular basis when compared with other methods of freezing and typically achieve a temperature of about –50°C.

They are very convenient to use since they can be transported easily and are therefore available to freeze tissues in places a long way from the laboratory. The sprays are a good reserve or emergency method of freezing since they can be stored without deteriorating. They can damage the outside of the block if sprayed directly on to the specimen since they are initially liquid and therefore act as solvents, but they do not penetrate beyond the very edge of the block so the inner regions are not affected. Currently their use is not encouraged in the interests of reducing the impact on the ozone layer.

Cooling with carbon dioxide

This method of cooling relies on the sudden expansion of high-pressure liquid CO_2 from a cylinder that rapidly expands adiabatically into a gas and absorbs heat in the expansion phase. The same effect is seen with CO_2 fire extinguishers, which produce a jet that appears white because of the production of cold CO_2 snow. Also like fire extinguishers, CO_2 cooling is very noisy. When the CO_2 is released, it expands rapidly, causing a very loud hissing noise.

CO_2 cylinders are reasonably convenient in the laboratory where they are only occasionally moved, but are more difficult to transport to distant sites. CO_2 cylinders are reasonably priced and readily available as they are also used for pumping (for example, beer in pubs). The cylinders must be used so that they run the liquid CO_2 out of the cylinder and not just the gas. This requires either a special siphoned cylinder, which can be used upright, or an ordinary one, which can be used inverted (with the outlet at the bottom). CO_2 blasts can reach temperatures as low as –70°C.

Cooling using solid carbon dioxide

Solid carbon dioxide is also known as 'cardice' or dry ice. It can be bought as large blocks and can also be prepared from liquid CO_2 by trapping the CO_2 snow produced when the liquid expands, as explained above. Solid CO_2 can be used as a dry material by packing it around the microtome chuck to cool it. Alternatively, solid CO_2 can be mixed with acetone to produce a freezing mixture in which chucks or tissues can be immersed. Both methods achieve a temperature of about –70°C. Buying blocks of CO_2 can be expensive and inconvenient as solid carbon dioxide must be delivered daily.

Cooling with liquid nitrogen

This liquefied gas is very cold (–196°C). Liquid nitrogen can be cheap if produced on site or bought in bulk. The very low temperature allows the possibility of very rapid freezing but there are some problems.

Liquid nitrogen has a tendency to boil when it comes in contact with warmer objects (see **Box 6.4**). The mechanical effects of boiling may damage some delicate tissues and the boiling slows down the rate of freezing. This slowing is due to the boiling at the surface of the block producing a layer of nitrogen gas around the block that acts as a layer of insulation and inhibits the transfer of heat.

Various techniques are used to get around this problem. Some workers simply use liquid nitrogen to cool less volatile liquids such as isopentane, which do not boil, and then use the cooled isopentane to cool the actual block. The isopentane becomes quite viscous at these low temperatures and the viscosity is used as an indication of when the cooling has reached the correct temperature. Other workers use metal foil to support the sample and enhance the rate of cooling. The speed of cooling can also be improved with powdered starch ('glove powder') as a dusting on the surface of the block. The fine powder reduces the boiling effect in the same way as boiling beads do in a beaker of heated liquid.

Box 6.4 // Liquid nitrogen is safe if handled carefully

Although liquid nitrogen is very cold, it is still reasonably safe because of the boiling effect – liquid nitrogen will simply run off anything warm like skin with the drops of liquid nitrogen held aloft like a miniature hovercraft on a cushion of nitrogen gas. It is a party trick of some conjurors and exhibitionists to pour liquid nitrogen on their hands. This is not my idea of a safe trick and liquid nitrogen does become extremely dangerous if it becomes trapped and cannot roll away. So folds in skin or clothing become dangerous traps for the supercold liquid. A particular risk is the wearing of jewellery, especially rings because the ability of metals to conduct away heat will guarantee extremely efficient freezing of the underlying skin. If treated with caution liquid nitrogen is safe but mishandled it can be a disaster. N.B. Often overlooked is the fact that nitrogen acts as an asphyxiant by displacing oxygen and should not be used in confined spaces.

Rather than cooling a liquid, liquid nitrogen can be used to cool large-blade metal forceps, which are then applied to the tissue. This technique can be used to freeze tissues whilst they are still in the body to eliminate the time between removal and freezing. This method only works well with very small pieces of tissue and special forceps that have extra-heavy jaws, which act as large heat sinks to draw the heat out of the tissue without themselves becoming warmed significantly.

Cutting sections using a freezing microtome

The freezing microtome technique uses a microtome with some convenient method of freezing the tissue on to the chuck (e.g. carbon dioxide blast or Peltier module) but only the specimen is kept cold. The rest of the microtome and the air around the microtome are at room temperature.

Tissue should be fixed first

The tissue usually needs to be fixed in formaldehyde to harden it and enable thin sections to be cut, as it is common for the sections to thaw out during the cutting process. Fixation in formaldehyde is best as it does not interfere with the freezing process and leaves the tissue fairly flexible. Mercury-, chromium- and osmium-based fixatives overharden the tissue, making thin sectioning difficult, and the salts can depress the freezing point of the tissue. Alcohol-based fixatives will not freeze since the freezing point of alcohol is very low (−112°C). Unfixed sections can be cut using a freezing microtome, but they will often disintegrate during staining.

The tissue is frozen on to the chuck

Tissue can be attached to the microtome chuck simply using water or saline as an adhesive, although some workers prefer to use a gum syrup, which has a thicker consistency. A piece of filter paper soaked in water placed on top of the chuck is often useful as it makes it easier to remove the tissue cleanly after sectioning.

Tissue can be frozen solid by one of the methods discussed above; the thermoelectric module and CO_2 blasting are the most common. Once frozen, they are cut using a conventional microtome and knife, usually a non-disposable knife, since ice is somewhat harder than wax.

Collection of free-floating sections

If the sections are cut very quickly, they can be made to jump off the knife directly into a bath of fluid, although this is difficult to control. If the sections are cut more slowly, then they melt on the knife and can be removed using a brush moistened with water. Although it is possible to attach the sections on to a slide and then air-dry them to achieve adhesion, they are usually easily washed off, even when using an adhesive such as gelatine. More often the sections are left free-floating and transferred from one reagent to another by draping them over a bent glass rod ('glass hockey stick'). This is more difficult than using a section attached to a slide and the final stained section still needs to be attached to a slide before mounting under a coverslip.

Cold knife technique

The section can be prevented from melting by cooling the knife and this also helps to reduce crumpling. The knife can be cooled using solid CO_2 held on the knife or by repeatedly blasting the knife with CO_2, or even by clamping a Peltier module to the knife.

> **Box 6.5 // Use of freezing microtomes in neurohistology**
>
> The freezing microtome has often been more popular with neurohistologists than with routine histologists. It is often quicker and simpler to cut large blocks of brain with a large freezing microtome than it is to process them through to other embedding media. The thicker sectioning associated with the freezing microtome is not a major problem. The sections needed for many neurological methods are often much thicker than for the rest of the body since the intention is to follow the nerve-cell processes.
>
> However, staining the free-floating sections is always more difficult than handling ones attached to slides but does have the advantage of two surfaces from which dyes and reagents can penetrate.

The ideal situation is to keep the knife well below freezing at between −10 and −20°C. The sections still have a tendency to roll and must be teased out during cutting using a fine brush. In the full cold knife technique, the tissues are placed on to a slide and into fixative before thawing occurs. This is difficult in practice and most workers simply use a cooled knife to give better sections.

Although widely used at one time, most of the freezing microtome techniques are rarely used for most tissues, although they are still used in neurological histology.

Cryostats

The cryostat or cold microtome technique involves keeping all steps in the cutting method at low temperature, with the block, microtome knife and indeed the whole microtome as well as the final section being kept below freezing. The cryostat is therefore a specially designed deep-freeze cabinet into which a microtome fits and with connections to allow the common controls to be operated from outside the cabinet without the microtomist having to put his hands into the cold chamber. This not only keeps the operator's hands warm but also stops the cabinet temperature from rising too much. A light is usually fitted inside so that it is easy to see what is happening on the microtome. There is generally a transparent lid to allow viewing whilst still keeping the chamber cold. There is also usually an anti-misting system and an automatic defrost system.

Good quality sections from unfixed tissues

Sections produced in the cryostat are generally of a much better quality than those produced using a freezing microtome. In the cryostat, even unfixed tissue will section well. In fact, fresh tissue generally cuts better and is easier to handle than fixed tissue. It is much more usual to fix after cutting than before and since the section always remains frozen there is no need to thaw the sections, which can be fixed whilst still frozen. It is also very common not to fix the tissues until after staining. Unfixed tissues are widely used in histochemistry. Variations on this technique are the most common methods for preparing frozen sections.

Freezing is done outside the cryostat

Tissues are usually attached to the microtome chuck outside the cryostat and then quickly transferred into the cold chamber of the cryostat. The attachment to the microtome chuck is fairly straightforward. The metal microtome chuck has a piece of moist filter paper placed on top and the fresh tissue is placed on top of the filter paper and the chuck. The tissue is then frozen by one of the rapid freezing techniques.

Tissues can be supported

If it is difficult to keep the tissue in position, it can be surrounded and supported by a gum solution or one of the commercial products (OCT or Cryo-M-Bed) containing polyvinyl derivatives, which are inert, water soluble and colourless. These are thick viscous solutions that allow awkward-shaped pieces of tissue to be held temporarily in place whilst freezing occurs. Once they are frozen, these support media cut smoothly and evenly with a similar texture to soft tissues.

Cutting cryostat sections

The microtome fitted into the cryostat can be of any type, but sliding and complex rotary microtomes can be difficult to lubricate at such low temperatures, so modified rocking or simple rotary microtomes are often fitted instead. Sections cut at low temperature have a tendency to roll up, so most cryostats use an anti-roll plate to keep them flat (see **Box 6.6**). This is simply a transparent plate of Perspex arranged in such a way that it leaves a narrow gap between the plate and the knife. This gap is wide enough to allow the section to slide down but narrow enough to prevent rolling. The correct adjustment of the anti-roll plate is one of the most important factors in producing good sections. The anti-roll plate is also one of the reasons why even laboratories that regularly use disposable knives for their wax sectioning still use solid non-disposable knives in the cryostat. It is much more difficult to arrange the anti-roll plate with a disposable knife holder than with a solid knife.

Box 6.6 // Improvements in anti-roll plates

Anti-roll plates have evolved over the years as various workers have tried to simplify the adjustments. The original technique was to wrap a piece of adhesive tape around each edge of the anti-roll plate as a rough indication of the gap needed. Adjustment screws were then added so the gap could be finely adjusted depending on the section thickness. The usual way of setting the thickness was to use a piece of newspaper as a 'feeler gauge'. The Times newspaper seemed to work best but I've no idea why! Teflon plates were then used to act as non-stick surfaces to ease the section down the knife more easily but were often opaque and so lost the visual feedback of how the section was cutting and seem to have vanished from use. Currently, anti-roll plates are made with a raised section at each end and this seems to work well – at least until the next innovation!

Picking up sections on to a slide

Sections can be gently removed from the knife using a pair of forceps and, whilst still frozen, placed in a container of fixative or other reagent without being attached to slides. More commonly, however, sections are picked up on to warm (room temperature) slides or coverslips by gently placing the slide or coverslip on top of the cold section where it rests on the knife. With unfixed tissue the adhesion is good and some of the protein from the unfixed tissues seems to act as its own adhesive as it is coagulated by the sudden temperature change. Adhesion can be improved further by briefly air-drying the section or by brief fixation.

If the tissue has been fixed before cutting, adhesion is more difficult and the slides may need to have an adhesive applied before picking up the sections. Fixed tissues also cut less easily than unfixed tissues and the cutting can sometimes be improved by treating the section with a sucrose solution before freezing. It is always advisable to extract fixative from fixed samples by immersing in running tap water for several hours before freezing to avoid any adverse effects.

Individual component temperatures

The cabinet temperature, knife temperature and block temperature can all be varied, sometimes independently of each other. Most sectioning is done with all of these at the same temperature (isothermic cryotomy), but a colder knife is useful for thin sections. The temperature of the chamber in particular can have a large effect on the ease of cutting and the optimum temperature is not identical for all specimens. Materials such as liver and kidney often cut best at –10 to –15°C, whilst skin and prostate often cut better at –15 to –25°C. Undecalcified bone usually requires a very low temperature of –30 to –40°C.

Impregnation techniques

Some tissues are not able to fully support themselves as frozen sections and may require extra support; for example, the large air spaces in lung tissue offer no support during sectioning. Also, some specimens need a surrounding medium to keep them together and make a reasonable-sized block; this occurs, for example, with very small fragments of curettings or bone marrow fragments that are too small to be used individually as a block. These difficult specimens can first be impregnated with a water-soluble support and then frozen as a single block.

Impregnating media

Many media have been tried and all can be useful, but most have one or more disadvantages. Materials that have been tried include gelatine, which has a tendency to absorb many dyes, agar, which shows a strong metachromatic effect, methyl cellulose (used in Polycell wallpaper paste), which does not set well and is messy, and soaps and detergents.

Gelatine is probably the most popular medium. The gelatine is made up as a 10–20% solution and kept molten at 37°C. The tissue is immersed in this warm, molten solution. If impregnation is likely to be a long process, it is usually better to fix the tissue in formaldehyde before impregnation to prevent any post-mortem changes. However, the fixative needs to be thoroughly washed out or it will also fix the gelatine and spoil the impregnation. Once impregnated, the tissue is then blocked out in a mould and this is then placed in a refrigerator to harden the gelatine. A block is cut from the solid gelatine and can be frozen immediately or can be fixed in formaldehyde to harden the block further before being frozen.

Giant sections (Gough and Wentworth sections)

This impregnation technique forms the basis of the method for cutting giant sections of whole lung (see *Fig. 6.2*). It is much easier to impregnate the lung with water-soluble gelatine than it is to process a whole lung through to paraffin wax. The fixed lung is impregnated with a weak solution of gelatine and then frozen in a deep-freeze cabinet. The small defects caused by slow freezing are less of a problem in this case, since the sections are not used with a microscope and are only used to give a cross-section for macroscopic viewing. The frozen lung is then sliced into sections, which are much thicker than is usual for microscopic sections (usually more than 300 μm thick) on a large microtome. The sections are then pressed and flattened on to absorbent paper, dried carefully and glazed with a layer of gelatine (*Fig. 6.2*). The final section can be used to show the amount of macroscopic lung damage, for example, silicosis in coal miners. This technique has been modified for many other embedding media including Polycell wallpaper paste and Persil detergent. The original technique was described by Gough and Wentworth in 1948; hence the sections are often still called Gough and Wentworth sections, even when the technique has been heavily modified.

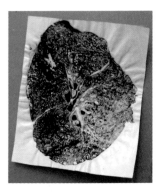

Figure 6.2
Large Gough and Wentworth section of a whole human lung. The tissue was frozen for support, sectioned (100 micrometres) using a large sledge microtome and then mounted on paper. The section shows large areas of black coal dust typical of a miner's lung. This type of section was used to determine lung damage in the assessment of compensation for pneumoconiosis in miners.

Box 6.7 // Visible Human Project

A variation on giant sectioning was co-ordinated by the US National Library of Medicine. This involved complete vertical sectioning of a male and a female cadaver at small regular intervals (male at 1 mm, female at 0.33 mm).

The technique involved embalming with formaldehyde, embedding in gelatine and freezing. The cross-sections were achieved by grinding away the surface, rather than true cryotomy, so there are no archival sections. The archiving was achieved by photographing each surface as it was exposed. The data are now available online as complete serial sections.

Both of the subjects had volunteered but there are still ethical concerns about the man as he was a convicted murderer who was judicially executed.

Freeze-drying

Another technique that uses freezing is the freeze-drying method. In this case freezing temporarily preserves and stabilizes the tissues as an alternative to fixing them. The water in the tissue is then removed by sublimation and the dry tissue can be impregnated directly with molten wax or resin without the need for any intermediate reagents. The advantage of this technique is that the tissues are not disturbed by being treated with a solvent, so water-soluble materials are preserved. It has proved useful for many materials such as glycogen that are disturbed by other preparative methods, but the single most important use is to demonstrate the amine neurotransmitter materials. These amines (often called biogenic amines) include adrenaline, noradrenaline, dopamine and 5-hydroxytryptamine (serotonin). These amines can be made fluorescent by exposing them to aldehyde fixatives (often called formalin-induced fluorescence) once they are dry.

Sublimation rather than evaporation

The important characteristic of the freeze-drying technique is sublimation of the water. Sublimation is the conversion of solid ice directly into a vapour without passing through the liquid phase. Sublimation only occurs if the ice is put into a very high vacuum and maintained at the triple phase eutectic point. Sublimation removes the water as vapour without damaging the tissue since it avoids the surface tension effects seen when a liquid evaporates.

Tissue is frozen and then warmed in a vacuum

The tissue is first very quickly frozen, usually referred to as **quenching** to distinguish it from slow freezing. This is best done using isopentane or a similar solvent with a high thermal conductivity cooled with liquid nitrogen. The frozen tissue is then put into a chamber with a Peltier module set to provide cooling and the tissue is kept at a low temperature to maintain it in a frozen state. The chamber is evacuated to a high vacuum (less than 1 torr or 133 mPa) to reduce the vapour pressure below that of ice and then it is slowly warmed up. Since a Peltier module is used for the cooling, the heat is applied simply by reversing the direction of the current and hence reversing the direction of heat flow, gently warming the tissue.

Sublimation keeps the tissue frozen
The ice sublimes as heat is applied; however, the tissue remains frozen and cold, even though heat is being applied continuously. The reason for this is that as the ice undergoes sublimation it absorbs heat. The vapour pressure in the chamber is kept very low either by using a chemical desiccant (phosphorus pentoxide is the usual drying agent) or by having an even colder region in the chamber such as a metal tube filled with liquid nitrogen. Water vapour will condense on to the cold spot, even in the high vacuum, and this creates a vapour gradient between the tissue and the trap.

Impregnation with wax
The drying process is very slow and takes several hours to reach completion, even for very small pieces of tissue, as up to 80% of the tissue is water. When it is completely dried, the tissue is allowed to return to room temperature and can then be fixed using formaldehyde vapour or can be immediately impregnated with wax or resin and made into a block for sectioning.

Suggested further reading

Bancroft, J., Layton, C. and Suvarna, S. (2012) Chapter 7: Microtomy FFPE and frozen sections. In: *Bancroft's Theory and practice of Histological Technique*, 7th edn. Elsevier Health Sciences.
Culling, C.F.A. (1975) *Handbook of Histopathological and Histochemical Techniques*. London: Butterworths.

Suggested website

http://library.med.utah.edu/WebPath/webpath.html

Self-assessment questions

1. Why does slow freezing damage tissues?
2. Name the reasons why frozen sections are used in (a) diagnosis of malignancy; (b) lipid staining; (c) enzyme techniques.
3. Outline the difference between a cryostat and a freezing microtome. Which is better?
4. What is the significance of liquid nitrogen boiling when it comes in contact with warm tissue?
5. Why do tissues still deteriorate even when stored at –20°C?

07 Staining theory

Learning objectives

After studying this chapter you should confidently be able to:

- **Describe why sections need to be coloured with dyes**
  Staining is needed to give contrast between different components of the tissues and allow examination by light microscopy.

- **Describe how dyes bind to tissues**
  Dyes bind by forming bonds with tissue components. Ionic and hydrogen bonding and van der Waals forces are probably all involved. Ionic staining is the most important and distinguishes between basophilic and acidophilic tissue components. Hydrogen bonding and van der Waals forces are less important but probably play a role in selectivity.

- **Describe the use of mordants in staining**
  Mordants are metal salts that help bind some dyes to tissues. Haematoxylin is the most important mordanted dye.

- **Define metachromasia and give examples of its use**
  Metachromasia produces a different colour in a tissue component to the colour of the dye solution. Toluidine blue is blue in solution but stains mast cell granules red.

- **Describe the main properties of haematoxylin and staining of tissues using haematoxylin solutions**
  Haematoxylin is a natural dye that requires oxidation to haematein before use as a stain. Haematoxylin is a mordanted dye that can stain many different elements in tissue depending on the mordant used. Using different mordants it can be used to stain nuclei, connective tissue fibres, nerve cells, muscle striations and mitochondria. It is usually used regressively.

- **Describe the use of silver as an impregnating metal**
  Silver solutions are easily reduced producing a dense black deposit and this reduction is autocatalytic. In argentaffin reactions, no extra reducing agent is needed, but argyrophil reactions require the addition of a reducing agent.

- **Describe the reasons for mounting tissues and outline the types of mounting media**
  Mounting media and coverslips not only protect the specimen but also make it translucent, making examination easier. Mounting media may be resinous (organic-based) or water-based solvents.

If sections of human tissue are examined under the microscope immediately after sectioning, they appear very dull and uninteresting. The tissue lacks contrast because all of the fixed materials have a similar refractive index and a similar colour so that a dull grey colour is all that can be seen. To bring out the structure of the tissues, it is essential to stain the cells to see the different parts in contrasting colours.

Staining is not simply random colouring of the sections but depends on using differences in the chemistry of the tissue to show the various components in different colours. This is most commonly done using dyes that can bind to the tissue in a selective way. Thus, the colours that are seen reflect the nature of the tissue and are not just a pretty picture. By using two or more dyes, it is possible to bring out the different cell and tissue components in several contrasting colours.

The commonest stain in use is the haematoxylin and eosin (H&E) stain, which colours the nuclei a dark blue or purple and stains the cytoplasm and connective tissue in shades of pink (see *Fig. 7.1*).

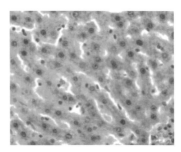

Figure 7.1
Simple haematoxylin and eosin-stained liver. The nuclei are visible as purple/blue round structures, the cytoplasm of the liver cells is pink and blood cells in the sinusoids have a more orange tinge.

7.1 Staining mechanisms

The binding of dyes to tissues is no different to any other chemical bonding and the mechanisms rely on the same binding forces that occur in all other organic compounds. The dye must form some type of bond or link to the tissue or they will simply be rinsed out of the tissue when the section is washed in another reagent. The usual forms of bonding can be involved. Each type has its own characteristics and bond strengths.

Bond type	Strength (kcal mole^{-1})
Ionic bonds	40–110
Hydrogen bonds	2–7
Van der Waals forces	1–2
Covalent bonds	35–212
Hydrophobic interactions	4–8

Ionic bonding

Ionic bonding involves electrostatic attraction between oppositely charged ions. One ion is a fixed ion in the tissue section and the other is the dye ion. Anionic (negatively charged) dyes will bind to cations (positively charged) in the tissue, and cationic dyes will bind to tissue anions.

Ionic bonding is the single most important form of bonding in most histological staining. Almost all simple staining can be understood and controlled by understanding the ionic charges involved.

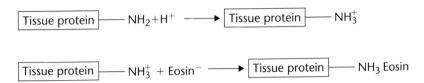

Figure 7.2
The ionization of tissue amines and subsequent binding of eosin

Negatively charged eosin ions will stain positively charged tissue ions

Eosin is an example of an anionic dye and is attracted to protein groups that are positively charged (cations) such as amino groups. First, the amino groups in the protein become ionized by binding to a hydrogen ion and this charged group then attracts the eosinic ion (see *Fig. 7.2*).

Eosin is usually sold as a salt such as sodium eosinate, which is readily soluble in water. Anionic dyes are also called acid dyes in histology because they are derived from coloured acids (in this case, eosinic acid) and not because of the pH of the solution. Anything that will stain with an acid dye is called **acidophilic**. In the case of materials staining with eosin, they could also be termed **eosinophilic**. Materials that are acidophilic include collagen, red blood cells and the cytoplasm of many cells.

Positively charged methylene blue ions will stain negatively charged tissue ions

Methylene blue is an example of a cationic dye and will bind to tissue anions such as carboxylic acid, sulphuric acid and phosphoric acid groups. These groups need to be ionized to bind the dyes (see *Fig. 7.3*).

Box 7.1 // Controlling pH allows dyes to be removed

Sensitivity to pH is exploited in the removal of dyes. This is used in differentiation and is also important in getting dye splashes off work surfaces, skin and clothing. Basic dyes can be removed by using acids (usually acid alcohol), whilst acidic dyes are most easily removed with a solution of ammonia in alcohol. These should always be very dilute solutions and used with extreme care. The solutions are extremely painful on the skin, especially cuts or sensitive areas such as around the eyes. It is much better to avoid getting the dye on your skin in the first place.

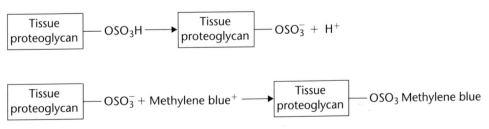

Figure 7.3
The ionization of sulphate groups in mucins and their subsequent binding of methylene blue

Cationic dyes are commonly called basic dyes and so substances staining with such dyes are called **basophilic**. Substances that bind basic dyes include nucleic acids and acid mucins.

Binding of dyes depends on tissue ionization

Proteins normally contain both acidic and basic amino acids and so it might be expected that proteins would take up both dyes. In practice this does not occur because most staining is done at a neutral or slightly acid pH. At these acid pH levels, the carboxyl groups of most amino acids are not ionized. Dyes will only bind to tissue groups when they are ionized; if the groups are non-ionized they will not attract the dye ions and will remain unstained.

Acid pH levels favour staining with anionic dyes

Ionic bonding to dyes and therefore staining is pH sensitive since the ionization of tissue groups is affected by pH. At an acidic pH, the high concentration of hydrogen ions favours the ionization of amino groups and results in strong staining of proteins by eosin, as described above. However, the same acidic pH will have the opposite effect on staining by methylene blue since weak acids, such as the carboxylic acids found in proteins, will be inhibited from ionizing by the high concentrations of hydrogen ions. Stronger acids, such as the phosphate groups found in nucleic acids and sulphate groups found in mucins, are less easily inhibited and will still ionize at the pH levels generally used in staining. This means that in slightly acid solutions methylene blue will act as a differential stain, picking out the nuclei but leaving the proteins unstained. Altering the pH can inhibit dyes from ionizing, but total inhibition of ionization of the salt forms of dyes will only occur at extreme pH levels. The ionization of dyes can be assumed to be complete at normal staining pH levels.

Alkaline conditions favour staining with cationic dyes

Alkaline solutions will have the opposite effect. The lack of hydrogen ions will allow the weakly acidic groups in proteins to ionize and methylene blue will stain both the cytoplasm and the nucleus. To get maximal staining with methylene blue, it is best to use Löffler's formula, which uses potassium hydroxide to raise the pH. Methylene blue will then stain all of the proteins and the nucleic acids so the whole of the tissue will appear blue and there will no longer be any differential staining. Eosin staining is depressed at high pH since the amino groups are now unionized and no longer attract and bind the eosin.

By careful selection of the pH, it is possible to get highly selective staining of individual components. This is most apparent in the staining of mucins where pH is used to control the binding of alcian blue to the weak carboxylic acid-containing mucins and the strong sulphated mucins (see **Chapter 8**).

Ionic interactions are long-range forces and can attract dyes to tissues over relatively large distances. Since there are two different charges, they can repel as well as attract (see **Fig. 7.4**).

Salt concentrations affect dye staining

Ionic binding can be inhibited by high salt concentrations, although weak salt solutions can act to increase the staining. This paradoxical effect of both enhancing and inhibiting dyeing can be understood by the effects of the ions. In strong salt solutions, there will be competition between the dye and the salt ions so staining will be inhibited. In many ways

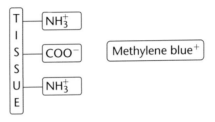

Figure 7.4
Effect of opposite charges. Although there is a negatively charged carboxyl group in the tissue that could bind the dye, the dye will not be attracted to it because of the two nearby positive charges. The net charge on the surface of the protein is positive, so the methylene blue ion is repelled

this resembles the effect of acids on basic dyes where the hydrogen ion is competing for binding with the dye molecule. Salt will affect both acid and basic dyes. Low concentrations of salt will be less inhibitory since there will be more dye than salt and once a dye has bound, the binding of the dye will be stronger than the binding of the salt. The salt may enhance staining slightly since both ions will be interacting, so the dye may be able to bind to sites to which it was previously not attracted due to the salt ions masking a repelling charge (see *Fig. 7.5*).

The critical electrolyte concentration (CEC) method of Scott utilizes magnesium chloride solutions of different molarities, in conjunction with alcian blue stain, to differentially stain acid mucins. Sulphated mucins will stain blue in solutions at 0.8 M and above, carboxylated mucins at 0.4 M and above.

Hydrogen bonding

Hydrogen bonding differs from other uses of the word 'bond' since it is a force of attraction between a hydrogen atom in one molecule and a small atom of high electronegativity in another molecule. Thus, it is an intermolecular force, not an intramolecular force as in the common use of the word bond. The hydrogen bond has a very limited range and will only form if the two interacting groups are brought sufficiently close together.

Hydrogen bonding is not affected by pH or salt concentration but is affected by strong hydrogen-bonding agents including urea and water. Hydrogen bonds are highly selective as they can occur only between certain groups (one must act as a donor and the other as

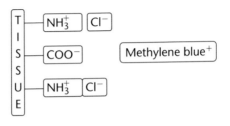

Figure 7.5
Effect of 'salting on'. Compare this with *Fig. 7.4*. The methylene blue is now able to bind to the carboxyl group, since the repelling charges on the amino groups are temporarily neutralized with chloride ions. The net charge on the surface of the protein is negative

a recipient). They probably play a role in the selectivity of dyes, but are usually secondary to ionic bonds unless special conditions are arranged to inhibit ionic interactions. This inhibition of ionic bonds can be achieved by using a non-aqueous solvent (water inhibits hydrogen bonding and favours ionization), a high salt concentration (which competes with the dye for tissue ion-binding sites) and an extreme pH (which is chosen to inhibit the ionization of tissue groups).

One or two instances of staining involve hydrogen bonding rather than ionic bonding. In the staining of elastin fibres hydrogen bonds are probably more important than ionic forces. More controversially the staining of amyloid by Congo red has been considered to be dominated by hydrogen-bond staining, which becomes effective when the dye intercalates (interleaves) neatly between the turns of the DNA helix, permitting the weak but close proximity hydrogen bonding to be effective.

Van der Waals forces

These are short-range forces and will only have an effect if the two atoms are between about 0.12 and 0.2 nm apart. If they are further apart, then there is no effective bonding force. Van der Waals forces can occur between any two atoms and are not specific for any atom or group. If the surface shape of the tissue protein and the shape of the dye match, then many van der Waals bonds can be formed. Thus, although they are individually very weak, they may add up to a significant binding force (especially important in staining amyloid with Congo red stain) (see **Box 7.2**) if the dye and protein have complementary molecular surfaces (see **Fig. 7.6**).

Van der Waals forces are believed to have a role in selectivity but probably only play a minor role in actual binding of the dye to the tissue. The ability to form many van der Waals bonds is one explanation of the finding that larger dyes will bind more strongly than small dyes, even though they may have the same number of ionizable groups. Van der Waals bonds are unaffected by pH, ions and hydrogen-bonding agents.

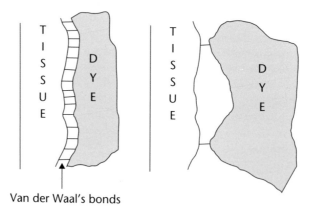

Van der Waal's bonds

Figure 7.6
Van der Waals forces and dye binding. If the dye and tissue have similar shapes, more atoms can be brought close enough together to form van der Waals forces and this can add up to a significant binding force. If the dye and tissue shapes are not complementary, only a few atoms will be close enough to form van der Waals forces and the dye will be less strongly bound

> ### Box 7.2 // Van der Waals forces are important in section adhesion
>
> Van der Waals forces are important in staining but in a quite different way to the binding of the dye. The adhesion of the section to the slide involves van der Waals interactions between the section and the glass. As the water evaporates from under the section, the lower surface of the section comes into contact with the flat smooth surface of the slide. Millions of van der Waals interactions are brought about and the section becomes firmly adherent.
>
> With the use of silane-treated slides, there is the addition of charge to the surface resulting in ionic bonding and the increased strength of adhesion may well reflect the difference in strength of the two bonds.

Covalent bonds

These are very strong bonds and are not easily broken once formed. They do not seem to be important in most staining reactions. They are important in some histochemical techniques e.g. periodic acid–Schiff, and in the attachment of dyes to antibodies in immunofluorescence. The so-called **reactive dyes** use covalent bonds to bind but are not used much in histology.

Hydrophobic interactions

Although they are sometimes called hydrophobic bonds, the forces are not chemical bonds in the conventional sense since they hold dyes in tissues by the exclusion of water from the regions of hydrophobic groups. The exclusion of water stabilizes the two groups involved by entropy/enthalpy changes. Hydrophobic interactions again are short range and are unaffected by hydrogen-bonding agents or salts. Altering the pH may change a particular group from a hydrophilic to a hydrophobic form by altering its ionization and this will alter the staining with hydrophobic dyes. Hydrophobic interactions are important in selectivity and play a major role in the staining of lipids, e.g. with oil red O.

7.2   Dye structure

Dyes are coloured organic compounds that can selectively bind to tissues. Most modern dyes are synthesized from simpler organic molecules, usually benzene or one of its derivatives (see **Box 7.3**). The modification of these compounds into dyes is a huge industry and the chemistry of dye synthesis can be complex, but a simple example will show the general nature of dye structure.

Chromophores

Most simple organic compounds such as alkanes, benzene and alcohols are colourless to the human eye but will absorb light outside the visible spectrum. Benzene, for example, absorbs strongly in the UV region of the spectrum but appears water-white to the human eye. Benzene must be altered so that it will absorb visible light and so become a visible coloured compound that can be a useful as a dye. Any group that makes an organic compound coloured is called a **chromophore**. Benzene can be made to absorb visible light by adding a suitable chromophore. In the example below (see **Fig. 7.7**), the chromophore used is the nitro group. Adding a single nitro group gives nitrobenzene, which is a pale yellow colour; adding a second and third group intensifies the yellow colour and trinitrobenzene is a strong yellow colour.

Box 7.3 // Synthetic dye discovery

Until the discovery of the first synthetic dye by W.H. Perkin in the 19th century, the only dyes available were natural ones. These tended to be somewhat drab and faded easily. The dye Tyrian purple was therefore a major discovery. Legend has it that this dye was discovered by the demigod Hercules. It is extracted from molluscs and by ancient standards was bright and did not fade quickly. It became a symbol of power and wealth in the ancient world and the dictionary still lists purple as meaning wealth and power (e.g. having a 'purple patch').

   The discovery of the first synthetic dye, Mauve, was entirely accidental. W.H. Perkin was a chemistry student trying to make the drug quinine from coal tar when he found some bright purple crystals. He borrowed some money and set up a factory to produce the dye and was rich enough to retire at 35. His dye "mauvein" was the height of chic in Parisian fashion houses for several years and admired by Queen Victoria.

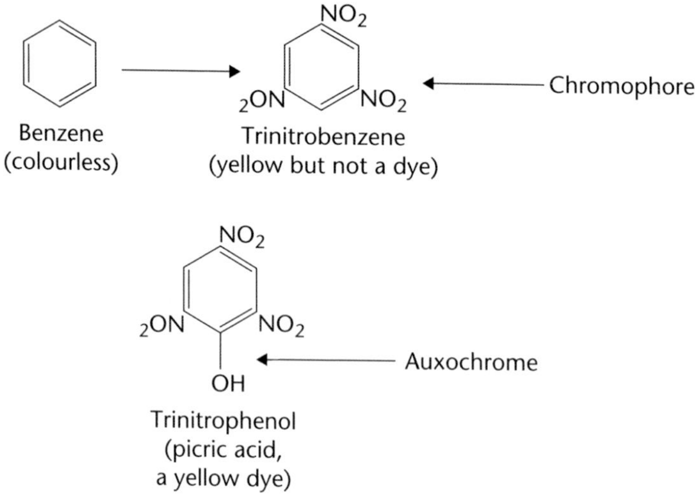

Figure 7.7
Conversion of benzene into a dye by the addition of a chromophore and an auxochrome

---

Auxochromes

Trinitrobenzene, although coloured, is still not a dye, as it will not bind to tissues. Treating the section with trinitrobenzene will temporarily colour it yellow in the same way that a plastic sponge appears coloured when it is soaked in a coloured liquid but the colour will wash out as soon as the tissue is rinsed in a solvent. To turn a coloured compound into a dye requires the addition of an ionizable group that will allow binding to the tissues. Such binding groups are called **auxochromes**. The addition of an ionizable OH group turns trinitrobenzene into the dye trinitrophenol, which is more commonly called picric acid in histology. Picric acid is an acid dye (the OH group is phenolic and ionizes by losing a hydrogen ion) and is very useful in histology (see *Box 7.4*); it is an essential part of the popular van Gieson counterstain.

   The most important chromophoric group in dye structure is not the nitro group but the quininoid arrangement of the benzene ring (see *Fig. 7.8*). This has two double bonds at either end of the ring and two double bonds on either side. This arrangement strongly absorbs parts of the visible light spectrum.

> **Box 7.4 // Picric acid**
>
> Picric acid is a powerful explosive and needs to be kept moist. It was known as Lyddite after the explosive works at Lydd where it was manufactured. It was originally discovered by oxidizing indigo, a natural dye from an Asiatic plant. Picric acid is an unusual material in histological terms since it is a good fixative and forms the basis of the highly rated Bouin's fixative, and it is also a useful dye, especially in trichrome stains, because it has a low molecular weight. It is used in histochemical tests to dissolve and thus identify the pigments found in malarial cells and following formalin fixation.

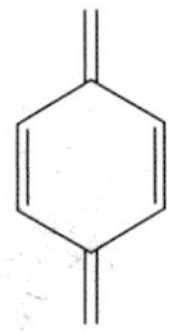

Figure 7.8
Quininoid structure, an important chromophore

Changing substituent groups

The exact colour of a dye is highly dependent on many other features of the molecule including molecular size and other substituents on the ring. Electron resonance through alternating single and double bonds, often in complex cyclic structures, has a significant effect in determining perceived colour. Most auxochromes alter the colour of the dye slightly (hence their name), as well as allowing the dye to bind to the tissue. Other groups or atoms are introduced solely to alter the colour and these are called modifier groups. Modifiers do not greatly alter the staining characteristics of the molecule, they simply alter the shade of colour.

Use of common dye names

Dyes are produced mainly for industrial uses such as textile dyeing, so a wide variety of different dyes have been synthesized to give a large range of colours. Dye manufacturers usually give the dyes they produce common names such as eosin or Congo red rather than their full chemical name and some of these names are copyrighted.

The dye structure shown in **Fig. 7.9** illustrates the complexity of many of these dyes. The full chemical name of this structure is 3,3'-((biphenyl)-4,4'-diylbis(azo))-bis(4-

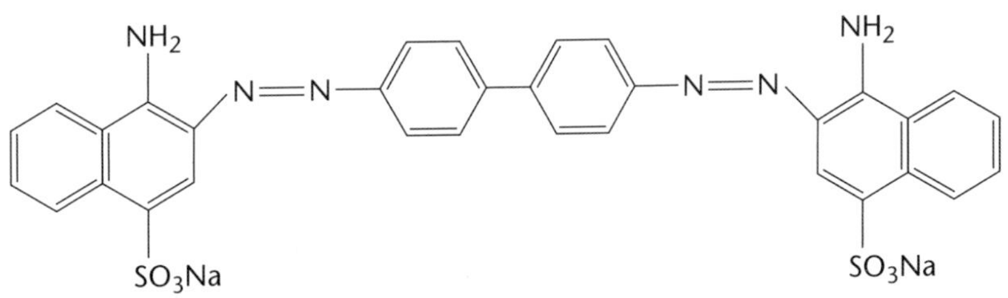

Figure 7.9
Structure of Congo red

amino-1-naphthalenesulphonic acid) disodium salt, whilst its common name is Congo red. The common name is easier to remember and to say, so most histologists stick to using these.

Different manufacturers may have different names for the same compound and this can be very confusing. If you ask for a dye by one name and get a bottle back with a different name on the label, you tend to think there has been a mistake. But if you order trypan blue you could get a bottle called chlorazol blue, which is the same dye by a different name. Eosin Y (yellowish eosin) has the following alternative names: acid red 87, bromoacid J, bromoacid S, bromoacid TS, bromoacid XL, bromoacid XX, bromofluorescein and bronze bromo.

Although a dye may have more than one name, it is usually easy to check with the supplier who will be able to put your mind at rest. More of a problem is the fact that different dyes can be sold by different manufacturers under the same name. For example, a dye called light green is usually considered an acid dye in histology and used for staining connective tissue, but the term light green is also used by some manufacturers for some basic dyes that will stain the nucleus and not the connective tissues. Buying the wrong dye can totally alter the results of a staining method.

The common dye names are derived from their industrial use rather than their histological use, e.g. fast green FCF (**for Colouring Food**) and brown FK (**for Kippers**). In industrial terms, an acid dye is one that would be used from an acidic solution and not necessarily one that would be anionic. Sometimes the names coincide with histological properties, e.g. basic fuchsin is a basic dye, but they are sometimes misleading, e.g. neutral red is a basic dye in histological terminology.

The Colour Index

To overcome all of the confusion there is a standard list of all dyes, their synonyms and their structures. This is called the Colour Index (CI). This is a monumental work of reference produced by The Society of Dyers and Colourists and each dye is given an individual number and listed along with its name(s) and properties. Since each dye on the list has a unique number to identify it, this list is the most reliable way of identifying a dye. When naming a dye in the description of a technique, the CI number should be given to avoid ambiguity, e.g. eosin Y (CI 45380).

CI numbers are arranged according to their structure, with the most important feature being their chromophoric group. For example, all nitroso dyes have numbers between 10000 and 10299, nitro dyes have numbers between 10300 and 10999, monoazo dyes have numbers between 11000 and 19999, and so on. There are 31 groups in all, with CI numbers up to 78000. Not all of these groups include important histological dyes; a few of the more important groups are listed below with examples of histological dyes from the group.

Nitro dyes. These have the nitro group -NO$_2$ as the chromophore, e.g. picric acid, martius yellow.

Azo dyes. These have the -N=N- group (azo) as the chromophore, e.g. orange G.

Triaryl methane dyes. These include the quininoid arrangement as the actual chromophore. The quininoid ring is shown as the one on the left in the diagram below, but since all three benzene rings are equivalent there can be rearrangement of the bonds and any of the benzene rings could take up this arrangement. There are a large number of dyes used in histology that fall into this category; a few examples are fuchsins, methyl violet, methyl blue and aniline blue. Final colour also depends on methylation of the benzene rings.

Anthraquinone. Here the quininoid ring is seen as the middle of the three fused rings. Examples are alizarin and carmine.

Xanthene. Here the quininoid ring is the right hand one of the three fused rings and the ring is tilted compared with the previous example. Examples include eosin and xanthene.

Thiazine. This is very similar to the previous example in overall structure, but the middle ring now has S and N as constituent atoms. This group contains many important metachromatic dyes, such as toluidine blue, methylene blue and azure A.

Histological classification

In histology it is often more useful to classify dyes by their action on tissues and hence their uses in histology. Two dyes within the same chemical group may have quite different uses in histology. For example, the two anthraquinone dyes in the list above are used quite differently. Carmine is an important nuclear stain, whilst alizarin is most commonly used to detect calcium in tissues. Also, dyes that are from totally different groups may quite easily be exchanged in histological techniques. The histological classification is only a broad guide to how a dye will work in practice, since the actual binding relies on many properties and not just the simple ionic nature.

Basic dyes are cationic and will stain anionic or acidic materials such as carboxylates, sulphates (many complex carbohydrates are sulphated) and phosphates (particularly the phosphates in nucleic acids). Most are used as nuclear stains and staining of cytoplasmic carboxyl groups is deliberately suppressed by using a slightly acid pH. Acidic substances that stain with basic dyes are termed basophilic.

Acidic dyes are anionic and will stain cationic or basic groups in tissues such as amino groups. Most are used to stain proteins in the cytoplasm and connective tissues. Substances that stain with acid dyes are called acidophilic.

Neutral dyes are simply compounds of basic and acid dyes. In this case, both ions are coloured. Such dye complexes will stain both nucleus and cytoplasm from a single dye bath. Romanowsky stains are neutral dyes made from more complex mixtures, usually with buffered eosin and methylene blue. These are the commonest dyes used in haematology. They are less common in histology but still very useful and include Giemsa, Leishman and Wright's stains.

Amphoteric dyes also have both anionic and cationic groups, but these are on the same ion. Such dyes can stain either the nucleus or the cytoplasm if conditions are appropriate.

Leuco compounds are dyes that have been rendered colourless by destroying their chromophores. They can be recolourised by tissues and are sometimes useful in histochemistry, e.g. Schiff reagent.

Natural dyes are simply dye substances extracted from natural sources. Although the main source of dyes for early microscopists, they have largely been replaced by synthetic dyes, which are usually more reliable, cheaper and can be supplied more readily. Natural dyes still in use include haematoxylin, carmine, orcein and litmus, although synthetic varieties are also available for some of these.

7.3   Non-dye constituents of staining solutions

As well as dyes, most staining solutions contain other components to improve the staining.

Mordants

Mordanting is the use of a non-dyeing compound to improve the binding of the dye, with the mordant involved being able to mediate a dye–tissue interaction. Mordanting of dyes has a long history and was crucial in early textile dyeing to fix the stain to the fabric and make it into a fast dye. Fast in this sense does not mean rapid but resistant to washing out or fading, and both of these properties are critical in the dyeing of textiles. However, the term mordant was very vague in its original usage and covered a number of mechanisms of binding dyes. The term has been adopted for some histological staining, but its use in

> **Box 7.5 // Traditional mordants for textile dyes**
>
> The methods used to make dyes stable and resistant to washing were often quite elaborate. Alums were commonly used as mordants, so their transfer to histology is hardly surprising. Organic materials were also used including tannins from wood bark, gallic acids from tree galls and ammonia from urine. The dye woad was extracted from the leaves of *Isatis tinctoria* and is identical to indigo. The importation of indigo was strongly resisted by the dyers in Britain and the last woad mills closed only in the 1930s in Lincolnshire. This was despite woad processing having the reputation of being the smelliest of industrial processes.

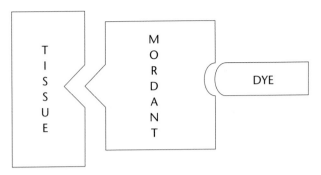

Figure 7.10
Mordanting. The dye can only bind strongly to the tissue when the mordant acts as a link between the two

histology is more restricted. It is usually only applied to conditions where the mordant acts as a link between the dye and the tissue and where the mordant is a metal salt (see *Fig. 7.10*).

The mechanism by which the mordant binds to the tissue is not certain but one likely mechanism is a dative covalency. The link to the dye would involve more than one such dative bond resulting in a chelate that was stable. The dye and mordant complex is sometimes called a **dye lake**. The groups on the dye forming the dative bonds are mainly oxygen-containing (e.g. in phenols, carboxyls and quinones) or nitrogen-containing (in amine, azo and nitro groups).

Since it is the mordant that binds to the tissue, the selectivity of the dye is controlled by selecting the mordant not the dye. The mordant gives greater stability to the stain and is not easily removed by water, alcohols or weak acids (i.e. it is a fast dye) and this makes it ideal when other stains are to be used afterwards, as the stain resists decolorization by the later reagents. Staining is commonly done with the dye and mordant present in the same solution, thus forming the dye lake in the stain before being applied to the tissue (e.g. Harris's haematoxylin and carmalum). The dye and mordant can also be used in two separate steps (e.g. Heidenhain's haematoxylin) and one or two techniques have used post-mordanting in which the dye is applied first and the mordant added afterwards.

Regressive use of mordanted dyes

It is also common to use mordanted dyes regressively (see *Box 7.6*). The differentiation is done by using strong acids (e.g. hydrochloric acid, often in alcoholic solution). Differentiation can also be done using excess mordant; for example, the iron alum in Heidenhain's haematoxylin can be used to slowly remove the excess haematoxylin. The excess mordant acts by displacing the dye lake and replacing it with a mordant with no attached dye. Theoretically it should be possible to devise a stain in which the balance of mordant to dye gives self-differentiation. A self-differentiating haematoxylin was described by Baker in 1962 but requires a constant and reliable dye, which is generally not available since most mordanting is done with natural dyes that vary in their composition from one year to the next.

Trapping agents

These differ from mordants in that they are always applied after the dye. They form large aggregates with the dye and result in the dye precipitating in the tissue. The large

> **Box 7.6 // Progressive and regressive staining**
>
> The way in which dyes are used can differ. One distinction is between **progressive** and **regressive** staining. Progressive staining is the simplest, with the dye being applied to the section until the desired density of colour is reached. Regressive staining involves overstaining the tissue so it is darker than is needed and then removing the excess to bring the colour down to the required level. The removal of the excess dye is termed **differentiation**.
>
> Regressive staining is often a better method of using stains. The reason is that dyes are rarely specific and will not only stain the structure being demonstrated but will also slightly colour the background, albeit less than the required structure. By removing some dye, the background can be cleared, since the background binding is usually weaker so the dye will be removed more readily. As long as the object being viewed has more dye than is required, differentiation will simply bring the colour down to the optimal level.

precipitate is more difficult to remove. The best-known example is the use of iodine to trap the violet dye inside the relatively impermeable wall of Gram-positive bacteria, whilst it can be removed from the more permeable Gram-negative organisms.

Accentuators and accelerators

Accentuators and accelerators are materials added to staining solutions to improve the staining reaction. Accentuators are generally simply used to control pH, e.g. potassium hydroxide in Löffler's methylene blue and phenol in carbol fuchsin. Accelerators are found in neurological techniques and are often hypnotic drugs such as barbiturates or chloral hydrate; their mechanism of enhancement is not known.

7.4 Metachromatic dyes and metachromasia

The term metachromasia is used when a dye stains a tissue component a different colour to the dye solution. For example, toluidine blue is a strong basic blue dye that stains nuclei a deep blue colour; however, it will also stain mast cell granules a pink colour. This colour shift that occurs with mast cells is called **metachromasia**, whilst the usual blue staining is called orthochromasia. Many dyes can show metachromasia but the thiazine group dyes are especially good for this type of staining.

Metachromasia is important as it is highly selective and only certain tissue structures can stain metachromatically. Substances that can be stained in this metachromatic way are called **chromotropes** and they include mucins, especially the sulphated mucins.

Mechanism of colour shift in metachromasia

The colour shift is always from a blue or violet dye to yellow or red staining. This means that the colour absorption shifts to shorter wavelengths, leaving only the longer wavelengths to be seen. This is believed to represent polymerization of the dye. The greater the degree of polymerization, the stronger the metachromasia. For example, toluidine blue will stain hyaluronic acid a blue colour, pectic acid (found in plants) a purple colour and mast cell granules a definite red colour. The difference is in the spacing of the acid groups, as shown in *Table 7.1* and *Fig. 7.11*.

Metachromasia requires water between the dye molecules to form the polymers and does not usually survive dehydration and clearing.

Table 7.1 Metachromasia and the spacing of acidic groups

Target	Distance apart of acidic groups (nm)	Staining
Hyaluronic acid	1.03	Blue (orthochromatic)
Pectic acid	0.5	Blue/purple (weakly metachromatic)
Mast cell granules	<0.4	Red (strongly metachromatic)

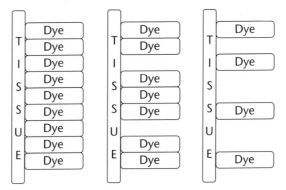

Figure 7.11
Metachromasia. The tissue on the left would be metachromatic as the dye has formed a polymeric form; the middle tissue would be weakly metachromatic as the polymeric forms are only a few molecules in size. The right-hand tissue would be orthochromatic as the dye molecules are widely spaced

7.5 Examples of important dyes and their uses in histology stains

Nuclear stains

Nuclear stains are very important in histology as the structure of the nucleus is often altered in disease. It is also easier to recognize tissue structure when only the nuclei are stained rather than when other structures are stained but the nucleus is unstained. Nuclear stains are important not only for looking at nuclear structure but also as counterstains for many staining techniques, since it is easier to recognize the location of the material in the tissue if a nuclear stain is used. For simple nuclear structure, a blue haematoxylin stain is ideal. However, for counterstaining this may not be the best stain to use, since a counterstain should be a different colour to the main technique so that it does not confuse identification. The blue/purple colours of haematoxylin often overlap with the colour of the main technique, so other nuclear counterstains are needed.

Haematoxylin

Haematoxylin is a natural product extracted from the heartwood of the tree *Haematoxylum campechianum*, which was named after the Campeche state in Mexico where it was originally found. It is now cultivated in the West Indies. The logwood of the tree is first extracted with hot water and the dye is then purified by precipitation with urea. The dry powder is usually quite pure (about 95%) but is not actually a dye. Haematoxylin is soluble in both water and alcohol but dissolves faster in alcohol, so stock solutions are often made by first dissolving the powder in alcohol and then diluting the alcoholic solution.

The formula of haematoxylin is shown below and it can be seen that there is no obvious chromophore and a solution of haematoxylin is not highly coloured.

Thus, haematoxylin itself is not a dye and for staining it must first be oxidized to haematein. Haematein has two fewer hydrogen atoms and the rearrangement of bonds introduces the quininoid ring structure and hence colour, as can be seen from the formula below:

Haematein is less soluble in water and alcohol than haematoxylin, but is soluble in ethylene glycol and glycerol. Haematein is only a weak acid dye, imparting a yellowish colour to the tissues, but when combined with a suitable mordant, haematein becomes probably the most widely used nuclear dye.

Haematein itself can be oxidized further to oxyhaematein, which is a weak acid dye but has no mordant dye capability:

Most working solutions do not completely oxidize the haematoxylin and the unoxidized part gradually oxidizes to haematein at the same time as some of the haematein oxidizes to oxyhaematein. This replenishes the working solution and greatly lengthens the life of the reagent. However, eventually all solutions will lose their strength and become useless. Oxidation is slower in acid conditions, so many solutions are deliberately kept acidic. During the natural oxidation, many haematoxylin solutions produce precipitates that must be removed by filtration before using the solution.

The oxidation of haematoxylin to haematein can occur in atmospheric oxygen, a process called ripening. This is a slow process and can take months, especially in cold and dark conditions, so that ripening is slower in the cold dark winter months than in the bright warm summer sunshine. This ripening is considered to give a longer shelf life but it is inconvenient if supplies run out, since it may take months to prepare a new batch.

Oxidation can also be carried out using oxidizing agents such as sodium iodate (200 mg per gram of haematoxylin), potassium permanganate (177 mg per gram of haematoxylin) or mercuric oxide (500 mg per gram of haematoxylin, although using smaller amounts than these traditional quantities will prolong the shelf life as explained above.

Box 7.7 // Brazilin

Brazilin, extracted from trees of the genus *Caesalpinia*, is very similar to haematoxylin (only one OH different) and can be oxidized to brazilein. It can be used in the same way as haematoxylin but has never become as popular in histology. It is from this dye that the country of Brazil took its name.

Brazilin   Brazilein

Mordants for haematoxylin

Haematoxylin is a very versatile stain and can be used to demonstrate many different tissue components in a highly selective way. The type of mordant used alters the specificity and colour of the stain.

Aluminium salts: haemalum. These are the commonest haematoxylin solutions and there are many different formulae but they all have similar results. Typical formulations include Harris's, Mayer's, Ehrlich's and Gill's haematoxylins. The mordant is usually either aluminium potassium sulphate (potash alum) or aluminium ammonium sulphate (ammonium alum). Because of their use of alum salts as mordants, these staining solutions are referred to as **haemalum** solutions. In acid solutions, the alum dye lakes are quite soluble and have a strong red colour. In alkaline conditions, the dye lakes are less soluble and have a strong blue colour. The dyeing bath is usually acidified and once staining is complete the section is rinsed in an alkaline solution. In hard water areas, the tap water is alkaline and simply rinsing in tap water will 'blue' the section. If the water is soft, then an alkaline solution can be prepared, e.g. lithium carbonate or tap-water substitutes. The haemalums are used regressively with a controlled differentiation in acid alcohol (1% HCl in 70% alcohol) followed by reblueing in water.

Ferric salts: iron haematoxylin. The ferric salts used are either ferric chloride or ferric ammonium sulphate (iron alum). The resulting stain is blacker and more intense, and will resist acidic counterstains such as van Gieson's better than haemalum. The ferric salts are oxidizing agents and will accelerate oxidation of the haematoxylin to haematein, which may result in overoxidation and loss of staining. The mordant is therefore either used separately (Heidenhain's) or the mordant and haematoxylin are mixed just before use (e.g. Weigert's). Differentiation is often done with excess mordant and requires microscopic control. There is no need to 'blue' the sections in an alkaline solution since the mordant produces an intense black colour regardless of the pH. The use of iron haematoxylins has declined following the introduction of the celestin blue–haemalum sequence, which also resists acid decoloration. Celestin blue solution is in fact a combined mordant, containing 0.5% dissolved Celestin Blue in 5% iron alum.

Other mordants that can be used to selectively stain specific tissue components are shown in *Table 7.2*.

Table 7.2   Other mordants that can be used with haematoxylin

Tissue element demonstrated	Mordant
Nuclei	Al or Fe
Myelin	Cr or Cu
Elastic fibres	Fe
Collagen	Mo
Neuroglia	W
Axis cylinders	Pb
Mucin	Al
Fibrin	W
Mitochondria	Fe
Heavy metals (Pb, Cu) and Ca	None (the metal in the tissue acts as the mordant and binds the dye)

Carmine and carminic acid

Carmine is a natural dye extracted from the red pigment cochineal, which is used in cooking. Cochineal itself is extracted from the bodies of the scale insect *Dactylopius coccus*. Carmine is actually a complex of aluminium and carminic acid rather than just the dye molecule. Commercial carmine powder is quite variable in its composition and in addition to the dye–aluminium complex also contains protein, calcium and other ions. Carminic acid is a glycoside with a glucose derivative joined to an anthraquinone structure (see below).

Carminic acid is the pure dye and is only slightly soluble in water but dissolves much better in solutions of an aluminium salt, when it forms the carmine complex. Solutions of the complex are not stable and significant deterioration occurs after only a few weeks of storage. For precise staining, it is better if the carmine solution is prepared from purified carminic acid.

For many years carmine was a major stain with its main advantage being its permanence when compared with other dyes. The problem with many other stains was that when they were mounted in Canada balsam, the acidity of the mountant caused significant fading in just a few months. At the height of its popularity, carmine was used for many different techniques and a great many methods were devised using carmine. The unreliability of the dye supply and the rising cost of the natural product have led to it becoming much less popular. The loss of popularity occurred at the same time as laboratories switched to the use of modern synthetic mounting media, which cause much less fading of other dyes so there was less need for a stable nuclear stain.

A good carmine stain is easy for inexperienced workers to use, as overstaining is difficult. Any excess dye can easily be removed with 1% HCl. The most popular stains were the carmalum techniques (e.g. Mayer's, Grenacher's) and acetocarmine.

Neutral red and safranine

These are popular red nuclear stains mainly used as counterstains to blue staining methods such as Perl's iron staining method. Both dyes are easily soluble in water and alcohol.

Neutral red can also be used as a vital stain when used at a very dilute (10^{-5}) concentration. Neutral red can act as an indicator, changing colour at pH 6.8–7.0 (turning yellowish in alkali). Neutral red stains nuclei red and cytoplasm pale yellow. Its use has been extensively replaced with nuclear fast red (kernechtrot or 'red kernel/nucleus' stain).

Methylene blue

This is a very widely used simple blue stain that does not require a mordant. It gives a quick and simple nuclear counterstain for red primary stains. It is readily soluble in water and alcohol.

Methylene blue was a major component of Romanowsky stains used in the staining of blood smears and bone marrow specimens. In the preparation of the Romanowsky dyes, it was used as 'polychrome methylene blue' (see ***Box 7.8***). The polychroming produces a range of dyes from the original methylene blue of which azure B is probably the most important. Modern Romanowsky stains generally use mixtures of pure dyes rather than the empirical polychromed methylene blue.

Methyl green

This blue/green nuclear stain is a useful nuclear counterstain and is also an important part of many techniques that differentiate between DNA and RNA in tissues. It is often contaminated with methyl violet, but this can be removed by washing with chloroform.

Box 7.8 // Polychrome methylene blue

One of the methods used to produce polychromed methylene blue was to allow fungi to grow in a solution of it. The metabolic actions of the fungi converted some of the dye to other compounds. The relationship between dyes and micro-organisms can be quite interesting. Many dyes will allow micro-organisms to grow in them and this changes their staining characteristics. Other dyes have been used as antibacterial agents and the selectivity of dyes led many people to believe that dyes might prove to be the 'magic bullets' that would kill selected bacteria and save humanity from infection.

Cytoplasmic stains

Cytoplasmic stains are often used as counterstains but can also be important to identify tissue components. Most techniques are also used to distinguish connective tissue fibres and other protein materials. The cytoplasmic stains should produce several different shades of colour so that the tissues can easily be distinguished. Most of the stains are acidic (anionic) dyes but can be used in mixtures to improve the contrast between different components.

Eosin

This is not a single dye but a variety of related dyes. All are derived from fluorescein, which is a useful fluorescent dye widely used to label antibodies but is useless for ordinary light microscopy. By substituting halogens or nitro groups for some hydrogens, a variety of shades of red can be produced from yellowish to bluish e.g. eosin Y (yellowish) changes to eosin B (bluish) if the bromine groups on positions 2′ and 7′ are changed to nitro groups.

The dyes are also fluorescent but are solely used as red dyes, although the parent dye fluorescein is widely used as a labelling compound in immunofluorescence. The sodium salts of the dyes are all freely soluble in water and fairly soluble in alcohol but will precipitate as eosinic acid if the pH is very low. However, adding dilute acids will improve eosin staining but may overdifferentiate the nuclear stain.

Eosin is a very good cytoplasmic stain as it gives several shades to the tissue. The range of shades can be extended even further if more than one dye is used in the solution. Some workers claim that up to seven different shades can be distinguished, although I have always found it difficult to distinguish more than about four.

Eosin solutions keep reasonably well unless they become contaminated by fungi, when they will develop significant growth. This growth can be inhibited by adding a small amount of thymol to the solution and this acidic material also enhances the staining.

Ethyl eosin is an ester rather than the more usual sodium salt and is only slightly soluble in water. It is used when eosin staining is needed from alcoholic solution. It must be differentiated in alcohol.

Eosin is also an important component of Romanowsky stains, which are all eosinates of azure dyes. Its pre-eminent role in staining is shown by the fact that many structures are referred to as eosinophilic when they will stain equally well with other acid dyes.

Eosin gives a good red cytoplasmic counterstain but if other colours are required then other dyes must be used.

Methyl blue and aniline blue

These are widely used blue anionic dyes with similar staining properties. Both are water soluble but insoluble in alcohol. They are often confused as both are also known as soluble blue and water blue. Both are quite large dye structures and are frequently used to stain connective tissue fibres.

Fast green FCF and light green SF

These are green anionic dyes similar to the blue dyes above and are frequently used as counterstains to red dyes. Fast green FCF is less prone to fading than light green SF.

Orange G, picric acid (trinitrophenol), metanil yellow and martius yellow

These are very pale-coloured dyes ideal for faint background staining or in conjunction with other acid dyes. Orange G is soluble in water but less so in alcohol and is a major component of the Papanicolaou stain used in cervical cytology.

Picric acid is a valuable stain in multiple acid dye techniques because of its small size, which permits the staining of red blood cells an intense yellow.

Connective tissue methods

Connective tissue consists mainly of collagen fibres, elastic fibres, glycosaminoglycans and cells. The main way of distinguishing the fibres and cells is by using a combination of acid dyes to stain different structures in differing colours. There is still uncertainty about the exact mechanisms of these techniques but they seem to depend on differences in dye size and differing permeabilities of tissues.

Acid dye combinations

The differing molecular weights and sizes of dyes affect their diffusion rate and their ability to permeate into small spaces in the tissue. The larger dyes will also be able to form more van der Waals forces. Thus, when two acid dyes compete for binding to tissue, the larger dye will generally tend to displace the smaller dye. In addition, the smaller dyes tend to be paler colours (yellowish), whilst the larger dyes are dense colours. These effects combine so that smaller paler dyes are overwhelmed by larger denser dyes when they compete directly.

Tissue permeability is related to the amount of protein that is present and the amount of water between the proteins. Loose collagenous (areolar) tissue has many minute fluid spaces and is very permeable, whilst erythrocytes are packed full of haemoglobin and are much denser. Most other cell types, including muscle, lie between these two. The concept of the differential acid dyeing technique is that only the small dye will penetrate into the dense red cells. The red blood cells should thus be stained with the smallest dye. In the less-dense collagenous tissue, the large and small dyes will be in competition. The larger dye will dominate and the collagen will appear stained only with the largest dye.

Van Gieson's stain

This stain uses two acid dyes (acid fuchsin and picric acid) to distinguish between acidophilic materials. Each dye, if used alone, would stain all cytoplasm and connective tissue. By combining them in a single solution, the tissue differences can be exploited. The open texture of the collagen, allowing free and rapid access to both dyes, stains red, whilst muscle and erythrocytes, which restrict access of larger dyes, stain yellow. Both dyes are mixed into a single solution along with hydrochloric acid to give a pH of 1–2.

Permeability and dye size considerations would suggest that the small dye will rapidly penetrate both the dense red blood cells and the looser connective tissues. The larger fuchsin molecules will penetrate into the connective tissues quite readily but will penetrate the denser red blood cells only slowly. Where both dyes are present, the fuchsin will displace or mask the paler picric acid with the result that the connective tissue will stain red but in the red blood cells the picric acid will not be displaced or masked and the red cells will stain yellow (see *Fig. 7.12*).

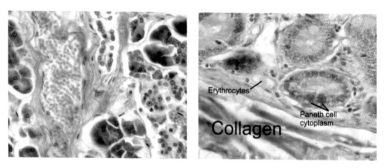

Figure 7.12
Van Gieson and trichrome methods. The left-hand photograph shows a van Gieson stain with red collagen and yellow erythrocytes. The Masson trichrome on the right shows yellow erythrocytes, red cytoplasm, especially of the Paneth cells, and blue collagen fibres.

Trichrome stains

These take the differential staining a stage further and use three different-sized dyes to selectively stain the three tissue densities. For example, picric acid (formula weight (FW) 229), acid fuchsin (FW 578) and methyl blue (FW 800) can be used as a trichrome mixture. The red blood cells are the densest tissue and stain with the smallest dye, the intermediate cytoplasm and muscle cells are stained red by the intermediate-sized dye and the collagenous tissue stains with the largest dye. In each case, only the largest dye of the competing dyes does the staining.

Heteropolyacids aid trichrome staining

Trichromes differ from van Gieson's stain in that an extra reagent is used in the form of one of the heteropolyacids. The heteropolyacids are either phosphomolybdic (sometimes called molybdophosphoric) acid or phosphotungstic (tungstophosphoric) acid. These improve the staining but whether they simply act as colourless dyes or have a more active role in some form of mordanting is still unresolved. Trichromes also differ in that the three dyes are usually used separately and sequentially, rather than in a single mixed reagent as in van Gieson's stain.

Trichromes can to some extent be 'tuned' to differentiate between tissue fibres by selecting dyes of appropriate sizes and by controlling the size of tissue spaces. Alcoholic solutions seem to affect penetration by allowing dyes to permeate more freely, possibly by increasing the size of tissue spaces. This makes the molecules act as if they were a slightly smaller size.

Molecular size and permeability: not the full story

Although the explanation given here accounts nicely for much of the staining with multiple acid dyes (trichromes and van Gieson's), there are anomalies and the exact mechanisms are still very much undetermined. In particular, if the dyes are used alone they will readily stain all of the tissue. Thus, the fuchsin dye in van Gieson's stain will stain red blood cells, showing that it is able to penetrate these structures.

There is the possibility that the timing is crucial and that by using a limited time the red dye would not have long enough to penetrate into the cells. Even on theoretical grounds this seems unlikely; a red blood cell is less than 8 μm across at its widest point and less than 3 μm thick. For diffusion across such small distances to take more than

2 min (which is a typical staining time for van Gieson's stain) would suggest an extremely dense material. If van Gieson's staining is extended to 30 min or more, there is no real major difference in the result.

Similar anomalies can also be seen when different combinations of dyes are used. It is always the smallest dye that stains the red blood cells and the largest that stains the collagen, but the same dye can stain both under different conditions. If acid fuchsin is used in combination with other dyes, it will stain erythrocytes if the other dye is larger, but will stain collagen and not erythrocytes if the other dye is smaller. The situation is more complex than the simple dye size and permeability would suggest, yet the concept does seem to hold in most practical applications and several good trichrome methods have been produced on the basis of this theory.

Dyes and quality control

As mentioned earlier most dyes are not produced for histologists but for textile dyers. The important property for textiles is a reliable final colour rather than chemical purity. Dye manufacturers therefore adjust their products to give consistent dyeing of fabrics rather than histological reliability.

Staining and reproducibility

Consternation can be experienced when a method fails to work and it is worth pausing here to consider the most likely cause or causes for an established method to fail. Essentially this is a fault finding or 'spot the variable' exercise. Try to understand the significance of the following important questions.

1. Has a control been used and is the control negative?
2. Has the section been fully dewaxed?
3. Have fresh reagents been tried?
4. Have the correct dyes been used?
5. Has an inappropriate solution been used for rinsing (e.g. alcohol instead of tap water)?
6. What is the quality of the laboratory's distilled water?
7. Have the correct timings been used?
8. Are all solutions at the correct pH?
9. Are all solutions of the correct strength / dilution?
10. Is there significant variation in laboratory temperature?

Answers to any or all of the above may be highly significant in resolving the problem.

Dye samples are impure and often unreliable

Dyes, unlike most biochemical reagents, are often impure substances and may contain significant amounts of other materials such as salts, dextrans and even other dyes. The actual content of the named dye rarely exceeds 95% and may be as little as 25% of the total weight. Different batches of dye will differ in their dye and contaminant content, which makes quality control in the histological laboratory difficult.

The non-dye constituents are often very important and may grossly affect the staining. To try to combat this problem, some laboratory suppliers offer **certified dyes** that have been tested biologically for their stated uses. Such dyes are more expensive but should match their stated uses reliably.

> **Box 7.9 // Contaminated dyes**
>
> Dyes have always been impure so it became important to have a good source of dyes. Dyestuffs are quite expensive to manufacture, so it was not unknown for dyes to be 'cut' with less expensive materials to make them more profitable. Some older samples of dye certainly seemed to have an insoluble residue left after preparing the staining solution. One manufacturer, however, became famous for the quality of his stains and if you read the old textbooks you will find his dyes being recommended time after time in techniques as being the best available. Nobody thought his dyes were purer, just better. It was said that 'not only does Herr Grübler have the best dyes, he also has the best impurities'. Grübler dyes lost their leading role following the Second World War when importing of dyes from Germany became impossible and laboratories had to find other sources.

It is also worth repeating that some dyes have many names and it should always be made clear which dye is needed by using CI numbers; otherwise the dye may be completely different. When a staining method suddenly stops staining as expected, it is worth checking that you have not got a different batch of dye to the usual one.

There is a growing tendency for laboratories to buy in many reagents in a ready-prepared form rather than making up stains from the original ingredients. This leads to more consistency in the laboratory as the scale of industrial production can be controlled more carefully than small irregularly prepared batches in the laboratory.

Checking dyes in a histology laboratory

Quality control of dyes within the laboratory is difficult, as many of the techniques used in quality control require complex equipment to analyse the dye samples (e.g. infrared spectroscopy, high-performance liquid chromatography), but some simple tests can usually be performed.

1. Chromatography. This will detect coloured contaminants of dyes and can be a sensitive way of comparing two dye batches. Simple paper chromatography using filter paper is often enough to pick out impure dye samples.
2. Measurement of absorption (including a full spectrum if a suitable spectrophotometer is available) can be used to determine the amount of dye in a sample and may also show contaminants.
3. Testing with standard dyeing techniques to determine whether the dye is suitable or needs altered staining times/conditions. Some dye batches may be suitable for one stain but not for others; for example, fuchsin samples may be good for use in Ziehl–Neelsen staining for mycobacterium but not for preparing Schiff's reagent.

Once the dye has been made up into a solution it may not be permanently stable. Dyes can alter due to oxidation by the air, bleaching by light, contamination by micro-organisms growing in the solution or chemical reactions between constituents of the dye solutions. Reagent bottles should be clearly labelled with the date of preparation and renewed at regular intervals or sooner if the staining seems to be suffering. If light accelerates the deterioration, then storage of the reagent in brown bottles to prevent light reaching the dye may help, although the dark glass will also mask any contamination and precipitation, so care must still be taken. Most techniques using reagents that need special storage (e.g. refrigeration) will usually give details.

7.6 Silver impregnation

Metallic impregnation is an alternative way of increasing the contrast in tissues. The commonest metal to use in light microscopy is silver, which produces a dense, black, fine deposit of silver and silver oxide where the silver ions have been reduced. Silver impregnation is also called silver staining, but the mechanism is quite different to the effects of dyes and the structures are actually plated with the silver rather than the silver being reversibly bound to the section.

Advantages

Silver impregnation has a number of advantages compared with dyeing techniques and has a number of very common applications. The main advantages of silver techniques are:

1. They are stable and do not fade. The end product is metallic silver, which if properly fixed and washed is effectively permanent. The silver deposit in black and white photographs is similar to the material produced by silver impregnation and photographs from 150 years ago are still in excellent condition. Dyed sections rarely last more than 10 years without some signs of fading.
2. The silver deposit is densely black, which gives good contrast and is excellent for taking photographs.
3. Silver techniques are very sensitive methods and will detect many materials that are difficult to demonstrate by dyeing. These materials include reticulin fibres (see *Fig. 7.13*), which are difficult to observe with haematoxylin and eosin staining but can be readily demonstrated with silver impregnation. Metal impregnation methods are more common in neurological methods, e.g. for axons, motor end plates and astroglia (see *Fig. 7.14*).
4. Slender objects are thickened because they become silver-plated. This can be useful for fine fibres such as reticulin or for slender bacteria such as spirochetes.

Disadvantages

1. The techniques can be unreliable and capricious. They will sometimes work well and other times will not work at all. This can extend to different workers. There sometimes seems to be one person in the laboratory who can get a technique to work perfectly, whilst everyone else struggles, even when using the same reagents. Staining times can vary tremendously from one day to the next when a fresh batch of silver solution is prepared.

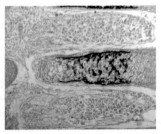 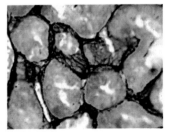 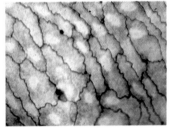

Figure 7.13
Silver impregnation techniques. The photograph on the left is a von Kossa stain on developing bone showing mineralisation of the central bone; the bones at the top and bottom are not yet mineralised. The middle photograph is kidney stained to show reticulin fibres surrounding the convoluted tubules. The right-hand photograph shows cell borders in a simple squamous pavement epithelium.

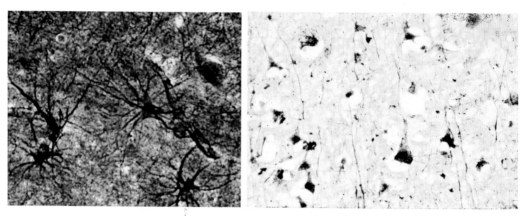

Figure 7.14
The use of silver impregnation to show astrocytes (left-hand image) and neurones (right-hand image).

2.   The silver solutions are often very alkaline. Strong alkaline solutions have a tendency to strip sections off the glass slides so extra care and adhesives are needed.
3.   Silver techniques are so sensitive that they can sometimes give non-specific background deposits ('dirty preparations').
4.   The techniques have a tendency to stain everything they come into contact with (hands, laboratory coats, benches, glassware, etc.). Silver nitrate produces nitric acid and oxidized silver on decomposition in a damp environment. This will cause burns to skin, hence the name 'lunar caustic' and its early medical use as a wart remover. Silver is very difficult to remove without using dangerous reagents, so clothing is often permanently stained. Silver solutions are easy to wash out if they are caught early enough, but as they look just like water it is not always obvious that there has been a spillage. Once reduced, the safest way to remove silver deposits is by using an iodine solution, which converts the silver to silver iodide, which is then soluble in sodium thiosulphate solutions.
5.   Some silver solutions have a tendency to become explosive if stored for more than 24h, as can the silver 'mirror' that forms on glassware.
6.   Silver is expensive; at the time of writing, the price for analytical grade (99% purity) is £102.50 for 25 g.
7.   Silver cannot be discarded into the drains as it is a heavy metal poison.

Use of silver

Silver is not the only metal that can be used for impregnations but is the most useful as it is easily reduced and any reduced silver acts as a catalyst for the reduction of more silver. This autocatalytic activity makes silver useful in many fields other than histology. The use of silver is widespread in photography and the chemistry of photography and the chemistry of silver impregnation are very closely related.

Silver solutions are reduced during the impregnation, so silver techniques are primarily methods for reducing materials. There are three different ways of producing silver deposits. These are the argentaffin reaction, the argyrophil reaction and ion-exchange reactions.

The argentaffin reaction

In the argentaffin reaction, the tissue contains reducing groups that are sufficiently strong and present in sufficient quantity to give a visible deposit without added reducing agents. These groups are often aldehyde groups and silver solutions can be used to replace the Schiff's reagent in the periodic acid–Schiff technique (see **Chapter 8**) to give periodic acid–silver. The argentaffin reaction occurs particularly with reducing pigments and is strongest with the pigment of enterochromaffin cells, which derives its alternative name (argentaffin pigment) from the reaction. The strong reaction in this case is due to phenolic components (5-hydroxytryptamine, or serotonin). The reaction only needs the addition of the silver solution, such as in the Masson–Fontana technique, but tends to be very slow and may take up to 24 h to give a deposit.

The argyrophil reaction

Many tissue groups are able to adsorb silver, possibly by ionic mechanisms as for dyeing. The silver is mainly adsorbed as silver ions but small amounts are reduced to silver atoms. These silver atoms are deposited at the site of reduction. The initial reduction reaction with silver only deposits submicroscopic atoms of silver at particularly reactive sites. Probably only a few, perhaps as few as two, atoms are deposited in this initial stage and these are too small to be visible, even with high-power microscopy. These silver atoms then act as catalytic sites where more silver can be deposited by the reducing action of a developer (see **Fig. 7.15**), e.g. formaldehyde or hydroquinone (quinol). In this case the developer does the main reduction and the tissue simply provides places where there are silver atoms to catalyse the reduction. This type of reaction where an external reducer or developer is added is called an argyrophilic reaction.

Ion-exchange reactions

Ion exchange can also deposit silver and this is used to detect mineralization of bone using the von Kossa technique. The section is treated with silver solution (silver nitrate)

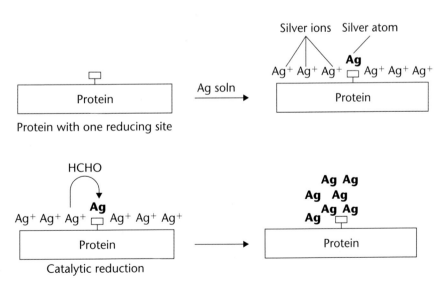

Figure 7.15
Silver deposition and reduction in the argyrophil reaction

and the phosphates and carbonates in the mineralized bone form insoluble silver salts. The silver salts are then blackened by UV light or hydroquinone solutions (see *Fig. 7.13*). Although often said to demonstrate calcification of bone, the method actually detects carbonates and phosphates.

$$CaCO_3 + 2AgNO_3 \rightarrow Ag_2CO_3 + Ca(NO_3)_2$$

$$Ag_2CO_3 \text{ (UV treated)} \rightarrow Ag_2O + CO_2$$

Black

Silver solutions

Ammoniacal silver solutions are used as they are easily reduced. Silver solutions always need careful preparation and some diamine silver solutions can become explosive if kept for more than 24 h. If they are being used in a glass container, then a simple safety precaution is to wrap them up with adhesive tape (Sellotape); if an explosion occurs the glass fragments will be held by the sticky tape. Claims have been made that a designated glass container with an accumulated silver mirror should be used to minimize contamination and produce a 'clean' reaction. Unfortunately, there is a significant risk of them exploding with the slightest tap against the glass. It is important always to use distilled water in any silver method, as tap water will react with the silver salt.

Several silver solutions can be used in silver techniques but they are not directly interchangeable as they differ in their sensitivity to reduction.

Silver nitrate

This is the commonest form of silver salt used in the preparation of silver solutions. Simple silver nitrate solutions are sometimes used, e.g. von Kossa's solution, or as sensitizer solutions, e.g. Beilschowsky's method for nerve fibres, but for most techniques a more readily reduced form is needed.

Silver diamine

Silver diamine solutions are prepared by precipitating the silver with a hydroxide solution and then redissolving in a minimum amount of ammonium hydroxide. These solutions are very alkaline and this makes sections more liable to detach during staining, so an adhesive is often advisable. The final solution can be explosive if it is stored for more than 24 h, but has the advantage of being very sensitive.

$2Ag^+ + 2OH^- \rightarrow Ag_2O + H_2O$ Precipitation of silver oxide

$Ag_2O + 4NH_3 + H_2O \rightarrow 2[Ag(NH_3)_2]^+ + 2OH^-$ Dissolving to form silver diamine

Silver carbonate

Silver carbonate solutions are prepared by precipitating the silver using either lithium or sodium carbonate solution. The precipitate is filtered and washed. This removal of the precipitating salt is different to the previous example of silver diamine where the hydroxide is left in the solution. The precipitate is then dissolved using strong ammonia as for the diamine solution. Silver carbonate solutions are claimed to be even more sensitive than diamine solutions.

Hexamine silver solutions

These use hexamine (methenamine or hexamethylenetetramine). When mixed with silver nitrate, this produces a white precipitate that immediately redissolves without the need to titrate with strong ammonia.

Background deposits

Silver techniques often produce a non-specific deposit due to contaminants. Very small deposits can often be reduced by **toning**. This involves using 'gold chloride' (sodium chloroaurate):

$$3Ag + (AuCl_4)^- \rightarrow Au + 3AgCl + Cl^-$$

Thus, three silver atoms are replaced by one gold atom. For very small deposits this will result in a great reduction in size (thus reducing the background staining) but the large deposits of the impregnated tissue will hardly be affected. Gold toning also alters the colour from an intense black to a warmer brown/black colour.

Following completion of the technique, the sections are usually treated with 'hypo' (sodium thiosulphate, previously called hyposulphate). This is a photographic fixer that dissolves excess silver ions and prevents them later depositing as background. It is probably not necessary in histological preparations, as all of the silver is usually completely reduced so there is little risk of further reduction, but it is always done 'just in case'.

Silver techniques

Reticulin can be demonstrated using silver impregnation and the following is a fairly typical silver staining technique based on the method proposed by Laidlaw in 1929.

First the reticulin is oxidized to give aldehyde groups:

Then the silver solution oxidizes the aldehydes to acids and in the process is itself reduced to silver atoms that precipitate at the site of reduction:

Aldehydes are one of the commoner reducing groups in tissues and silver solutions can often be used to detect the presence of aldehydes.

Silver techniques vary quite widely in their conditions

There are many variations on silver techniques that seem to give good results. It is largely a matter of preference which technique works best in a particular laboratory. There are probably differences between the laboratories that are not particularly mentioned or even controlled that make one method more suitable for one laboratory than another. These variations include tissue fixation and processing, water quality (both tap and distilled or deionized water), ambient temperature and ambient light. The complexed ammonium ion concentration is highly significant and will decrease if old ammonia solutions are used to make up a working solution. Long term storage of small volumes in large containers is not good practice since ammonia will evaporate into the air space and escape on opening. There will then be less than expected available ammonia per given volume.

The actual concentrations of silver vary quite markedly from 1 g per 100 ml (Foot method) to 10 g per 100 ml (Laidlaw method). Times and temperatures also vary from 30 s (Gordon and Sweet method) to 60 min (Perdrau method) and temperatures from room temperature of 20°C up to temperatures of 70°C (Lillie method).

This wide variation might suggest that the technique is quite insensitive to conditions and would work reliably, regardless of any slight technical errors, but this is not the case. Silver techniques are more difficult to get exactly right than most staining methods and require care, patience and experience to get an even impregnation and lack of non-specific background. The wide variation and the number of modified methods is actually a reflection of this. Many people have tried, and largely failed, to get an automatic and reliable technique.

7.7 General treatment of sections during staining

After drying to achieve adhesion, paraffin wax sections are still not ready to be stained, as they are totally impregnated with wax, which forms a waterproof coating and prevents dye access to the proteins.

Section rehydration

For staining, the wax must be removed and the section rehydrated. This is conveniently done using stainless steel racks that hold a number of slides and flat staining dishes. The dry labelled slides are placed in the rack and the wax is removed. Xylene is still the most commonly used reagent for this process. Xylene is less commonly used for processing tissues because of its tendency to cause shrinkage and hardening, but there is no problem with shrinkage at this stage because the tissues are firmly attached to a rigid slide, and hardening is no longer a difficulty as there is no further sectioning to be done. Removal of the wax needs to be complete; if any wax remains it will result in uneven staining. Treatment with xylene for 5 min is usually sufficient.

The sections can then be transferred through a series of graded alcohols (typically 100%, 95%, 70%) and finally into distilled water. They do not need prolonged times in any of these baths since penetration is very rapid through the thin sections; 30 s with gentle agitation will usually be enough.

This process of returning a paraffin section to water is usually called either dewaxing, or 'taking the section to water' (see **Box 7.10**) or occasionally by the somewhat grander phrase of 'deceration'. One of these phrases will be found at the beginning of most staining schedules and must always be done when paraffin sections are used.

> **Box 7.10 // Dewaxing or 'taking to water'**
>
> If the section is frozen then the phrase 'take the section to water' can be ignored as the section is already in water. Often students do not think and try to process the tissue as if it was a paraffin section, which can totally ruin the section. For this reason, personally I prefer the phrase 'take the section to water' to 'dewaxing', as it is less prescriptive and suggests that the histologist needs to get the section into water from the medium the section is currently embedded in.
>
> Sections are best dewaxed immediately before being stained. Once rehydrated, they will slowly deteriorate, especially if kept in water. Deterioration is not rapid and sections can be kept in water, alcohol or xylene for quite long periods, although they do gradually lose adhesion. This is similar to the 'soaking' of pans with dried-on food residue. Since soaking will remove dried-on food, it is hardly surprising that it will also remove dried-on sections. It is also bad practice to allow sections to dry out at any stage in the staining, as this can introduce minute air bubbles into the protein mesh, which is the cellular structure. These air bubbles may remain and not be removed when the tissue is subsequently placed into another reagent and will end up in the final preparation. There are occasionally instances when sections need to be dried, but these are the exception.

Once fully rehydrated, the sections can be stained in aqueous reagents until they are ready to be mounted.

Automated staining

Automated processing of tissues is widely accepted and a similar automation is possible with staining. The same general principles apply to both situations. Automation frees staff from a routine task that is relatively straightforward and allows them to do more demanding tasks. The use of an absolutely regular procedure ensures that there is little variation in results, so that direct comparisons are valid from one batch of stained sections to the next. This accuracy and reproducibility are crucial in some applications such as diagnostic and exfoliative cytology (see **Chapter 13**) where the colour of the cytoplasm is an important diagnostic feature.

The disadvantage is that there is less flexibility. All of the sections will be given the same treatment, regardless of their requirements. It is also only feasible for techniques that are carried out for a large number of samples. Machines are fine for doing hundreds of haematoxylin and eosin stains, but it is not reasonable to use a machine for stains where the technique is only required for two or three slides each day. It also does not lend itself to situations where different results are needed; for example, when photographing at low magnifications, an overstained section will give better results than the usual staining intensity. An ordinary stain will give insufficient contrast for the film's recording capabilities but a more-intense stain will give stronger differences between the tissue components.

Automated staining also demands reproducible reagents. If there is a change in a reagent's staining properties, the machine will not recognize this or compensate for the change in the way that a person would. Most histologists can easily compensate for gradual changes in reagents as they age or for sudden alterations from a new batch of stain without too many problems. Machines only follow the program and cannot tell that there is any need to change. Any alterations result in machines needing to be reprogrammed, for example, if a different batch of reagent is prepared. This inflexibility may also result in reagents being discarded sooner than they would be for manual staining in order to maintain a standard program.

Automated staining machines are also less flexible in producing single stains, even when they are already programmed for that stain. Thus, producing a single slide may hold up some types of machine; these machines must go through the full cycle before another section can even begin since the steps are uneven. These machines are inefficient for staining single sections. An alternative strategy is to have all the steps the same length (e.g. 1 min) so that sections can be added at any time and will follow the same path. The difficulty here is that, if a longer time is needed, then several baths of the same reagent are required. These machines often cannot cope with large numbers of sections in a short space of time.

Automated staining machines are very useful for absolute regularity with large numbers of sections needing the same treatment at the same time. They have found a significant role in two main areas:

1. Haematoxylin and eosin staining in histology, Papanicolaou staining in cytology and blood-film staining in haematology. This is because the sheer numbers of slides needing staining make it worthwhile.
2. Immunohistochemistry, nucleic acid hybridization and similar techniques. Here the actual numbers are smaller but the need for absolute consistency is greater, so these techniques have moved to more automation.

The use of automatic coverslipping machines is often linked to automated staining. The process of mounting sections is very mundane, so automation is possible. There is less requirement for variety in mounting, so provided they are working well these machines are a useful addition to the laboratory.

7.8   Section mounting

Use of coverslips

Mounting of sections under a coverslip is essential to get the best and clearest view of the specimen. You only need to compare an unmounted wet section at the end of staining with a properly mounted section to see the difference. The microscope manufacturers usually assume that the specimen will be mounted in a medium with a high refractive index and covered with a thin glass coverslip and calculate all of their optical corrections on that basis. The difference can be seen in **Fig. 7.16** where the optical paths through a wet section and a mounted section are compared. The unmounted section has twice as many refracting surfaces and the opaque tissue will transmit much less light.

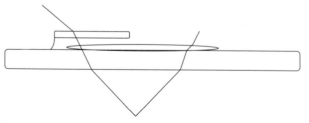

Figure 7.16
Effect of a coverslip on viewing sections. On the left, a section mounted in a high refractive medium has only two refractions, whilst a section in water has up to four refracting surfaces

The coverslip should have a thickness of 0.17 mm (No. 1 coverslip) for the best results, as this is the thickness used in optical calculations. Thicker coverslips such as a No. 2 coverslip will interfere marginally with the clarity and very thick coverslips may even prevent the oil-immersion lens being used as they can have a greater thickness than the normal working distance of the oil-immersion lens.

Mounting media

The mounting medium should have a high refractive index (RI). Most tissues have an RI of between 1.5 and 1.55, so a mounting medium with an RI in this range will give maximum clarity. There is no single mounting medium that is suitable for all specimens and stains. There are two major types of mounting media used and the difference is in the solvent. The commonest types are the resinous mounting media, which are based on hydrophobic organic solvents, usually xylene, and which need the section to be dehydrated and cleared before mounting. Water-based mounting media will accept tissues straight from distilled water and are used when a xylene-based medium would not be appropriate, e.g. if the dye or histochemical reaction product is soluble in xylene or dehydrating alcohols.

The properties that need to be considered in a mounting medium are:

1. Refractive index. If the RI is much lower than 1.5, then tissues will not be completely transparent and diffraction will occur. This is usually a disadvantage as it reduces clarity but it can sometimes be an advantage as it will give some contrast to even unstained tissues.
2. Clarity under normal conditions of use. Some media can become opaque as they dry out and are not suitable for long-term preservation.
3. Effects on the stain itself. Some mounting media will cause fading. This is most common with acidic mounting materials, which will cause significant fading, especially in the light. Some media may also act as solvents for the dyes and as a consequence the dye diffuses or leaches out into the mountant. This will gradually obscure the tissues.
4. Fluorescence. This is really only critical for fluorescence microscopy but it is generally a useful characteristic for a general mounting medium since it eliminates the need to use a special mountant when fluorescence is being used.
5. Setting. The ability of a mountant to dry or set quickly and hold the coverslip in place is very useful. Many aqueous-based media fail to harden sufficiently and the coverslip will need 'ringing' to preserve the section.

Resinous mounting media

Canada balsam. This was the original resinous mounting medium used in histology. Canada balsam is derived from the *Abies balsamea* fir tree and is available as a dried, brittle, yellow solid. It will melt at high temperature and is soluble in xylene. Approximately 60 g in 100 ml of xylene gives a good working mountant, although it takes a few days to dissolve completely. The yellow colour of the mountant hardly seems to matter when viewed through the microscope. The mountant is usually significantly acid and will cause fading, especially of basic dyes. It is relatively expensive and is mainly of historical importance rather than being a common mountant.

DPX. This is a synthetic polystyrene resin that is dissolved in xylene and has some plasticizer added. The initials come from the components: Distrene 80 (a commercial polystyrene), plasticizer (e.g. dibutyl phthalate) and xylene. It is a water-white clear solution and is one of the more popular mountants in use. It has very little tendency to fade dyes and hardens in about 24 h. The specimens do not need ringing.

Other synthetic media are available such as Permount or Entellan, which are commercial brand names.

When a section comes to the end of normal staining and is ready for mounting in a resinous mounting medium, it needs to be dehydrated and cleared in xylene before finally being mounted. This is simply the reverse of the dewaxing, but it is better not to use the same reagents as they gradually become contaminated with the other reagents, e.g. dewaxing will leave wax in the xylene, which can interfere with the mounting medium.

Aqueous mounting media

There is no fully satisfactory aqueous medium and several different ones are used for different purposes. They differ in the way in which the RI of water (1.33) is raised sufficiently to give a clear image. Most are best considered as temporary mounts and need ringing to hold the coverslip in place and prevent drying out. Tissues do not need any treatment before mounting and can be mounted directly from water or buffer.

Glycerol. Glycerol is a trihydric alcohol with a high RI. It can be used alone or with the addition of a buffer to control the pH. It is a useful medium for fluorescent staining, for example, for immunofluorescent antibody techniques. The addition of *p*-phenylenediamine is said to retard the fading of fluorescence. It neither hardens nor dries out and is usually used as a very short-term mountant, although it can be ringed for slightly longer use.

Glycerol jelly. This uses the addition of gelatine (up to 12% in some formulations) to allow the medium to set. The usual formulation has a lower RI (1.42) than most mounting media, so the clarity is reduced and some unstained structures will be visible. It is solid at room temperature and needs to be melted in a waterbath before use. It is very easy to get air bubbles trapped in this medium, so it is convenient to melt it and get rid of any air bubbles by warming it in a vacuum-embedding oven. Glycerol jelly is quite a good growth medium for some bacteria and fungi, so there is usually an antibacterial additive (e.g. phenol), but it still does not keep well. Sections may also allow the growth of organisms in storage, so it is best thought of as a temporary mount.

Apathy's medium. This uses a gum (gum arabic or gum acacia) and sucrose to raise the RI. It has an RI of around 1.5, so it can give nicely transparent preparations. It has a tendency to crystallize in storage and can set by drying but this is quite slow. Again, it may need the addition of an antibacterial agent to help preserve it.

Polyvinyl alcohol or polyvinylpyrollidone media. These are synthetic and less liable to bacterial contamination than the organic-based mountants, although the addition of phenol is still advisable. They dissolve in water or buffer but need constant stirring. They solidify slowly by evaporation but specimens can be ringed to prevent this. These are more permanent than the other water-based mounting media, but are still not as good as a resinous medium.

Temporary mounts need ringing

Ringing is the term used for sealing the edges of a coverslip when the mounting medium does not set. Ringing was originally so called because the coverslips were round and so there was a ring of the sealant round the coverslip. Ringing was done on a turntable to give a nice neat finish. Originally it used a gold size followed by a black asphaltum varnish. This produced a very neat finish and some commercial suppliers of prepared slides still finish many of their preparations in a similar way as it looks good. Most laboratories have dropped this and ringing is now just a temporary expedient rather than an aesthetic requirement.

Good temporary ringing can be achieved in a number of ways. Ordinary nail varnish works quite well and comes in a bottle with its own brush, which makes it convenient and simple. The only drawback is that it is dissolved in acetone, which may affect some materials, although I have never found this to be a problem.

Many styrene-based cements can also be used and again are convenient as they come in tubes ready to squeeze out around the coverslip. Again the solvent is a theoretical problem but I have not had problems. These cements can often be semi-permanent. Paraffin wax can also be used. A piece of warmed metal (such as the flat end of a broad spatula) is used to apply a layer of molten wax, which immediately sets. Provided the slide is dry, this is quick and easy but the seal can be easily broken and will not store well.

Storage of slides

Mounted slides should always be carefully labelled and stored horizontally until fully dry and set when they can be stored on their edge or end. Stained slides should be stored away from light as the dyes will fade even in the best mountant.

Suggested further reading

Bancroft, J., Layton, C. and Suvarna, S. (2012) Chapter 9: How histological stains work. In: *Bancroft's Theory and Practice of Histological Technique*, 7th edn. Elsevier Health Sciences.

Kiernan, J.A. (2008) *Histological and Histochemical Methods: theory and practice*, 4th edn. Scion Publishing.

Suggested websites

http://library.med.utah.edu/WebPath/webpath.html
www.histology.leeds.ac.uk/
http://stainsfile.info/StainsFile/jindex.html

Self-assessment questions

1. What is the role of chromophores and auxochromes in dye structure?
2. How do basic and acidic dyes bring out the structure of tissues? Name one acidic and one basic dye.
3. How do pH and salt concentration alter dye binding?
4. A small amount of mordant causes staining but an excess of mordant removes the staining. Explain this oddity.
5. Why does haematoxylin mordanted with aluminium salts stain nuclei but other mordants cause haematoxylin to stain connective tissues or nerve fibres?

6. Toluidine blue will stain mast cell granules red. What is the name of this phenomenon? Why does the colour change occur?

7. Name one red and one blue nuclear stain. When would you use a red nuclear stain and when would you use a blue one?

8. Why do some haematoxylin solutions initially improve with keeping and then deteriorate?

9. Outline why permeability and dye size might explain trichrome staining with three acid dyes.

10. Distinguish between argentaffin and argyrophil silver impregnation.

11. Why is silver the best metal for metallic impregnation techniques?

12. Why might a lipid-staining technique recommend mounting in glycerol jelly instead of DPX?

13. Why do most laboratories routinely use a resinous mounting medium?

08 Histochemistry: general considerations and the histochemistry of carbohydrates and lipids

Learning objectives

After studying this chapter you should confidently be able to:

- **Outline the problems associated with identifying and localizing chemicals in tissue sections**
  Histochemistry is limited by the need to have an insoluble coloured product to localize the site of reaction. Tissue preparation must retain the materials in their original site. Careful controls are needed to ensure that methods are sufficiently sensitive and selective.

- **Describe the periodic acid–Schiff (PAS) reaction and be able to outline the problems in interpreting the PAS reaction**
  The PAS reaction detects materials containing sugars, including polysaccharides, mucins and glycoproteins. In the PAS reaction, the periodic acid oxidizes the adjacent hydroxyl groups to aldehydes, which then react with the Schiff's reagent. Difficulties include non-specific staining due to pre-existing aldehydes in tissues and identifying the source of a positive reaction from a wide range of materials that are PAS-positive.

- **Describe how to identify and localize mucins in cells and tissues**
  Mucins can be classified into major groups using PAS, alcian blue and aldehyde fuchsin in combination. Further identification can be achieved using enzymes and controlled electrolyte concentration to inhibit staining.

- **Describe the histological classification of mucins**
  Mucins are classed into neutral mucins, which have no acidic groups on the carbohydrate, and acidic mucins, which have acidic groups that react with alcian blue. The acidic mucins are further subdivided depending on the type and number of acidic groups into carboxylated, strongly sulphated and weakly sulphated types.

- **Describe how lipids can be demonstrated in tissues**
  Many lipids can be demonstrated by their hydrophobic nature and the ability to concentrate fat-soluble dyes from an aqueous solvent (lysochromy or sudanophilia). Some lipids require special treatment before showing sudanophilia. Special techniques can be applied to identify subgroups of lipids.

Histochemistry is a method of investigating tissue structure that, instead of using relatively crude dyeing of the tissues, uses reasonably specific chemical tests to identify and pinpoint the location of chemicals within the tissues. In this regard, histochemistry is a modified form of chemistry or biochemistry.

131

There are a number of fairly specialized subdisciplines within histochemistry, each of which has its own problems and solutions. These include the histochemical identification of: carbohydrates and lipids (discussed in this chapter), nucleic acids, proteins and enzymes (see **Chapter 9**), and pigments and minerals (see **Chapter 10**).

8.1   Limitations of histochemistry

Histochemistry may have a lot in common with biochemistry but it has certain extra problems as biochemists investigating tissues usually have some quite significant advantages over histochemists.

- Histochemistry must produce a coloured final result to make it visible under the microscope. Biochemists can certainly employ methods that produce coloured final reaction products, but they can also quite happily employ methods that use materials only 'coloured' in the UV region and such results are invisible to the human eye and useless for microscopy.
- Histochemistry must produce a material that is insoluble so that the position of the chemical reaction in the tissue can be located accurately. Slight solubility of the final material is a problem as it can allow the coloured material to move from its true site to a nearby location and give a false localization.
- Histochemistry must be highly sensitive as it must accurately detect the material present within a single cell and in some cases within a single organelle such as a mitochondrion. Biochemists can often use several kilograms of tissue to make a single preparation and concentrate the material they are investigating.
- Histochemistry must never allow the material being investigated to be separated from its environment, as pinpointing its position in the tissue is the rationale for using histochemistry. The biochemist can separate and purify the material using a range of techniques such as chromatography, electrophoresis and dialysis. Indeed, the speed of movement of the material in such separation techniques is often the best method of identifying closely related materials.

Histochemistry is therefore often more difficult to achieve in practice and is often less certain in its identification of unknowns than the equivalent biochemical detection. It is also much more difficult to get an accurate quantification (see **Section 18.2**). Thus, it is best to think of biochemistry and histochemistry as complementary techniques that supplement each other rather than competing with each other; they give different information and, when used together, give a more complete picture of the chemistry of the cell than either could alone. Biochemistry can determine the types and quantities of material present in a tissue, whilst histochemistry can identify the position and distribution of the materials. Histochemistry also relies on the knowledge gained from biochemistry to enable satisfactory histochemical techniques to be developed.

8.2   Use of controls in histochemistry

Controls are important in histochemistry and should include a minimum of a **positive control** and a **negative control**. Without controls the methods are little more than demonstration techniques, which, although still useful, cannot be considered as rigorous chemical testing.

Positive controls

The purpose of a positive control is to check that the technique is working satisfactorily. For this purpose, you need a weakly reacting material. The reason for using a positive control is to ensure that any material in the test section will be detected. If a strongly reacting control is used, it is possible that the test is working for a strong reaction but is insensitive to a weak reaction. A strongly reacting control may also be included to give a good positive result for comparison, but it should be in addition to a weak control, not a replacement. Where a material may have several forms that react differently (such a case would be amyloid, which has a number of different types), it may be necessary to have more than one positive control material (see **Box 8.1**). Thus, for amyloid the controls may include both the common AA and AL amyloid types (see **Chapter 11**) and perhaps even some of the less-common types if they are likely to occur in one or more of the test sections.

Where a technique is being used quantitatively, the use of a range of controls is needed to cover the range of results expected and to act as a standard for the quantitation.

The positive control should be as close as possible to the material being looked for in the test section. For example, using the periodic acid–Schiff (PAS) technique, it would be possible to check that the technique is working by using glycogen, a mucin, reticulin fibres or several other materials. If you are looking for glycoproteins in the pituitary, the control should be glycoprotein in the pituitary and not glycogen in the liver. The use of a slightly different substance is not as good a control as it may react differently.

The positive controls should be carried out at the same time and in the same way as the test sections. It is not enough to control the technique once at the beginning of the day and then use the result for several different batches of tests run at different times. A single positive control may be sufficient for a batch of sections that are tested at the same time, provided the test sections are similar in type and in the reaction expected.

Box 8.1 // Control blocks can include many tissues

If many different tissues are being examined together, then the number of control sections can become tedious. In some cases, it is possible to have a large number of control tissues yet only one control section. The method is to pack a large number of test tissues into one block. The tissues may be blocked out separately in one mould or small rods of tissues can be cut and assembled into a single checkerboard or composite block. For control purposes, you only need a very small sample of tissue to see if the technique is working as expected, so as little as 1 mm^2 of tissue will suffice. A single checkerboard may contain 64 separate tissue samples in a single section. This is an effective way of reducing the number of control slides without compromising the number of control tissues. The only disadvantage is the time and effort needed to make the composite blocks.

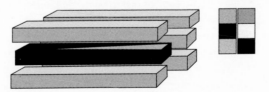

Rods of tissue are assembled into a solid block and when sectioned appear as a checkerboard of tissues.

Negative controls

Negative controls are used to test the specificity of the reaction and ensure there are no false-positive results. The tissue should again be as close as possible to the test section in type and in expected reaction. In many cases it is possible to use another slide of the test section as a control by destroying the reactivity of the material being tested (see section on Treated tissues below.) It may be necessary to have a negative control for each section if there is some background colour that might be confused with the positive result. Only by comparing the negative and test sections can you be sure that the result is a genuine positive result.

Control results

The controls should always be examined first. If the results are satisfactory, then the test slides can be examined and the results analysed.

If there is any sign of a positive result in the negative control, then the method may not be specific and some factor other than the test material may be reacting. If the positive control is negative, or unexpectedly weakly reacting, then the method may not have been sufficiently sensitive or may have been performed wrongly. In both cases, the tests will need to be repeated. The controls may need to be investigated first to ensure that the results can be relied on before repeating the tests.

A simple rerun of the controls is a useful starting point, taking extra care to ensure that all the steps are carried out correctly. This will detect whether the results were due to a simple error in technique. The rerun can be done using fresh reagents to ensure that it is not a reagent error and different control slides to ensure that it is not a control section error. If the rerun fails to give the expected results, then the whole technique needs to be reassessed. Each step may need to be controlled separately to detect which step is going awry. This may result in a large number of controls but is the only way to ensure reliable results.

Controls can be of various types

The source of controls is important and it is sometimes necessary to use different types of control tissues especially when investigating unexpected results.

Control tissues

Positive control. This involves having tissues that you know contain the material you are looking for. The material can be a sample from a previous test that came up positive. Most laboratories are constantly looking for good positive-control materials to use in subsequent tests. The use of stock controls assumes that the material is stable enough to be kept. Some control materials do deteriorate during storage and may need regular replacement.

If the material cannot be stored or if there is no stock control available, it may be necessary to use fresh animal tissue. This would be the case for enzyme techniques where it may not be possible to store tissues as the enzymes may rapidly deteriorate. In this case, it is essential to use a tissue that you are certain contains the material being detected.

Negative control. Here it is essential to choose a tissue that does not contain the material being demonstrated to check for non-specific reactions. Ideally, the negative-control tissue should be from the same organ and the same species as the test sections.

Treated tissues

These are an excellent negative control since the actual substance being tested is treated to remove or destroy its staining ability. This can be used not only to ensure that the method is specific but also when the method is known to be non-specific. For example, the PAS technique is useful but stains many different materials and is not specific for glycogen. It can be used to demonstrate glycogen specifically by using a control in which the glycogen is removed by enzyme digestion with amylase. Other PAS-positive materials such as mucins, glycoproteins and reticulin will be unaffected. Thus, a material that stains in both the test and treated tissue control is not glycogen. Any staining that is positive in the test but missing from the amylase-treated tissue is glycogen. These controls are not as easy to examine and interpret as the use of sections that are naturally negative, but this is a vital method in many histochemical investigations.

The reactivity of the material can be destroyed in a number of ways:

1. Enzyme digestion, e.g. amylase for glycogen, collagenase for collagen, ribonuclease for RNA.
2. Extraction with a solvent, e.g. chloroform/methanol will dissolve lipids.
3. Use of a specific inhibitor, e.g. oxaloacetic acid will specifically inhibit succinic dehydrogenase but not other enzymes.
4. Heating to destroy enzyme activity, e.g. acid phosphatase using Gomori's lead phosphate technique; any coloration will be non-specific binding of the lead salts.
5. Blocking of reactive groups. This may involve reacting the group with reagent, e.g. acetylation of hydroxyl groups (using acetic anhydride in pyridine). The term is also loosely applied to complete removal of the group, e.g. deamination with nitrous acid (see *Box 8.2* for further comments).

Altered technique

It is possible to ensure that there is no non-specific reaction simply by omitting an essential step in the technique. For example, an easy negative control for an enzyme is to omit the substrate from the incubating solution. The absence of substrate means that the enzyme cannot catalyse its normal reaction so any positive result is non-specific. Similarly, omitting the periodate oxidation in PAS acts as a good test for any preformed aldehydes.

These types of control are very useful for investigating unexpected positive results. By omitting different steps in the procedure it is often possible to identify the source of the unexpected reaction. This control of individual steps is used in immunological and related techniques where there may be six or seven steps that could be the source of the false-positive reaction.

Box 8.2 // Blocking controls

The blocking methods are also used to prevent background reactions by removing the reactive material before performing the test. Thus, preformed aldehydes in tissues can be blocked by reduction with borohydride. This is useful, for example, after glutaraldehyde fixation when there are many free aldehydes from the glutaraldehyde. The use of any method using Schiff's reagent on glutaraldehyde-fixed tissues is therefore difficult unless these abnormal aldehydes from fixation are first blocked.

The use of blocking is also important in immunoperoxidase methods to destroy endogenous peroxidases (see *Chapter 12*).

8.3   Use of histochemical investigations

Histochemical tests as simple staining methods

Many histochemical techniques are used in routine histology not so much for the chemical information they can provide as for the fact that they are often better at showing a particular structure than simple dyes could ever be. Examples include the use of acetyl cholinesterase techniques to demonstrate motor end plates, the PAS technique to visualize goblet cells, nucleic acid techniques to identify plasma cells and lipid techniques to show myelin degeneration. In these cases, the histochemical controls are often omitted since the technique is not being used in its most critical way. This should not be taken to mean that the controls are never necessary, just that they can be omitted for simple demonstration methods; they would still be needed for more rigorous histochemical testing.

Staining techniques in conjunction with histochemical investigations

It is also true that many staining techniques that are not strictly histochemical are often used in histochemical investigations where they may confirm the results of another test. Two or more non-specific tests may between them give results that are quite acceptable as histochemical identification, even though individually they are not specific. Thus, alcian blue will stain many forms of mucin and aldehyde fuchsin will stain many materials including sulphated mucins, elastic fibres and some cell inclusions. If a material stains with alcian blue and with aldehyde fuchsin (in separate sections, not in the same individual section), then it can safely be assumed that the material is a sulphated mucin. Individually the two techniques are not specific but their combined results are at the very least highly selective.

Whether and how to fix

In all histochemistry, as indeed in all histological investigations, the early treatment of the tissue is vitally important. Improper treatment of the tissue, such as inappropriate fixation, will result in the material under investigation being altered, moved or even completely lost. As histochemistry involves precise chemical investigation, it is often done on unfixed tissue. This means that the tissues are unstable and must be used quickly. Once the histochemical reaction has occurred and the final insoluble product is formed, it is often possible to post-fix the tissues to make them more permanent.

Box 8.3 // Histochemistry is not always difficult

Histochemistry is sometimes perceived as being difficult and obscure. This is not true. Histochemistry is not always difficult or complex. The Perls technique for iron is one of the most sensitive of the histochemical techniques. In this case, ordinary paraffin wax sections will suffice since the material being identified is protein linked and therefore retained in paraffin sections. The reagents are not complex or critical and the Perls method can be used for first-year degree practical classes. It works reliably for the students and they can observe the iron present within the spleen cells. Individual macrophages can be distinguished easily in their preparations. Any technique that can be done first time every time by inexperienced workers and still detect the minute amounts of iron present inside a single macrophage can hardly be called difficult.

Not all histochemistry is this easy, but a surprising amount can be performed with little difficulty given reasonable care.

8.4 Carbohydrates

The carbohydrates found in tissues are mainly sugars and their derivatives. Simple sugars such as glucose and fructose are extremely difficult to preserve in tissue sections as they are very soluble in water. Distinguishing between simple sugars by chemical tests is quite difficult and glucose is also almost ubiquitous in cells and tissues. The identification of free monosaccharides is rarely attempted in tissues since the sugars are likely to have been lost or moved from their original site and any positive reactions are likely to be the glucose that is found everywhere.

Sugars can, however, be bound to a number of other materials to give a range of sugar-containing materials that are of interest to histochemists. The presence of any sugar can usually be identified by the periodic acid–Schiff (PAS) technique. This is the single most important technique in carbohydrate identification as it identifies most neutral sugar- and sialic acid-containing materials. It does not, however, identify the type of sugar; it simply indicates that a saccharide is present.

The PAS reaction

The PAS reaction involves a two-step chemical reaction in which there is an oxidation step in periodic acid to generate aldehyde groups from the carbohydrate, followed by the addition of a reagent sensitive for aldehydes (Schiff's reagent), which stains the site of the sugars. The method is widely quoted in descriptions of cells indicating whether the cell is PAS-positive or PAS-negative (see *Box 8.4*).

Box 8.4 // Negative results with the PAS technique can still be useful

The PAS reaction is such a useful indicator that it is one of the few stains where a negative result is often quoted and many cellular components will be described in books as PAS-negative. The negative result means that no sugars are present, so it acts as a method of distinguishing glycoprotein hormones from simple proteins, e.g. in the pituitary.

This use of the negative result is quite unusual and means that it is used in situations where you expect no reaction. This means that controls need to be more stringent than normal since a negative result is expected and poor technique will give a negative result.

Other techniques that have this distinction include Sudan staining and the Perls method. It is also the case with the Gram bacterial stain where Gram-negative indicates a great deal about the structure of the bacterium.

Periodate oxidation

The oxidation reaction acts on the glycol (hydroxyls on adjacent carbons) part of the sugar and produces two aldehydes and at the same time splits the bond between the two carbons:

$$
\begin{array}{c}
H-\overset{|}{C}-OH \\
| \\
H-\overset{|}{C}-OH \\
|
\end{array}
\quad +IO_4^- \longrightarrow \quad
\begin{array}{c}
H-\overset{|}{C}=O \\
\\
H-\overset{|}{C}=O \\
|
\end{array}
\quad +IO_3^- +H_2O
$$

Reaction with Schiff's reagent

The aldehydes then react with Schiff's reagent to produce a product that is a strong red/purple colour. The reaction can also occur with some derivatives of sugars and glycols, e.g.

amino and alkylamino sugars. Most materials that contain sugars, such as glycoproteins, will be PAS-positive. Strongly sulphated mucins in connective tissue and hyaluronic acid are negative with the standard PAS technique, despite containing large numbers of sugar groups, but other mucins are PAS-positive.

Oxidation agents

Periodic acid (HIO_4) is the most commonly used oxidant. It is made up as a 0.5 or 1% aqueous solution (0.044 M) and oxidation is for 5–10 min at room temperature. A longer oxidation may be needed for some materials to become Schiff-reactive (e.g. sulphated mucopolysaccharides). Adding extra steps to the technique may also increase the range of materials that can be demonstrated (e.g. borohydride treatment). Alcoholic periodic acid can be used instead of the aqueous solution if there is the possibility of the material dissolving in water (e.g. R.D. Hotchkiss, in 1948, recommended alcoholic periodate oxidation for glycogen).

Other oxidants can be used to give the same reaction, but most will continue oxidizing after the production of aldehydes, and this further oxidation of the aldehydes to acids results in loss of staining. This further oxidation does not happen with periodic acid and it is this that makes periodic acid the most popular method. Other oxidants that can be used include chromic acid, potassium permanganate, lead tetra-acetate and sodium bismuthate. These oxidants need more precise timing to obtain maximal staining and it is this critical timing that makes them much less popular. However, they are occasionally included in other staining methods; for example, chromic acid oxidation is used in the Gridley technique for fungi, and also the Grocott method for *Pneumocystis*.

Schiff's reagent

Schiff's reagent is a sensitive reagent for detecting aldehydes and stains them a pink/purple colour. It is quite specific for aldehydes. Even the related ketones do not react with the PAS method. Schiff's reagent consists of a solution of basic fuchsin (pararosaniline) that has been decolorized with sulphurous acid. The sulphurous acid adds an extra sulphurous group to the central carbon of the dye. This disrupts the chromophore group and its associated electron resonance with consequent loss of colour.

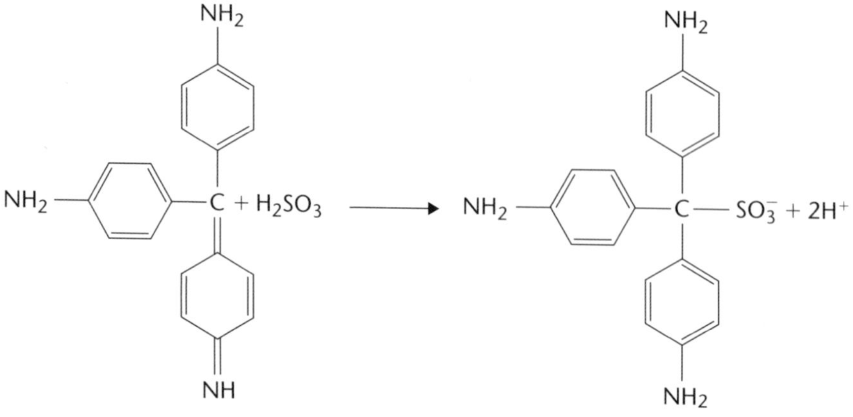

Pararosaniline with
the quininoid chromophore
as the bottom ring

Schiff's reagent
has no quininoid ring
and is colourless

On reacting with the tissue, it seems likely that the alkylsulphonic derivative of the dye is formed. This entails a rearrangement of the sulphonic group and restores the chromophoric group, but this is not simple restoration of the original dye colour since the product is a significantly different colour to the original dye.

The quininoid ring chromophore is restored
when the Schiff's reagent reacts with an aldehyde

To prepare Schiff's reagent therefore requires a solution of basic fuchsin or pararosaniline. The fuchsin needs to be specifically for the preparation of Schiff's reagent as not all 'basic fuchsins' are suitable.

Varieties of Schiff's reagent
The sulphurous acid needed to decolorize the dye can be produced using the following methods:

- Sulphite or metabisulphite and HCl at 50°C (de Thomasi method). This is one of the most popular recipes.
- Metabisulphite and HCl at room temperature with prolonged shaking (Lillie method). This 'cold Schiff' is more acidic than the previous recipe.
- Thionyl chloride (Barger and Lamarter method).
- Bubbling sulphur dioxide through the solution (Itikawa and Oguru method).
- Sodium dithionite (Alexander method).

Sensitivity and shelf-life
Although these different methods all produce effective reagents, there are differences in their sensitivity and shelf-life. The amount of sulphurous acid seems to be the main difference affecting the Schiff's reagent. If there is an excess of sulphurous acid, the Schiff's reagent will have better storage qualities but may have reduced sensitivity. Schiff's reagent should be stored at 0–4°C when it may keep for months.

After the sulphurous decolorization, Schiff's reagent is often a yellowish colour and this can be removed by treating the solution with activated charcoal. A slight yellowish tinge does not seem to matter and Schiff's reagent will still work fine, but if a pink colour is produced, this indicates degradation of the reagent and it should be discarded. Because of the problems involved in preparing Schiff's reagent, many laboratories now prefer to buy it from a commercial supplier.

Washing out
In the original techniques, the Schiff's reagent was washed out of the tissues using sulphurous reagents as a rinsing solution. For most routine purposes, the use of sulphurous

acid rinses after the Schiff's reagent has largely been dropped in favour of simply washing in running water for several minutes; however, sulphurous rinses may still be needed for critical applications. The water wash generally increases the intensity of the colour as any excess sulphurous acid is removed.

Alternative reagents

A PAS-stained section can be counterstained in many ways but the use of a nuclear stain such as haematoxylin is usually sufficient. The traditional Schiff's reagent prepared from fuchsin or pararosaniline can be replaced by reagents prepared with other dyes such as thionin. Instead of Schiff's reagent, the aldehydes can be detected with a silver reagent, usually methenamine (hexamine) silver. This technique is called the PA–silver method but is less popular.

Interpreting the results of a PAS reaction

Many materials are PAS-positive but if badly carried out the PAS technique can also give false-positive results, so controlling and interpreting the results are very important before relying on the PAS reaction as a method of identifying a material.

False-positive reactions. The first type of false-positive PAS reaction is given by an improper Schiff's reagent. If all of the dye has not been decolorized or if the Schiff's reagent is old, the section may develop a pink colour independent of the presence of aldehydes. This should not happen if the Schiff's reagent is prepared correctly, is reasonably fresh and has been stored correctly. Schiff's reagent should always be examined for signs of deterioration before it is used, but deterioration of the reagent is always a possible explanation if results seem unusual.

The second type of false-positive reaction occurs when there are preformed aldehydes within the tissue before the periodic acid oxidation. These aldehydes will react with the Schiff's reagent and give the same colour as if it was a PAS-positive substance. Preformed aldehydes can come from the use of an aldehyde fixative. Formaldehyde fixation is not normally a problem as the formaldehyde is effectively washed out of the tissue during processing, but other aldehydes, particularly glutaraldehyde, can cause difficulties.

Preformed aldehydes also occur naturally in small amounts in tissues, and particularly after mercuric chloride fixation they can appear as the so-called **plasmal** reaction. The plasmal reaction is mainly due to lipids ('plasmalogens') and so this is more likely in frozen sections, but it can occur sometimes in paraffin sections. Blocking of fixative-derived or preformed aldehydes is possible, e.g. using sodium borohydride, but is not a common procedure.

To ensure that non-specific reactions are not occurring, a second section can be carried through the technique but omitting the periodic acid oxidation. Any material or component stained in both sections is not PAS-positive but something else, usually a preformed aldehyde.

False-negative reactions. The cause of false-negative reactions may be a reagent that is not working effectively or inadequate treatment with one or more reagents. A positive-control section where you are sure that the material is present should be stained alongside the test section to ensure that the technique is working. The control should be a weak rather than a strong positive and should be as close in type and material to the test section as possible.

A negative result when a positive reaction was expected may also be due to poor technique; staining a slide upside down is one example. The test should be repeated ensuring that no errors occur.

The materials being detected may have been lost due to processing or delayed fixation. If this is the case, the only way to correct it is to get a new specimen.

PAS-positive materials

Some of the more commonly encountered materials that can be PAS-positive are shown below. This is not a comprehensive list but indicates the range of materials that need to be considered when a positive reaction is found.

Amyloid	Glycogen
Basement membranes	Hyaline membrane of neonatal lung
Cartilage	Lipochrome pigments
Cellulose	Pituitary basophil granules
Cerebroside lipids	Starch
Chitin	Thyroid colloid
Epithelial mucins	Zymogen granules in pancreas
Fungi (chitin)	

If PAS is being used for a specific confirmation, then the identity of the material will probably be obvious. The problem of identification can arise if there is an unexpected positive reaction or if the identification of a material needs to be confirmed beyond doubt.

There is no simple procedure to be certain of the identity of a PAS-positive material. Several techniques can be applied:

1. Simple appearance can often be useful. Some PAS-positive materials can be recognized simply by their shape (morphology), e.g. fungal hyphae are sufficiently distinctive for their morphological appearance to be enough for a positive identification.
2. Stain a second section with another technique that selectively stains the material that you believe is the source of the PAS reaction. For example, starch can be identified by treating it with iodine when it will develop a distinctive blue colour. Even if the other technique is not itself specific, the combination of PAS reactivity and staining with a second technique may be unique, e.g. lipochrome pigments also stain with the Ziehl–Neelson technique and Sudan black, so anything that stains positive with all three methods can be identified as lipofuscin.
3. Dissolving or destroying the material chemically, e.g. dissolving glycolipids in hydrophobic solvents shows that they are soluble in organic solvents and this is a strong indication of their lipid nature. This requires two sections to be stained together. One section is pretreated with a lipid solvent and after the extraction both sections are stained using the PAS technique. Materials that stain in both sections are not glycolipids since they are insoluble in the lipid solvent. Glycolipids will show positively in the section that was not extracted but will not stain in the treated section.
4. Enzyme digestion. Really this is the same as (3) but enzymes are such precise reagents that this can be used as a positive identification and is therefore worthy of separate consideration. This is useful for glycogen. Diastase (amylase) specifically digests glycogen and starch. Since starch is not a normal constituent of animal tissues, this makes it highly selective and starch can be eliminated by iodine staining. Predigesting one section with amylase and comparing it with an undigested section, both of which have been stained by PAS, allows identification of the glycogen. It also, of course, eliminates glycogen as a possible material when both sections are positive.

Specific carbohydrates found in tissues

Polysaccharides: glycogen and starch

Glycogen is the major form of carbohydrate found in human tissues and acts as the main energy store in many cells such as muscle and liver (see **Box 8.5**). Glycogen is one of the commonest materials giving a PAS-positive reaction in tissues. Glycogen is a polymeric form of glucose (the glucose is mainly α-1,4 linked in straight chains but there are some α-1,6 links, which give branched chains). It is related to starch, which has the same general chemical composition but is generally found as larger polymeric molecules. Since only one material (glucose) makes up these materials, they are termed homoglycans. Two forms can be identified by electron microscopy:

- α-glycogen, which forms distinctive rosettes of β particles;
- β-glycogen, in which the individual components of the rosettes are free as individual 20–40 nm particles.

Glycogen is found in many cells, with large amounts in liver and muscle cells.

Box 8.5 // The glycogen content of cells is variable

The amount of glycogen is quite variable in tissues as it is used continuously as an energy source by cells. The amount of glucose can alter quite markedly depending on diet and activity. Long-distance runners attempt to manipulate their glycogen stores by eating a high-protein/low-carbohydrate diet until shortly before a race. The lack of dietary carbohydrate depletes their liver and muscle cells of glycogen since protein cannot be converted easily into carbohydrate by the cell's metabolism. Then, for a few hours before the race, they eat a high-carbohydrate diet, which causes the cells to actively store as much glycogen as possible. This carbohydrate loading means that they can have maximum glycogen stores for the race. Glycogen is a much more readily available energy store, so the muscles will be that little bit more efficient than if they were running on lipids, the next available source of energy.

Glycogen relocation within the cell during fixation. Glycogen illustrates an important characteristic of histochemistry, namely the role of fixation in the successful demonstration of a tissue material. The preservation of glycogen is important. If it is not fixed properly, glycogen can be lost completely or its location within the tissue can be altered. Routine fixation with formaldehyde is not ideal as some of the glycogen is lost, so the amount left in the section is diminished. Glycogen is also relocated by 'streaming' in the cytoplasm and becomes polarized within the cell. In streaming, the glycogen is effectively washed to the far side of the cell by the fixative where it then becomes fixed. Using an alcohol-based fixative retains more glycogen and diminishes the streaming artefact (Gendre's fluid is recommended to give the best results). However, the best preservation is achieved by avoiding fixation altogether and using freeze-drying to prepare the tissue.

Starch contamination. The presence of starch in animal tissues usually means contamination of the specimen, the commonest source being the glove powder used in many rubber gloves to make them easier to put on (see **Box 8.6**). Starch can be distinguished from glycogen by staining with iodine (starch stains blue, glycogen stains brown); also starch is usually in the form of grains that show Maltese-cross birefringence with a polarizing microscope.

> **Box 8.6 // Contamination from rubber gloves**
>
> Starch was originally used to replace mineral talc as a glove powder. Lubrication is needed with latex gloves to allow them to slip on to the hands easily. Talc is a magnesium silicate and remained in the body, causing problems including 'talc granulomas'. Starch was used to perform the same task as it is more easily degraded in the body. Starch has now been found to have fewer although still significant problems and can delay wound healing. The Medical Devices Agency in the UK recommended that the use of powdered gloves should be discontinued. Gloves made of plastic, which is not as sticky as rubber latex, are now common in surgical practice. Non-surgical gloves may still use glove powder.

Cellulose

Cellulose is another homoglycan of glucose but is β-1,4 linked. It is usually found in plant materials but this can include cotton wool, gauze, filter paper and clothing, so cellulose fibres are one of the commonest contaminants of tissues. Cellulose fibres are usually easy to recognize: they are PAS-positive, birefringent, insoluble in all common reagents and resistant to enzyme digestion by amylase.

Chitin

Chitin is a homoglycan of *N*-acetylglucosamine and is β-1,4 linked. It is found in the exoskeleton of invertebrates. It is of only minor importance in mammalian histochemistry where it occurs as either a contaminant or as part of a complete invertebrate parasite.

Many PAS-positive substances are protein–sugar combinations

Many of the most important compounds containing sugars are mixtures of protein and sugars. These can be many and varied in form, structure and function. They are studied by many different disciplines and this has led to a variety of nomenclatures that do not always agree with each other. One simple distinction is between glycoproteins and proteoglycans. The major distinction here is between the dominant chemical in the combination:

Glycoproteins are proteins that have some sugar added as a post-translational modification but are predominantly protein. They are widespread as membrane proteins where the sugars act as 'stoppers' to prevent the proteins slipping back through the membrane. The sugars always occur on the outside of the cell and form the **glycocalyx** (the carbohydrate cell coat) where they act as antigens and cell-recognition structures. They also occur as serum proteins such as immunoglobulins, structural proteins such as collagen and as secretions, both exocrine and endocrine. Their histochemical properties are determined mainly by the protein but they do give a positive PAS reaction.

Proteoglycans are mainly carbohydrate in nature with a small protein core surrounded by large polysaccharide chains. It is this group that has confusing nomenclature. They are often known in histology by the imprecise terms of mucins or mucopolysaccharides. The histological classification of mucins that will be followed here is that of Cook (1974).

8.5 Mucins

Mucins are a group of substances with certain physical, chemical and staining characteristics in common. They all contain large amounts of sugars attached to a small amount of protein and they all stain PAS-positive. Many mucins react with the dye alcian blue at acidic pH. Many also show metachromasia and this forms the basis of the histological classification.

Identification of mucin type can be important in some diseases (see **Box 8.7**).

Histological classification of mucins

The histological classification of mucins is based on staining properties. Two major groups of mucins can be identified:

Neutral mucins

Neutral mucins are PAS-positive but alcian blue-negative and display no metachromasia with thiazine dyes such as toluidine blue.

Acid mucins

Acid mucins are PAS-positive and also stain with alcian blue; they demonstrate some degree of metachromasia. Acid mucins can be further subdivided based on their staining behaviour with regard to pH due to the type and number of the ionizable acidic groups:

Strongly sulphated acid mucins stain with alcian blue and are metachromatic, even at very low pH (less than pH 1). They are mostly connective tissue mucins, e.g. cartilage, chondroitin sulphates, keratan sulphates.

Weakly sulphated mucins stain with alcian blue and are metachromatic only when the pH is greater than 1, e.g. epithelial mucins in colonic goblet cells.

Weakly acidic (carboxylated) mucins stain with alcian blue and are metachromatic only when the pH is greater than 2.5, e.g. epithelial mucins of the salivary gland and the connective tissue hyaluronic acid. These can be further subdivided by the type of acid group (either uronic acid or sialic acid), which can be identified by enzyme digestion with hyaluronidase or sialidase followed by alcian blue staining.

Box 8.7 // Diseases of mucin metabolism

Mucins are normally very useful substances, but if the body loses the ability to destroy mucins, they can accumulate and cause severe disruption. This happens in a number of rare diseases called mucopolysaccharidoses. These are inherited gene defects in which the affected individuals accumulate mucinous substances inside and between cells. The accumulations occur because of an enzyme defect in the lysosomes. The severity of the disease and the type of accumulated material depends on which enzymes are defective or missing.

Affected babies often appear normal at birth and may develop normally at first. It is often only when the child seems to develop frequent infections that the disease is diagnosed. The effects are progressive and vary widely in their severity. Patients often develop coarse facial features (hence the derogatory and now defunct term 'gargoylism' for these diseases). Thick skin and unusual hair growth also disfigure these patients. Skeletal involvement often causes distortion of the body. Corneal abnormalities are not uncommon and can lead to impaired vision. Internal organs may enlarge, e.g. the liver (hepatomegaly) and spleen (splenomegaly). There will be mental retardation in severe cases. Their overall growth is usually stunted and their life expectancy is usually short.

There are several variants of the disease of which the best known are Hunter's disease and Hurler's disease.

Identification of mucins

Identification of mucins is most conveniently done by using several staining techniques. The routine stains such as PAS and alcian blue give information that is useful, but they can be even more useful if more than one technique is applied to the same section in a sequence so that the techniques interfere with each other and effectively compete for the carbohydrate.

PAS reaction

This is positive for all groups of mucins. However, if the mucin is stained first with another technique (e.g. alcian blue), the dye will prevent or block the PAS reaction and the mucin will not then stain PAS-positive.

Alcian blue reaction

This is a cationic dye with an unusual structure (it is a copper phthalocyanine dye) and if the pH is kept acidic it will only stain acid mucins. It does not stain nuclei, probably due to the size and shape of the dye and the conformation of the DNA. Although other cationic dyes can be used to stain mucins, it is this lack of nuclear reactivity combined with a strong colour that resists fading that has made alcian blue so popular in staining mucins. By varying the pH or salt concentration, alcian blue can identify the different types of mucin.

Aldehyde fuchsin reaction

This stains substances containing sulphur groups a strong purple colour. It will stain sulphated mucins very strongly. However, it will also stain many other tissue components purple (e.g. elastic fibres), even though they are not mucins. Thus, only components that stain with both alcian blue and aldehyde fuchsin can be classified as sulphated mucins. By staining with aldehyde fuchsin first, it is possible to inhibit or mask alcian blue staining. The mechanism for the affinity of aldehyde fuchsin for sulphated mucins is not known and the stain should be regarded as an empirical test (i.e. no rational reason); however, it seems to work reliably.

Critical electrolyte concentration

This technique uses differing concentrations of magnesium chloride (or different pH levels) to inhibit alcian blue staining in mucins. The dyes are made up in the appropriate concentration of the salt as shown below and then used to stain sections.

$MgCl_2$	pH	Type of mucin
<0.06 M	>2.5	All acid mucins are alcian blue-positive
0.2–0.3 M	1–2	All sulphated mucins are alcian blue-positive
>0.5 M	<1	Only strongly sulphated acid mucins are alcian blue-positive

If a group of sections is treated with the range of concentrations, then it follows that if a mucin is stained under all of the conditions, it is a strongly sulphated mucin. If a mucin stains in the first two solutions but not in the highest concentration, then it cannot be a strongly sulphated mucin and must be a weakly sulphated mucin. If it only stains alcian blue-positive in the weakest concentration of salt, then the mucin must be a carboxylated mucin.

Combining these techniques

By combining these techniques it is possible to identify the major groups of mucins.

Alcian blue/PAS. The alcian blue blocks the PAS reaction in acid mucins and so distinguishes between neutral and acid mucins. Acid mucins stain blue and neutral mucins stain purple from the PAS (see *Fig. 8.1*).

Aldehyde fuchsin/alcian blue. The aldehyde fuchsin stains the sulphated mucins dark purple, whilst the carboxylated mucins take up the alcian blue.

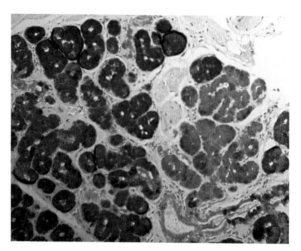

Figure 8.1
Salivary gland stained with alcian blue/PAS for mucins. The pink-stained cells contain neutral mucin, whilst the blue-stained ones contain acid mucin.

Finer distinctions can be made by using enzymes to differentiate between the carboxylated mucins and by using the critical electrolyte concentration technique to distinguish between the strongly sulphated and weakly sulphated acid mucins (see *Fig. 8.2*).

8.6   Fats and lipids

Lipids are usually defined by their solubility in fat solvents and their insolubility in water. Solubility is referred to by the terms:

- **hydrophobic**, which refers to materials that are insoluble in water but soluble in organic solvents;
- **hydrophilic**, which refers to materials that are soluble in water but not soluble in organic solvents; and
- **amphipathic**, which refers to materials that have parts of the molecule that are hydrophobic and parts that are hydrophilic.

All lipids are therefore at least partly hydrophobic but are otherwise often quite varied and they may have very little in common except their solubility. For example, the differences in structure, distribution and function between a steroid hormone such as testosterone

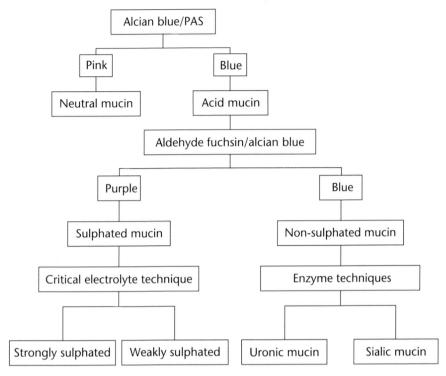

Figure 8.2
This scheme in *Fig. 8.2* looks easier to do than it is in practice. The colours can sometimes be vague and inconclusive, some mucins react poorly in practice and there are conflicting results in some cases. Nonetheless the scheme is useful for identifying mucins.

and the adipose fat found in the subcutaneous tissues is much greater than the similarities in their solubility. Nevertheless, in histological terms, the solubility of lipids has been of crucial importance. The identification and location of lipids is generally less well understood and less frequently investigated histologically because the traditional wax sectioning techniques remove the lipids (see *Box 8.8* for further comments). It is almost a truism to say that lipids are materials that cannot be investigated in paraffin wax sections. Because of this, historically most histologists have chosen to investigate other things.

Box 8.8 // Problems with processing fatty tissues

Even though fats are soluble in the solvents used for clearing, they can still cause problems in the cutting of wax blocks. Lipids will act as a waterproofing layer and will prevent the complete dehydration of blocks that are rich in fats. The fat itself will be dissolved gradually by the clearing agent and the block will then be left incompletely dehydrated. The water content may be very small, but unless the clearing agent is water-tolerant and can complete the dehydration, the final block will have some water trapped within the tissues. During storage, this water can evaporate causing the tissue to shrink, leaving a depressed block.

   If there is much fat, then it may not be completely dissolved by the clearing agent. The final block will be not be impregnated properly and will not section easily, if at all.

Types of fats and lipids

The classification and structure of lipids have been confused and different methods of grouping the various types have been used. Metabolically there are two types of lipid: these are the **steroids**, which have a complex ring structure as the hydrophobic part of the molecule, and **lipids**, in which the hydrophobic part of the molecule is an aliphatic fatty acid. In many biochemical textbooks, these two materials are dealt with quite separately since the metabolic pathways dealing with steroids are completely different to those dealing with fatty acids. Histology textbooks more commonly use a classification that is based on whether the lipid is conjugated or not. In many of the demonstration techniques used in histology, the critical difference is often whether the lipid is a hydrophobic lipid or an amphipathic lipid.

Conjugated and unconjugated lipids

The list below shows the different categories of conjugated (esters or amides) and unconjugated lipids.

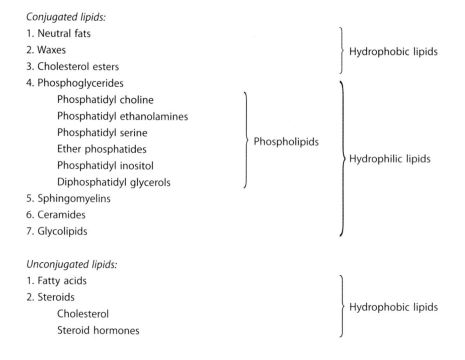

Conjugated lipids:
1. Neutral fats
2. Waxes
3. Cholesterol esters
4. Phosphoglycerides
    Phosphatidyl choline
    Phosphatidyl ethanolamines
    Phosphatidyl serine
    Ether phosphatides
    Phosphatidyl inositol
    Diphosphatidyl glycerols
5. Sphingomyelins
6. Ceramides
7. Glycolipids

Hydrophobic lipids

Phospholipids

Hydrophilic lipids

Unconjugated lipids:
1. Fatty acids
2. Steroids
    Cholesterol
    Steroid hormones

Hydrophobic lipids

Section preparation for lipids

The first thing that must be considered in investigating lipids is how they can be preserved and sectioned.

Fixation is difficult for lipids. Most fixatives used in histology are protein fixatives and do not fix lipids; at best they leave them unaffected (e.g. formaldehyde), at worst they may even destroy them (e.g. alcohol). Only two fixatives have any significant fixative action on lipids and neither is commonly used. The first is osmium tetroxide, which adds on to double bonds in tissues including those in unsaturated lipids.

$$ \mathrm{H-C} \underset{\mathrm{H-C}}{\overset{\mathrm{H-C}}{\Big\|}} +OsO_4 \longrightarrow \mathrm{H-C-O} \underset{\mathrm{H-C-O}}{\overset{}{}} Os \underset{O}{\overset{O}{}} $$

This fixes the lipid but also blackens it. This makes osmium-fixed lipid unsuitable for most histochemistry. Osmium tetroxide is, however, very important in electron microscopy since the fixation and blackening preserves and stains membranes, which are a vital part of the ultrastructure of cells.

The second fixative that can preserve lipids is potassium dichromate, which can preserve some phospholipids, although it has little effect on triglycerides or steroids. This has been used in histochemistry in the Marchi technique and the Baker technique for phospholipids, but otherwise the prolonged dichromate fixation needed to fix the phospholipids is not common.

The majority of lipid histochemistry is therefore done on unfixed lipid because, even if the sections have been fixed to preserve protein, the fixatives usually have no effect on the lipid. The sections are usually cut frozen. This means that the best fixative is formaldehyde, which allows good frozen sections and does not alter lipids. The addition of 10% calcium chloride is believed to help preserve phospholipids but should be kept short (<2 days) to prevent the formation of soaps with free fatty acids. Fixation is often delayed until after sectioning (or even after staining) when fixation times can be considerably reduced.

Sections are almost always prepared by a frozen section technique, with cryostat sections generally being preferred and fixation being done after cutting.

Neutral lipids
Simple neutral lipids can be localized very easily using **lysochrome** methods, which rely on the hydrophobic nature of the lipid. Lysochrome staining relies on using a dye that is very soluble in lipids but relatively insoluble in aqueous solvents.

If the dye is made up in an aqueous solvent, then when it is applied to the tissues the hydrophobic dye will become concentrated in the tissue lipids, even when used at only a low concentration in the watery phase. Such a solution of hydrophobic dye will rapidly colour the lipid by a simple solution mechanism, leaving the hydrophilic proteins and nucleic acids uncoloured, as it is much more soluble in the hydrophobic lipid than in the aqueous solvent (see **Figs 8.3** and **8.4**).

Box 8.9 // Lipids in accidents
The staining of lipids is one method of determining whether a person died instantaneously in an accident, even if they are not examined until many hours or days later. In an accident in which bones are broken, the fatty material in the bone marrow will be released into the bloodstream. The fatty material does not dissolve but remains as a solid fat embolism. The embolisms are carried around in the bloodstream until they reach a capillary where they lodge. The first capillary bed that the droplets encounter is usually in the lung. Thus, in a fatal accident, if a fresh section of lung is taken and stained by a lysochromy technique, then the finding of lipid droplets from the marrow shows that the heart went on beating after the bones were broken so the person did not die instantaneously.

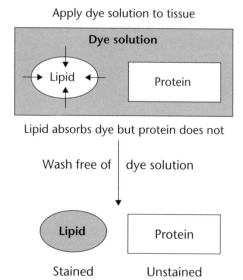

Figure 8.3
The lysochrome staining principle of differential solubility.

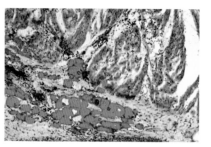

Figure 8.4
Heart stained with oil red O and counterstained with H&E. The hydrophobic lipid in the adipose tissue has stained strongly with the hydrophobic dye, whilst the cardiac muscle and fibrous connective tissue have failed to absorb the oil red O and instead are stained with the counterstain.

The first techniques used in this way relied on Sudan dyes (Sudan I, II, III and IV) and this type of reaction is still often called **sudanophilia**. Only Sudan III and IV were ever popular in histology and these have now been replaced by an even more highly coloured and hydrophobic dye, oil red O. In either case the hydrophobic dye must be dissolved at high concentration in an aqueous solvent in which it is not very soluble. This contradiction means that it is not usually a simple solution.

One of the earliest reliable techniques was that of Herxheimer who used acetone and alcohol. This has significant solvent ability for the dye but unfortunately it will also dissolve some lipids. It gives excellent demonstration of major fat deposits such as adipose tissue but may dissolve very small droplets of lipid such as might be found inside non-adipose cells. Since then, the lipid-dissolving ability of the solvent has been reduced in a variety of ways (see, for example, ***Box 8.10***).

Lillie and Ashburne introduced isopropanol as a solvent. The stock solution of the dye is a saturated solution in 100% isopropanol, but for use the stock solution is diluted

> ### Box 8.10 // Avoiding the use of solvents
> The dangers of dissolving the lipid by using an organic solvent as the dye carrier led to an attempt to avoid using a solvent. The gelatine method developed by Govan used gelatine as a carrier instead. In this method, the dye was first dissolved in acetone and this was mixed with a hot solution of gelatine. The acetone evaporated, leaving the dye trapped within the colloidal suspension of the gelatine. The method does work but the slight theoretical advantage of not having any solvent present did not compensate for a complex and messy technique and this method has not stood the test of time. A good theoretical idea does not always translate into a useful and workable method.

immediately before being used to give a final concentration of 60% isopropanol. This weaker alcoholic solution has much less solvent action on the lipid, but in diluting the stock a supersaturated solution is produced, which gives a high concentration of dye. The supersaturated solution is unstable and must be used immediately.

Chiffelle and Putt used propylene glycol and Gomori used triethyl phosphate, both of which are claimed not to dissolve lipid although they will dissolve the dyes.

Most laboratories use one of these three techniques, depending mainly on personal preference.

Since all of the solvents use the dye at close to saturation, it is important to filter the dye solutions to remove any precipitate and to prevent evaporation during staining, which would cause deposits to occur. Removal of excess dye solution is usually done with the same solvent that was used for the dye and this is then followed by washing in water. This rinse in solvent helps to prevent any precipitation of the dye and keeps the background clear. The most commonly used dyes are red and are usually counterstained with haematoxylin.

Temperature alters lysochrome dye solubility

Staining at room temperature will demonstrate neutral lipids but not most phospholipids, cholesterol or their derivatives, which are crystalline at room temperature. These can be rendered sudanophilic by heating to above their melting point during staining. Protein-bound lipid also may not be demonstrated by the simple technique as the protein restricts the entry of the hydrophobic dye. These protein-bound lipids are sometimes called masked lipids; they can still be demonstrated provided the lipid can be detached from the dye, e.g. by hydrolysis with acid.

Free fatty acids and phosphoglycerides, being more amphipathic, have a greater tendency to be extracted by the solvents used but their solubility can be reduced by bromination, which also converts any cholesterol to a compound that is liquid at room temperature and so will also stain.

Sudan black

One further Sudan dye used is Sudan black. This is a mixture of two main hydrophobic dyes plus several hydrophilic components. It is a more aggressive dye mixture and will often stain lipids that are partly hydrophilic (amphipathic), which would be left unstained by oil red O. Unfortunately, it will sometimes stain other tissue components as well as lipids, making interpretation more difficult. This unwanted staining increases as the dye solution ages and is less pronounced with fresh dye solutions.

A negative control for lysochrome techniques can be performed by using a section extracted with a 2:1 chloroform:methanol mixture for 1 h. The solvent will extract all of the lipids and so destroy any true sudanophilia. Any coloration after the extraction will be non-specific and not lipid.

Phospholipids

These are lipids that contain a phosphate group. They are mainly phosphoglycerides and are principally associated with membranes, including the myelin sheaths of nerves. The presence of the phosphate group renders them more hydrophilic (i.e. they are amphipathic) than the neutral fats, so they stain less readily with lysochrome techniques and are more rapidly and easily extracted than other lipids unless they are bound to protein.

Baker's acid haematein staining

This involves the reaction of phospholipids, especially choline-containing phospholipids, with potassium dichromate. The dichromate adds on to the lipids:

The chromium can then act as a mordant for the haematein dye and the phospholipids stain blue. Any lipid unaffected by the dichromate will be lost during dehydration and clearing, so the reaction is reasonably selective. It is usually controlled by using a second sample, which is treated with a simple solvent extraction before treatment with dichromate. The control differentiates phospholipids from tissue ions that may act as mordants. This technique, as well as being a histochemical technique, is also a good way of demonstrating mitochondria.

The original technique was a block-staining method and involved prolonged exposure to all of the reagents (a total of 3–5 days), but this can be reduced by using cryostat sections instead of blocks of tissue, which also makes controls more realistic as adjacent sections can be used for the test and the control.

Marchi techniques for myelin

This relies on the following properties:

1. Osmium tetroxide is soluble in both hydrophobic and amphipathic lipids.
2. Osmium tetroxide is an oxidizing agent that oxidizes and blackens lipids.
3. Hydrophilic oxidizing agents such as potassium dichromate or potassium chlorate are only able to penetrate amphipathic lipids.
4. The oxidizing and blackening effect of osmium tetroxide is inhibited by other oxidizing agents.

Thus, if a mixed reagent of osmium tetroxide and potassium dichromate (a hydrophilic oxidizing agent) is applied to tissues, the amphipathic lipids will absorb both the osmium tetroxide and the potassium dichromate. Since the dichromate will inhibit the blackening effect of the osmium tetroxide, the amphipathic lipids will not become blackened. The hydrophobic lipids will absorb the osmium tetroxide but not the potassium dichromate, so they will become blackened. The total effect, therefore, is that hydrophobic lipids will be blackened but more hydrophilic or amphipathic lipids will not.

This is one way of identifying degenerating myelin. Normal myelin consists of hydrophilic phospholipid, whilst degenerating myelin is converted to hydrophobic neutral lipid. With Marchi-type techniques, normal myelin is unstained and degenerate myelin is black. The original Marchi method used potassium dichromate, but the Swank–Davenport method of using potassium chlorate is now generally preferred.

Osmium tetroxide α-naphthylamine (OTAN)

This is similar to the Marchi-type techniques, but the osmium that is bound by phospholipids but not blackened is visualized using α-naphthylamine, which gives an orange/red or orange/brown complex with the osmium. The OTAN technique has lost popularity because its specificity has been questioned, and because hydrophobic lipids and some proteins may also be stained. In addition, α-naphthylamine can be carcinogenic, especially if contaminated with β-naphthylamine.

Steroids: cholesterol

Cholesterol is the only steroid that is present in large enough quantities to be detectable in tissue sections. Large amounts of cholesterol may form crystals within the tissues and when they dissolve out during processing, their presence is indicated by angular gaps in the tissues referred to as cholesterol clefts. Cholesterol is also associated with disease (see *Box 8.11*).

Cholesterol is not normally sudanophilic

Cholesterol will not be stained by the normal lysochrome technique as it has a high melting point and is crystalline at room temperature. By staining at a temperature above its melting point (150°C), it will become sudanophilic, but staining at this high temperature is inconvenient. It can also be rendered sudanophilic by treatment with bromine water (2.5% aqueous bromine), when it is converted into an oily liquid that will stain with Sudan black.

The PAN method

The best specific staining method for cholesterol is the perchloric acid/naphthaquinone reaction (PAN). The perchloric acid converts the cholesterol to a diene, which then reacts

---

Box 8.11 // Cholesterol in disease

Large amounts of cholesterol are found mainly in degenerating tissues, such as in necrosis. The cholesterol is a normal constituent of cell membranes but is difficult for the body to remove. As tissues degenerate, cholesterol accumulates. This occurs, for example, in atherosclerosis. The wall of an artery becomes damaged and as cells disintegrate cholesterol accumulates, partly from the degenerating cells and partly from **low-density lipoprotein** (LDL), which carries cholesterol in the blood and also gets trapped in the vessel wall. The lipid is ingested by macrophages, which develop a foamy-looking cytoplasm ('foam cells'). Repair mechanisms are activated and result in a fibrotic plaque. The lesion gets worse with time and can lead to narrowing of the arterial lumen and even complete blockage. The degeneration is slow and may even reverse if the cholesterol can be removed by **high-density lipoprotein** (HDL). The exact mechanism of this degeneration is still unresolved but high levels of LDL cholesterol are bad for the arteries and high levels of HDL are good for avoiding cardiovascular disease.

with 1,2-naphthaquinone to give a blue product. In practice, the cholesterol first needs to be oxidized, which can be done using ferric chloride. The reagent also includes ethanol and formaldehyde, but their exact roles are not completely clear. The reaction is usually done on a hotplate at 70°C and the reagent is constantly brushed on to the section to prevent drying.

Cholesterol esters can be distinguished from free cholesterol by treating a second section with digitonin. This precipitates free cholesterol as a digitonide, which is insoluble in acetone. The section is then treated with acetone. The acetone dissolves the cholesterol esters but leaves the cholesterol digitonide. The PAN technique is then applied to the extracted section. Material reacting in both sections is free cholesterol, whilst any material reacting in the straight PAN but absent from the digitonin-treated slide is a cholesterol ester.

Free fatty acids

Free fatty acids are unusual in normal tissues but can occur in acute pancreatitis when the release of lipases results in the fats being enzymically hydrolysed. Free fatty acids can be demonstrated by treating the section with copper salts. Any fatty acid will form insoluble copper soaps with the copper ions. The copper soaps can then be demonstrated with copper-detecting reagents such as rubeanic acid (dithio-oxamide).

In many cases, the free fatty acids will already have formed calcium soaps if they have been in contact with solutions containing calcium ions. This may occur in the body or if the tissues have been fixed in formal calcium. These calcium soaps need to be hydrolysed with hydrochloric acid before being treated with the copper ions.

Phosphatidyl ethanolamine-based lipids (plasmalogens)

Plasmalogens can be detected by the plasmal reaction (from which they get their name). This involves treatment with mercuric chloride, which reacts with the vinyl ether linkage found in plasmalogens. This releases an aldehyde, which can be detected using Schiff's reagent. The reaction may even occur without pretreatment of the tissues with mercuric chloride if the Schiff's reagent is highly acid and the treatment with Schiff's is prolonged, as the acidity can have a similar bond-splitting effect.

Unsaturated lipids

The double bonds in lipids can be detected in a number of ways. As mentioned above, osmium tetroxide adds across double bonds and blackens the lipid in the process. Double

---

Box 8.12 // Diseases of lipid metabolism

As with the mucopolysaccharidoses, if there is a defect in the destruction of lipids it can lead to a disease in which lipids build up. Tay–Sachs disease is a genetic disease in which the hexosaminidase enzyme of lysosomes is defective. This prevents the destruction of one type of glycolipid (gangliosides), which accumulates in the lysosomes. The childhood form of this disease is particularly associated with Jewish populations originating in Eastern Europe (Ashkenazi Jews). The Ashkenazi Jews have a much higher incidence of the gene in the population, with perhaps as many as 1 in 25 people being a carrier of the gene compared with 1 in 500 in the general population. Affected children are therefore much commoner in these Jewish populations. The disease causes swelling of neurones, which can lead to seizures, poor development and blindness. The disease begins to affect the child from about the age of 3 months and leads to death at around 4–5 years old.

bonds can also be oxidized to aldehydes, which can then be detected using Schiff's reagent. The oxidation can be achieved by using a powerful oxidizing agent such as performic acid (performic acid–Schiff, PAAS), but will occur slowly, even with atmospheric oxygen. This slow oxidation will result in aldehydes occurring in the section without any pretreatment and is called a pseudoplasmal reaction. This is a nuisance when using Schiff's reagent as it can give a non-specific background colour to the sections. The oxidation by air can be speeded up by irradiating the section with UV light and the UV Schiff technique is an alternative to using the PAAS and avoids the use of the highly corrosive performic acid.

Glycolipids

These will give a positive reaction with the PAS technique but require controls to distinguish them from mucins and other carbohydrates. The simplest is an extraction with chloroform/methanol. There can also be confusion with plasmalogens, which can give a positive reaction with Schiff's reagent and pseudoplasmal reactions.

Extraction techniques as controls and for identification

Lipids can be identified by their solubility. Extraction is a standard technique in lipid biochemistry and the difference in solubility of the different lipids forms the basis of chromatography, which is the best way of identifying individual lipids. In histology, the solubility of a lipid is determined by lipid staining of two equivalent sections, one of which has been extracted with a particular solvent. By using different solvents, it is possible to selectively remove different lipids. The example given in **Table 8.1** shows the method of Keilig.

Table 8.1 Selective removal of lipids by the Keilig method

Solvent	Lipids dissolved
Cold acetone	Glycerides, cholesterol and its esters and other steroids
Hot acetone	Cerebroside lipids
Hot ether	Lecithin and cephalins
Hot chloroform and methanol	All lipids

The table gives a false impression of the ease and accuracy of the method. For example, it is usually necessary to use a Soxhlet apparatus and continuous refluxing for long periods to ensure removal of the lipid. The problems with extraction are:

1. The lipid may relocalize rather than dissolve. This may be unrelated to the solvent since heating itself alters the lipid and it may move significantly during the treatment, giving the impression that it has dissolved whilst it has only flowed to a different area of the tissue.
2. The solubility is determined on pure lipids, but in tissues the lipids are often mixtures and this can affect their solubility in solvents.
3. Lipids in tissues are often intimately bound to proteins, which may alter their solubility.
4. Lipids in sections have often been exposed to fixatives, which may affect their solubility.

In practice, extraction techniques are rarely used as a major way of identifying individual lipids but they can be very useful control methods.

Suggested further reading

Bancroft, J.D. and Cook, H.C. (1994) *Manual of Histological Techniques and their Diagnostic Application.* Churchill Livingstone.

Bancroft, J., Layton, C. and Suvarna, S. (2012) Chapter 11: Connective and mesenchymal tissues and Chapter 12: Carbohydrates. In: *Bancroft's Theory and Practice of Histological Technique,* 7th edn. Elsevier Health Sciences.

Culling, C., Allison, R. and Barr, W. (1985) *Cellular Pathology Technique,* 4th edn. Butterworths.

Kiernan, J.A. (2008) *Histological and Histochemical Methods: theory and practice,* 4th edn. Scion Publishing.

Lamar Jones, M. (2002) Lipids. In: *Theory and Practice of Histological Techniques* (eds Bancroft, J.D. and Gamble, M.), 5th edn. Churchill Livingstone.

Totty, B.A. (2002) Mucins. In: *Theory and Practice of Histological Techniques* (eds Bancroft, J.D. and Gamble, M.), 5th edn. Churchill Livingstone.

Suggested websites

http://library.med.utah.edu/WebPath/webpath.html
http://stainsfile.info/StainsFile/jindex.html

Self-assessment questions

1. What are the extra difficulties involved in identifying chemicals in sections compared with in solution?
2. What types of material stain with the PAS technique?
3. Why is periodic acid preferred as the oxidizing agent compared with chromic acid or potassium permanganate?
4. List the main types of mucin.
5. Outline the principle of the critical electrolyte concentration technique of using alcian blue.
6. How could you distinguish between hyaluronic acid and sialic acid, both of which stain with alcian blue?
7. Why are lipid techniques unpopular with many histologists?
8. What is the principle of staining lipids with Sudan-type dyes?
9. Which, if any, fixatives can fix lipids?
10. Why does normal myelin react differently with lipid-staining techniques compared with degenerating myelin?

09 Histochemistry of nucleic acids, proteins and enzymes

Learning objectives

After studying this chapter you should confidently be able to:

- **Describe the methods of staining nucleic acids in sections**
  RNA and DNA can be stained using basic dyes or more selective techniques. DNA can be specifically demonstrated using the Feulgen technique or using fluorescent dyes such as ethidium bromide. To distinguish DNA from RNA, it is possible to use selective methods such as methyl green–pyronin or acridine orange or by selective destruction using enzymes.

- **Describe the main methods for identifying proteins in sections**
  Proteins can be stained with simple stains like eosin or by using more selective methods for particular proteins, e.g. acid orcein for elastin and martius/scarlet/blue for fibrin. Individual proteins are best identified by immunotechniques. Proteins with enzyme activity can be demonstrated by their action on a substrate to produce an insoluble coloured product.

- **Discuss the difficulties of retaining enzyme activity in sections**
  Enzymes are easily inactivated by fixation, processing or other handling of tissues or by simple exposure to room temperature. Even if they remain active, many enzymes will become delocalized or even completely lost from the tissue.

- **Outline the problems involved in quantifying enzyme activity in sections**
  Enzymes may be inactivated or may diffuse out of the tissues before quantitation can be done. The kinetics of enzymes in sections may not follow Beer–Lambert kinetics because of diffusion limitation.

9.1 Nucleic acids

The general structure of the nucleic acids is a backbone of alternating pentose sugar and phosphate groups. Each sugar has a nitrogenous base attached to it. The base can be either a purine or a pyrimidine. The backbone can interact with similar nucleic acid strands to form double strands by hydrogen bonding between the purine and pyrimidine bases. There are four different bases that can occur in any nucleic acid and the sequence of these bases is the basis of the genetic code. Nucleic acids include several quite different forms of material that differ in their chemical composition and structural forms.

DNA and RNA

Ribonucleic acid (RNA) has a simple ribose sugar in the main backbone of the molecule, whilst deoxyribonucleic acid (DNA) has the deoxyribose form of the sugar in its backbone. They differ slightly in their bases such that DNA includes the base thymine, whilst RNA uses uracil in the same position; in both forms the other three bases are the same (adenine, guanine and cytosine).

DNA is found almost exclusively in the nucleus (very small amounts are found in the mitochondria). Whilst all of the RNA is actually produced inside the nucleus, it is usually quickly transported into the cytoplasm, so that most of the RNA in a cell is found in the cytoplasm or in the nucleolus, which produces the ribosomal RNA.

Probably the most important differences in the properties of the nucleic acids, from a histochemical point of view, are the size of the molecule, the shape of the molecule and whether there is any attached protein.

DNA double helix

DNA forms a double helix and is strongly associated with histone proteins. This combination forms nucleoprotein and is the material usually referred to by histologists as chromatin. The DNA is wrapped around the histones and each structure forms a characteristic unit called a nucleosome. The DNA in a cell usually forms very large molecules with each chromosome being a single DNA molecule. Thus, an average human chromosome consists of around 1×10^8 nucleotides and the DNA helix would measure about 5 cm if straightened out completely. The wrapping around the histones reduces this to around 1 mm total length and more coiling and folding reduce it even further. Thus, the DNA within the cell is always highly coiled.

Active and inert DNA

Depending on how tightly packed, or condensed, it is, the DNA is described as being heterochromatic (darkly staining and very tightly packed) or euchromatic (lighter staining and less densely packed). Heterochromatin is inactive and not transcribed, whilst euchromatin is being actively used by the cell. The DNA may also be modified by being methylated. In methylation, some of the cytosine bases are changed to 5-methyl cytosine. This conversion is catalysed by enzymes. Methylation is believed to be a method of controlling gene activity as inactive genes tend to be more heavily methylated.

Different forms of RNA

RNA occurs in several different forms within the cell:

Ribosomal RNA (rRNA) does not form a double helix and is always associated with protein. This RNA forms part of the small structures called ribosomes. Although too small to be adequately resolved as individual structures by the light microscope, ribosomes can form larger aggregates (polyribosomes), which can be visible in the cytoplasm.

Transfer RNA (tRNA) does not form a double helix but forms a series of hairpin loops of RNA. Transfer RNA is not bound to protein.

Messenger RNA (mRNA) does not form a double helix but occurs as a single-stranded RNA and is not firmly bound to protein.

The differences in function of the three forms of RNA are of only minor importance in most histological investigations since all three forms of RNA are associated in different ways with the production of protein. Thus, for most histological purposes, they are all indicators of active protein synthesis and provided any one of them is retained, the loss of the other forms is less critical. It is only when mRNA is needed in hybridization studies (see **Chapter 16**) that it is essential to retain a particular kind of RNA.

Fixation affects either the physical or chemical structure of nucleic acids

In paraffin sections, only nucleic acid that is firmly attached to protein will be consistently retained since most fixatives act on proteins and not on nucleic acids. This means that only DNA and rRNA are retained by all fixatives and can always be demonstrated in tissues. Some fixatives, whilst not truly fixing the nucleic acid, may still alter it either chemically or physically.

Additive fixatives usually combine with and alter the reactive groups of nucleic acids, whilst non-additive fixatives alter the double helix regions of DNA and so disturb its physical form. There is no fixative that retains both the chemical and physical form of nucleic acids. Some methods rely on chemical reactivity, whilst others use physical form to trap the dyes. Whenever nucleic acids are to be investigated, it is essential to use a fixative that retains the particular characteristic of the nucleic acid that is used in the demonstration method.

Generally, fixatives containing acetic acid give better preservation and are preferred for DNA studies, but formaldehyde is satisfactory for DNA and rRNA.

Staining techniques

The commonest staining methods for nucleic acid are the basic dyes. Nucleic acids are strongly basophilic and will stain with almost any basic dye. This is good for morphological purposes but basophilia is not limited to nucleic acids and needs controlling with enzymes in order to get specific identification.

Feulgen technique

The Feulgen technique is specific for DNA and will not stain RNA or any other components of cells. The technique uses Schiff's reagent, which reacts with aldehydes (see **Chapter 8**). The DNA in tissues does not contain any free aldehydes, but aldehydes can be produced by acid hydrolysis. The hydrolysis removes the bases from the DNA leaving pentose sugars that contain aldehyde groups and therefore stain pink with Schiff's reagent.

Hydrolysis of DNA is usually done with 1 M HCl at 60°C. Care must be taken to get the timing of the hydrolysis correct or the DNA will be completely hydrolysed to soluble fragments of nucleic acid and these will wash out of the section and the staining will be reduced. Any RNA is completely hydrolysed by the usual hydrolysis conditions and does not stain.

The Feulgen technique is quantitative when performed correctly, and the amount of DNA in each cell can be measured by microdensitometry. The amount of DNA in an individual cell is usually constant in most human cells except when:

- The cell is actively dividing. Before division can occur, the cell must duplicate the DNA so that when the cell splits into two, both daughter cells will get a complete set of chromosomes.
- The cell has an unusual chromosome number.

Both of these can be associated with malignant cells, so investigations are currently underway to determine whether an automatic technique can be developed that will reliably identify malignant cells by measuring their DNA content.

The Feulgen hydrolysis step is dependent on the fixative used. Picric acid performs the same hydrolysis as HCl, so material fixed with picric acid is not suitable for the Feulgen technique nor is tissue decalcified in acid solutions.

The following are the optimum times for some fixatives and show the effects of fixation on the nucleic acid: formaldehyde, 8 min; Carnoy's, 8 min; Helly's, 5 min; Susa's, 18 min. If there is as little as 30 s error in the timing, the results will be less than optimal and will be non-quantitative.

Fluorescent stains

There are a number of very valuable fluorescent dyes that can be used to demonstrate nucleic acids. These include quinacrine, Hoechst 33258, ethidium bromide, acridine orange and methyl green–pyronin.

Intercalating dyes. Quinacrine, Hoechst 33258 and ethidium bromide act by intercalating into the double helix. Intercalation is where the dye slides between the stacks of bases and is dependent on the double helix structure. These dyes are only used to stain DNA and can be used as vital stains to show DNA in living cells. Quinacrine has been used to identify chromosome bands (so-called Q-banding) used in identifying chromosomal structure and diagnosing chromosome disorders. This is because it produces bright fluorescence in adenine/thymine-rich regions and less fluorescence in guanine/cytosine-rich regions. When quinacrine is used on the condensed chromosomes found in mitosis, it gives a banding pattern of bright and dim transverse bands. The non-fluorescent dark bands represent regions rich in guanine and cytosine, whilst the brighter fluorescent bands represent regions rich in adenine and thymine.

Acridine orange. Acridine orange is different to the previous dyes in that it can be used for both DNA and RNA. With acridine orange staining, double-stranded DNA fluoresces green, whilst single-stranded RNA fluoresces red. As staining depends on the conformation of the strands to distinguish the two types of nucleic acid, it follows that any treatment that alters the structure will destroy this differentiation. Formaldehyde-fixed and paraffin-processed materials do not work reliably with this technique but it has been used successfully on alcohol-fixed smears.

Methyl green–pyronin

The methyl green–pyronin technique (Unna–Pappenheim) uses two different basic dyes to differentiate between DNA and RNA. In some ways it is the equivalent of the van Gieson or trichrome method of using acid dyes. The dyes used are methyl green, a triaryl methane dye, and pyronin Y, a xanthene dye. The dyes differ in molecular weight, molecular size and molecular shape. When used at pH 4.6, they compete for binding to the nucleic acids and differentially stain the DNA a green colour and the RNA a red colour (see *Box 9.1*).

The mechanism of this differential staining is disputed. It seems likely that the larger methyl green stains by binding to the outside of the DNA helix and is bound by the phosphate groups and possibly by linking non-ionically to the protein and sugars. Pyronin, being a smaller, more planar molecule, can probably intercalate between the bases and stain nucleic acids in this way. The differences in staining probably reflect the differences in structure, with intercalation being more easily achieved with the more open structure of RNA than with the compact double helix of DNA.

This technique has many variants; this is probably because it is sometimes fickle and erratic in the results it gives. It is sensitive to pH and needs to be buffered at pH 4–5

> **Box 9.1 // Pyroninophilia**
>
> The identification of RNA by this technique is referred to as **pyroninophilia**. The term is applied to cells such as lymphoblasts and plasma cells that are active in the production of antibodies. Although the presence of nucleic acids in the cytoplasm can be indicated by cytoplasmic basophilia, the term pyroninophilia is more limited and only refers to nucleic acids, whereas basophilia refers to any acidic material in the cytoplasm.

(there is some dispute as to the best pH, with most methods using pH 4.8 but others recommending a pH as low as 4). The pyronin is rapidly extracted by ethanol dehydration, so the section is usually dehydrated in butanol instead. Dye purity seems to be a problem, with the methyl green usually being heavily contaminated, and repeated extraction with chloroform is recommended to remove the methyl violet. The method is also sensitive to dye strength. All these factors make it difficult to get uniform results in every laboratory on every occasion it is used.

Using specific enzymes and chemical extraction as controls

There is always doubt as to the nature of a stained material when using dyes for histochemical identification so it is important to control these methods using an extraction technique. This is done by selectively destroying one or both nucleic acids and comparing the digested section with an undigested one. Both sections are stained by the technique being used and only if the stain disappears from the digested section can it be reliably assumed that the material is the nucleic acid. If the colour is present in both sections, it is a non-specific reaction, probably due to the dye binding to a protein. This digestion is usually sufficiently specific that it can be used with a relatively non-selective basic stain such as methylene blue as a simple method of locating a specific nucleic acid.

Enzymatic digestion. This is highly specific if done correctly but needs careful control as enzymes are sensitive to fixation, pH and the presence of ions and the enzyme must be pure. The enzymes are readily available commercially and, providing pure forms of the enzymes are used rather than impure extracts, they are highly specific. Impure forms of enzymes prepared from materials such as pancreas can be contaminated with proteases. **Deoxyribonuclease** is usually prepared from pancreatic tissue and used as a dilute solution (0.05 mg ml^{-1}) in buffer at pH 7.6. Deoxyribonuclease requires magnesium ions to work effectively, so it is usually better to add magnesium chloride or sulphate to the buffer to ensure optimum activity. **Ribonuclease** is usually used as a simple solution in distilled water but at a higher concentration (0.5 mg ml^{-1}). Extraction is done by incubating at 37°C and then washing in water to remove the digestion products.

In situ hybridization

In addition to the general staining techniques, mRNA and DNA can be identified by hybridization with labelled probes. This method identifies specific sequences of DNA and will be covered in more detail in *Chapter 16*.

9.2   Proteins

Proteins are the most important structural component in most sections as they are better fixed and more easily retained than most other tissue materials. However, individual proteins are difficult to identify as all proteins are very similar chemically and differ mainly in size and shape and neither of these properties is easy to identify in tissue sections.

Several different approaches are available for the staining of proteins.

General protein stains

Proteins are usually acidophilic and will stain with a wide range of dyes. These include acid dyes such as eosin, ponceau red and Coomassie blue. Although these dyes are useful to indicate the total quantity of protein present within the tissues, they are not selective for particular proteins. These dyes are used as general counterstains to give general colour to tissues and are also widely used in electrophoresis to show the protein bands. If used correctly, they can bind to tissues stoichiometrically and can be used to measure the total amount of protein present accurately. The use of dyes as a quantitative measure is more commonly done in electrophoresis than in histology. The density of staining is used by most microscopists as a guide to the total amount of material present, but this is just used as a subjective guide and is not accurately measured.

Digestion techniques

These use specific enzymes to remove and thus identify proteins. By comparing two sections stained with a general protein stain, it should be possible to see whether a particular protein has disappeared. Collagen, for example, can be identified by collagenase. This resembles the use of deoxyribonuclease in identifying DNA but is less specific. Although the names of the enzymes, such as collagenase, often suggest a high degree of specificity, they are usually much less specific than might be expected and the name of the protease indicates only a differential rate, not an absolute specificity. Thus, collagenase will act more quickly on collagen than on other proteins, but will still have some proteolytic activity towards other proteins. Despite this, they are still useful as controls in other techniques.

Selective protein techniques

A few techniques are available that are claimed to be selective for specific proteins. Although these techniques can sometimes be highly selective and very useful for histological purposes, they are often not fully specific and other proteins or tissue components will also take up the stain, albeit to a lesser extent.

Elastic fibres and elastin stains

Elastic fibres are different to most proteins in that they have relatively few polar amino acids and many hydrophobic side chains. This unusual amino acid structure means that they stain rather poorly with the usual ionic dyeing techniques such as haematoxylin and eosin, van Gieson's and trichromes and instead need more specialized techniques (see *Fig. 9.1*). These possibly rely more on van der Waals bonds or hydrogen bonds than ionic bonding and many use an extreme pH that limits the staining of other fibres and proteins.

Relatively selective techniques include: Gomori's aldehyde fuchsin technique, Weigert's resorcinol fuchsin, Miller's, Verhoeff's haematoxylin technique and the Taenzer–Unna orcein technique.

Oxytalan fibres

These stain rather poorly by most techniques but after oxidation (e.g. with peracetic acid) they stain as for elastic fibres. They are found in tendons, periodontal membranes and ligaments.

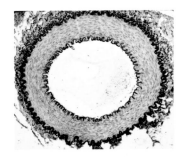

Figure 9.1
Cross-section of aorta stained with orcein to demonstrate the elastic fibres.

Box 9.2 // Elastosis

The term elastosis refers to a deterioration in the connective tissues of the skin. There is a general loss of connective tissue fibres, not just elastic fibres, and the fibres that remain are abnormal. The changes gradually occur with wear and tear on the skin of normal living but are accelerated by exposure to sunlight. This change can be seen by examining different areas of skin from the same person. Skin that is regularly exposed to sunlight is more affected than skin only a short distance away that is not subjected to the same amount of light (e.g. behind the ears). In elastosis the skin loses its elasticity, develops wrinkles and is more easily damaged.

Fibrin

Fibrin is produced from the precursor fibrinogen by the coagulation of blood and is also present in some inflammatory exudates (see *Fig. 9.2*). Fibrin is not stable in the body and gradually alters with age. Fibrin can be demonstrated by different techniques as these changes occur and as it is gradually replaced during healing (see *Fig. 9.3*). Fibrin that is less than 24 h old is difficult to stain specifically by any technique and this also means that fibrinogen in plasma will not stain by these selective methods.

The Gram staining technique is useful mainly because it is also used in staining bacteria. The use of the Gram stain in an inflammatory condition will therefore show fibrin at the same time as the bacteria causing the inflammation. The phosphotungstic

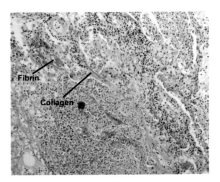

Figure 9.2
Inflamed lung stained by the martius/scarlet/blue method. Fibrin stains red, whilst collagen stains blue.

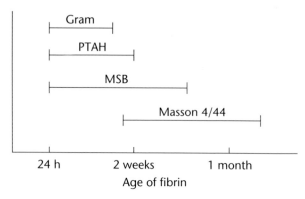

Figure 9.3
Staining techniques for fibrin as it ages

acid–haematoxylin (PTAH) technique is excellent for many structures and will show fibrin as a dark blue fibre. PTAH is not specific for fibrin but is a useful confirmatory stain. The martius/scarlet/blue (MSB) and Masson 4/44 methods are both trichrome techniques that were specially designed to be selective for fibrin, but again they are not specific and both need careful control and differentiation to achieve the desired result.

Fibrinoid
Fibrinoid is an amorphous material that stains similarly to fibrin but is associated with necrosis and disease rather than simple bleeding. It consists of fibrin (possibly aged or degrading), other plasma proteins and perhaps some mucopolysaccharide.

Amyloid
Amyloid is a pathological protein. Amyloid is formed in several diseases from a variety of precursor proteins. It stains selectively with congo red and thioflavine dyes. Amyloid is considered in more detail in **Chapter 11**.

Specific amino acid techniques
There are a range of biochemically based spot tests that will detect individual amino acids (e.g. Millon's for tyrosine) or specific groups associated with several amino acids (e.g. ninhydrin–Schiff for amino groups). The amino acid techniques demonstrate only protein-bound amino acids since free amino acids are readily lost into solvents including water.

In most cases, all proteins will react since all proteins contain a wide range of amino acids. This makes them somewhat pointless for general use as it is easier simply to use a general stain such as eosin (but see **Box 9.3**). A few examples of spot tests are given in *Table 9.1*.

Control of spot tests for amino acids is achieved either by selective removal, e.g. amino groups with nitrous acid:

$$\text{R-NH}_2 + \text{HNO}_2 \rightarrow \text{R-OH} + \text{N}_2 + \text{H}_2\text{O}$$

or by blocking, e.g. acetylation of amine groups with acetic anhydride:

$$\text{R-NH}_2 + \text{CH}_3\text{COOCOCH}_3 \rightarrow \text{R-NHCOCH}_3 + \text{CH}_3\text{COOH}$$

Box 9.3 // Tryptophan and amyloid

The p-dimethylaminobenzaldehyde (DMAB) method for tryptophan has been used as a method of demonstrating the pathological protein amyloid. This will work well with certain types of amyloid that have high levels of tryptophan in their polypeptides but is not a generally reliable method of identifying all amyloids. The low levels of tryptophan in most proteins make this method more useful than most of the other single amino acid techniques. Most proteins will barely react. It is only in cells such as Paneth cells and zymogen granules in the pancreas that significant reaction will be seen.

Table 9.1 Examples of spot tests for single amino acids

Group	Amino acid(s)	Technique
Amine	E.g. lysine	Ninhydrin–Schiff
Phenyl	Tyrosine	Millon
Disulphide/sulphydryl	E.g. cysteine	Performic acid–alcian blue
Indole	Tryptophan	DMAB–nitrite (see **Box 9.3**)
Guanidyl	Arginine	Sakaguchi

Single amino acids are rarely demonstrated in diagnosis but may sometimes be demonstrated in research. When performing these techniques, it must be remembered that the fixation and other treatments that are used in tissue preparation are likely to alter the reactivity of the proteins. Glutaraldehyde and to a lesser extent formaldehyde will bind to amino groups, mercury-containing fixatives will affect sulphur-containing groups and picric acid will form picrates with proteins. The best fixatives for the histochemistry of amino acids in proteins are the non-additive fixatives such as alcohol or Carnoy's.

Immunological techniques

These are the most specific and accurate but will only detect a particular protein. Immunological techniques are probably the best method of localizing individual proteins and have become very important in identifying cells, diseases and metabolic changes in cells. They are so important that they will be dealt with as a separate chapter (*Chapter 12*).

Specific action of the protein

Receptors and other binding proteins can be detected by using labelled ligands. These techniques usually resemble the immunotechniques in their methodology. The most important of this group of techniques are the enzyme techniques where the enzyme is located by its action on a substrate. These are less specific than immunological techniques; for example, acid phosphatase is a broad title for many enzymes. However, they are often useful as they indicate function rather than single protein species (see *Box 9.4*).

9.3 Enzymes

Enzyme histochemistry is the localization of enzymes in tissues by their specific action on a substrate.

> **Box 9.4 // Isoenzymes**
>
> Many enzymes have more than one type present within the body. These are isoenzymes or isozymes and often can be separated by electrophoresis. This technique is of little use in sections where localization of the enzyme is critical. However, many isoenzymes can also be identified by differences in their sensitivity to inhibitors or pH.
>
> Acid phosphatase is found in many cells but one form is resistant to inhibition by tartrate ions. Tartrate-resistant acid phosphatase (TRAP) is limited to the osteoclasts that resorb bone, so osteoclasts can be identified by using a TRAP technique.
>
> Similar inhibition methods are available for a range of enzymes and can identify subtypes of enzyme reasonably well.

Tissue preparation for enzyme techniques

Enzyme activity is an extremely labile phenomenon and most enzymes are rapidly inactivated by most tissue preparative methods. For this reason, paraffin wax- and plastic-embedded sections are almost entirely useless for demonstrating enzyme activity and cryostat sections are the more usual way of preparing the sections.

Fixation can sometimes be useful to help stop enzymes diffusing and this can be done either on the block before sectioning or immediately after cutting when the section can be rapidly fixed. Fixing the section rather than the block is often better as the length of fixation can be kept brief since penetration into a thin section is rapid compared with a large block of tissue.

For electron microscopy (EM), where the block needs to be embedded in plastic to obtain thin enough sections, the staining is done on the fresh tissue before processing ('block staining'). This means that only one method can be performed on each block of tissue and a separate block is needed for each enzyme being investigated. This is not necessarily as drastic as it sounds since EM blocks can be very small; indeed, a cryostat section for light microscopy can be used as an EM block. However, it does mean that each block has to be processed and sectioned separately. Once the enzyme technique is finished, the block can then be fixed as usual since there is no longer any difficulty if the enzyme is inactivated. The tissue is then processed and cut as for any other EM block.

Enzyme demonstration techniques

For histology it is usually more important to get accurate localization than precise quantitation and so the techniques are often quite different from biochemical techniques. In particular they must produce a visible (i.e. coloured) insoluble product. Various methods of localization have been tried with differing degrees of success.

Immunotechniques

Antibodies against the enzyme can be used to localize the enzyme by any of the usual immunotechniques. However, immunotechniques do not show enzyme activity, simply the presence of an antigenic protein. Thus, some inborn errors of metabolism may show high levels of enzyme even though there is no active enzyme present. In these cases, the cell may be producing an aberrant protein but the mutated enzyme has no catalytic activity.

Substrate film techniques

Substrate film techniques can detect digestion but are inaccurate and non-specific. The enzyme is detected by allowing it to diffuse out of the section into a thin film of substrate layered on to the slide (see *Fig. 9.4*). This can be done for proteases with a layer of protein or for nucleases with a film of DNA. The enzyme digests the film leaving a hole in the film. The film and section are then stained and any holes in the stained film indicate where the enzyme activity is located. This technique is not popular and has very limited application. It is only applicable to a few enzymes that break down a substrate that can be made into a suitable film. It gives relatively poor specificity and poor localization.

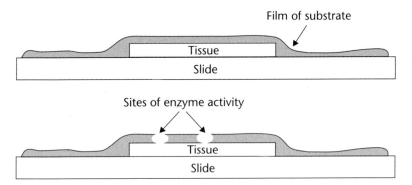

Figure 9.4
Substrate film technique. The upper diagram shows a layer of substrate before digestion and the lower diagram shows holes where the enzyme has digested the film

Dissolved substrate techniques

These are the most common techniques and provide the most useful way of detecting enzyme activity. They use dissolved reagents and by careful choice of reagents it is arranged so that one of the final products of the reaction is coloured and insoluble. Variations on the technique include the following:

- The natural reaction product is coloured and insoluble, e.g. DOPA oxidase.
- The reaction produces a product that, although not itself insoluble, can be precipitated as a coloured product. This is a very useful way of detecting many enzyme reactions, e.g. the azo dye technique for phosphatase.
- The reaction produces a product that, although not itself insoluble, can be precipitated. However, the precipitate is not coloured and must be treated further to produce the colour, e.g. the Gomori phosphatase technique.

Hydrolytic enzymes

Only two major types of enzyme can be easily and reliably identified in sections. These are **hydrolytic enzymes,** in which the substrate is split using water, and **oxidoreductase enzymes,** in which there is a change in redox potential. A few other individual enzymes can be demonstrated but these are the exception.

Hydrolytic enzymes were the first enzymes to be localized in tissue sections and the relevant techniques were introduced by Gomori. The basic idea of the technique has remained but the method has been modified to give more reliable results. The enzymes that were demonstrated by these initial techniques were phosphatases. Phosphatases split organic phosphate molecules releasing inorganic phosphate ions that are then precipitated by ions in the substrate solution and the precipitate is then visualized.

Gomori's acid phosphatase

The method uses β-glycerophosphate as the substrate and incorporates lead ions into the substrate solution to trap the phosphate ions. Lead phosphate is insoluble in acid solutions and so is precipitated at the site of enzyme activity. The lead phosphate is then visualized using a sulphide treatment to convert it to a black precipitate (see *Fig. 9.5*).

This method is easily adapted to other substrates that contain phosphates, e.g. ATP (for ATPases). It suffers from the disadvantage that lead ions can bind non-specifically to tissues (especially nuclei) and so requires careful technique and good controls to ensure that the sites of enzyme activity are correctly located. Lead is also an unnatural ion in cells and can act as an enzyme poison.

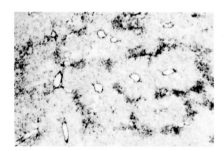

Figure 9.5
Liver stained by the Gomori method for phosphatase. The site of the enzyme appears black at the periphery of the liver lobules.

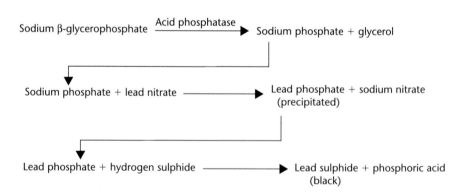

When these methods were first used to localize phosphatase enzymes, there was initially some doubt about their specificity and reliability. The lead ions, for example, could easily be absorbed on to other sites and give non-specific staining. Another problem was that the method assumes that, as the enzyme splits the glycerophosphate, precipitation of the phosphate ion will be instantaneous. This is not necessarily true. It was possible that the localization was wrong and that the phosphate ions diffused some distance before being finally precipitated. What was required was confirmation of the result using a totally different technique. This second technique was developed in the form of the azo-coupling technique.

Azo-coupling technique

This uses an artificial substrate, α-naphthol phosphate. When α-naphthol phosphate is hydrolysed by a phosphatase it releases phosphate ions and α-naphthol. The α-naphthol can react with a diazonium salt such as fast red B to form an insoluble azo dye at the site of the enzyme activity:

α-Naphthol phosphate + H₂O ⟶ α-Naphthol

α-Naphthol + N≡N⁺ Diazonium cation Fast red B NO₂ + OH⁻

Azo dye

When this technique was used, it was found to give similar results to the metal precipitation method of Gomori. Thus, the α-naphthol phosphate technique confirmed the results from the Gomori technique and the Gomori technique validated the α-naphthol phosphate method. With two techniques giving the same results, it is possible to have confidence in both. In practice, the α-naphthol phosphate method has some advantages over the Gomori technique and is now the more popular technique. It is very simple to perform (see **Box 9.5**) and gives strong reactions, especially using the later modified naphthol derivatives such as the naphthol AS derivatives developed by Burstone.

General techniques can be adapted for other enzymes

The α-naphthol method is quite adaptable and has been developed for many hydrolase enzymes as well as the phosphatases; for example, naphthol acetate can be used for esterase enzymes and naphthol glucuronate for glucuronidase enzymes. When hydrolysed,

> ## Box 9.5 // Identification of semen
>
> Acid phosphatase is found in prostatic fluid and can be used to localize and identify semen stains in cases of rape. If a semen stain is found in a case of rape, then the sperm can be used for DNA fingerprinting. The difficulty is in locating these stains once they have dried. If a piece of clothing is believed to have been contaminated with semen, it can be very tedious to search it visually to find minute traces of semen. If a damp sheet of paper is pressed against the clothing for a short period, then some of the protein in the semen will diffuse on to the paper, including the acid phosphatase enzyme. The paper can then be sprayed with acid phosphatase reagent. The resulting reaction will show exactly where the semen is, sperm can be located more quickly and the DNA testing can be carried out.
>
> This has also been taken up commercially for a less commendable purpose. A commercial spray was marketed to be used to spray the underclothes of a partner, the logic being that semen stains were an indication of marital infidelity. Unfortunately the logic is faulty. Men suffering from prostate problems will generate a positive reaction even when faithful. The manufacturers earned an 'Ig Nobel Prize' for their efforts.

these substrates release α-naphthol, which can couple with a diazonium salt as above. In general, provided the natural substrate of an enzyme can be coupled to α-naphthol without stopping the enzyme hydrolysing the compound, then this technique can be adapted for any hydrolase enzyme.

The metal trapping technique of Gomori can also be extended to other hydrolases; for example, lipases split fatty materials and release fatty acids that can form insoluble soaps with calcium ions in the substrate solutions. These insoluble calcium salts can then be visualized using the cobalt/sulphide technique of Gomori. Usually the synthetic detergent Tween is used as the lipid rather than simple soap.

Acetyl cholinesterase can be demonstrated using acetyl thiocholine. The esterase releases the thiocholine, which can be trapped with copper ions, and the copper thiocholine can be blackened with hydrogen sulphide.

Oxidoreductase enzymes

Oxidoreductases are enzymes that affect their substrates by changing the redox potential and often transfer electrons between the reactants. Several types of oxidoreductase enzyme can be demonstrated and they differ in the type of change they produce. Three types will be considered here: dehydrogenases, which transfer hydrogen from a substrate to a carrier molecule; oxidases, in which oxygen acts as the oxidizing agent; and peroxidases, which involve the use of peroxides.

Dehydrogenase enzymes

The action of dehydrogenase enzymes involves the exchange of hydrogen between two substances, the first of which is usually called the substrate and the other the hydrogen acceptor (or sometimes electron carrier). The hydrogen acceptor therefore changes from an oxidized state to a reduced state, whilst the substrate becomes oxidized (dehydrogenated). These changes are described for the forward reaction, but since most of these reactions are easily reversible the metabolism of the cell may well use the opposite direction.

The common hydrogen acceptors in the cell are NAD^+ ($NADH_2$), $NADP^+$ ($NADPH_2$) and FAD^+ ($FADH_2$) with the reduced form of the co-enzyme shown in parentheses.

In the detection of dehydrogenase enzymes, an artificial hydrogen acceptor replaces the natural NAD or FAD. The usual artificial acceptors are tetrazolium compounds, which

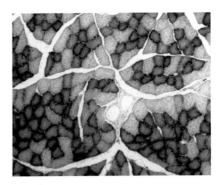

Figure 9.6
Cross-section of voluntary muscle stained for succinic dehydrogenase enzyme. The dark blue fibres are the red-muscle fibres and are rich in mitochondria. The paler fibres are the white-muscle fibres and have fewer mitochondria.

are converted from a soluble non-coloured form to a coloured insoluble precipitate, called formazan, when they become reduced (see *Fig. 9.6*).

There are several forms of tetrazolium, which differ in their redox potential and solubility of the formazan. The different redox potentials of the tetrazolium compounds affect which enzymes each form of tetrazolium will be best suited to demonstrate. The solubility of the formazan is only important in that it determines how the slides should be mounted.

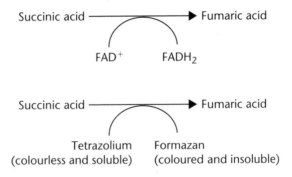

The technique for succinic dehydrogenase works quite well using only the tetrazolium salt (nitro blue tetrazolium) and enzyme substrate (sodium succinate) in a buffer since the enzyme incorporates a molecule of FAD. Many dehydrogenases, including lactic dehydrogenase, do not incorporate a natural hydrogen acceptor and need a soluble natural acceptor to be present in the substrate solution.

The lactic dehydrogenase transfers the hydrogens to NAD, which is its natural co-factor, and the NAD in turn transfers the hydrogens to the tetrazolium. This second transfer is carried out by another enzyme known as either a tetrazolium reductase or a **diaphorase**. These enzymes occur in the mitochondria and this results in lactic dehydrogenase being apparently located in the mitochondria. The reason is that there is a double reaction occurring. The lactic dehydrogenase transfers hydrogen from lactic acid to NAD in the cytosol and the NAD then diffuses into the mitochondria where it is acted on by diaphorases, which reduce the tetrazolium and produce the final colour.

Thus, lactic dehydrogenase techniques usually show deposits of blue in the mitochondria, even though the mitochondria do not contain any lactic dehydrogenase. Since diffusion is always going to be a significant factor in dehydrogenases, the location of the enzyme is best considered as accurate enough to identify cells containing the enzyme but not accurate enough to identify the subcellular compartment.

Preventing destruction of reduced coenzymes by the electron transport chain. In cases where soluble cofactors such as NAD (or NADP) are being used, there is usually a need to add extra NAD to the substrate solution. (The reagent dilutes the natural NAD found in the cell and so the concentration is too low to maintain the reaction at a reasonable rate.) On entering the mitochondria, the $NADH_2$ would normally be acted on by the electron transport chain (oxidative phosphorylation). The hydrogens are transferred along the chain and eventually form water by combining with atmospheric oxygen. If this natural reaction is allowed to occur, it would result in very little colour developing as very little of the $NADH_2$ would form the coloured formazan and most would simply produce water. To prevent this loss of $NADH_2$, the cytochromes of the electron transport chain can be poisoned by cyanide; thus, none of the $NADH_2$ is converted to water and it all forms formazan. Azide has a similar function and the same effect can be achieved by removing all of the oxygen and performing the reaction under anaerobic conditions.

Extra electron transport reagents are sometimes added to improve the transfer of electrons to the tetrazolium, e.g. Meldola's blue or phenazine methosulphate. These will speed up the reaction but in some cases of long incubation may give a non-specific deposit of formazan.

Oxidase enzymes

Oxidase enzymes catalyse oxidation reactions that involve oxygen as the hydrogen acceptor. Four atoms of hydrogen are oxidized by one molecule of oxygen. Thus, the reaction is often shown as one molecule of substrate (which loses two hydrogen atoms) and $\frac{1}{2}O_2$:

$$O_2 + 2(\text{Substrate}) \rightarrow 2H_2O + 2(\text{Product})$$

or

$$\frac{1}{2}O_2 + \text{Substrate} \rightarrow H_2O + \text{Product}$$

Although it is possible theoretically to use a tetrazolium salt as a hydrogen acceptor to replace oxygen, in practice the redox potentials for the reactions are not compatible with the redox potentials of the available tetrazolium salts. The methods used to demonstrate the oxidase enzymes therefore tend to be unique to individual reactions, rather than variations on a common theme. Two examples that are important in practice will be considered.

Catechol oxidases. These are a group of copper-containing enzymes that catalyse the conversion of tyrosine to dihydroxyphenylalanine (DOPA) and then the further oxidation of DOPA to DOPA quinone, which then spontaneously undergoes a series of reactions to form melanin. The enzyme may also be called tyrosine oxidase or DOPA oxidase depending on the reaction occurring.

DOPA quinone spontaneously oxidizes and decarboxylates to form indole-5,6-quinone and this polymerizes to melanin. The exact molecular nature of melanin is still uncertain but it is insoluble in all of the usual solvents, and is stable and very densely coloured (it appears black).

In tissues, the enzyme is usually demonstrated by using DOPA, thus demonstrating the second fast reaction rather than the slower first reaction. The sites of activity show up as black as the melanin develops. This histochemical reaction is unusual as it uses a natural substrate (DOPA) and natural product (melanin) rather than an artificial analogue (see *Box 9.6*).

Box 9.6 // DOPA in the brain

Melanin normally occurs as a protective pigment in the skin and eyes but it also occurs in the brain in a region called the substantia nigra. The melanin here is related not to protection from UV light but to the inhibitory neurotransmitter dopamine. In Parkinson's disease, there is a loss of these pigmented neurones and loss of dopamine secretion. The chemical L-DOPA is used as a treatment as it can cross the blood–brain barrier, whilst dopamine cannot.

Peroxidases. These enzymes catalyse the conversion of peroxide to water and oxygen. The active oxygen then oxidizes a substrate to a coloured insoluble product. These enzymes are found not only in animal tissue but also in plants such as horseradish (see *Box 9.7*).

$$H_2O_2 + \text{substrate} = \text{oxidized substrate} + 2H_2O$$

Box 9.7 // Using horseradish peroxidase as a reagent

Horseradish peroxidase has become a popular enzyme in research. It can be demonstrated using the technique outlined here and gives a brown deposit that can easily be seen in sections. It can be used as a tracer in several different ways.

As it is a protein, it can be used to identify where blood vessel permeability changes in experiments. If the blood vessel is impermeable, then the peroxidase cannot escape, but if the permeability increases, as in an inflammatory condition, then the peroxidase will escape and be trapped in the tissues. By sectioning the tissues and demonstrating the presence of the enzyme by histochemistry, the leaky sites can be seen. It is also used in antibody labelling techniques (see *Chapter 12*).

Diaminobenzidine $+$ H_2O_2 $\longrightarrow$ Quinone imine derivative $+$ $2H_2O$

The quinone imine then polymerizes to a brown product. The polymerization reaction may also be catalysed by peroxidase as it involves a further oxidation.

Examples of the use of redox and hydrolytic enzymes

An example of the use of redox and hydrolytic enzymes is in the recognition of different muscle types. The nature of a disease affecting muscle can often be recognized in a biopsy by changes in the distribution of the types of muscle cells present. By staining for the enzymes, the muscle types can be recognized (see *Table 9.2*). The myosin ATPase enzymes vary in their activity sensitivity to pre-incubation at different pH levels. The activity of the phosphorylase enzyme is not important in classifying muscle types but is useful in diagnosing McArdle's disease (phosphorylase deficiency).

Quantitation of enzyme activity

The apparent enzyme activity as observed by the colour of the final deposit is not always directly proportional to the actual amount of enzyme. This is due to the fact that the usual Beer–Lambert relationship is not necessarily applicable. In the absence of the Beer–Lambert relationship, it is difficult or impossible to relate the density of colour to the activity of the enzyme.

The problems associated with quantifying enzymes in tissue sections include:

1. Loss of enzyme activity during fixation, cutting, etc. All processing of tissues, even just preparing frozen sections, can involve significant loss of enzyme activity. The actual amount of enzyme activity in a section is always less than was present in the original tissue.

Table 9.2   Enzymes and muscle types

Muscle type	Succinic dehydrogenase	Myosin ATPase			Phosphorylase
		pH 9.4	pH 4.6	pH 4.3	
Type 1 (red)	+++	+	+++	+++	+
Type 2a	++	+++	0	0	+++
Type 2b	+	+++	+++	0	+++

2. The rate of an enzyme's activity in a section may be controlled by how rapidly the substrate diffuses into the tissue, rather than the amount of enzyme present. The diffusion of substrate can effectively be one-dimensional, with all of the substrate coming directly down on to the site of the enzyme. Diffusion from the side will always be slower and may be non-existent if there are adjacent enzyme sites, and of course diffusion from below is not possible since sections are normally attached to an impermeable glass slide or coverslip. If the rate of diffusion is slower than the maximal activity of the enzyme, then the amount of colour produced will not follow the usual Beer–Lambert law and enzyme activity will be underestimated.

3. Diffusion rate may be altered by the precipitation of insoluble product. The nature of histochemical demonstration requires the production of an insoluble product at the site of activity. This makes the diffusion process even worse. There may be a maximum colour that is able to develop where diffusion is completely blocked, so again Beer–Lambert relationships are not valid.

4. Section thickness may not be constant. This is a major problem as section thickness determines how much enzyme is present. If the thickness of sections from a microtome fluctuates by as little as 1 μm (1/1000 of a millimetre), it can alter the thickness of the section by as much as 20 or 25% (a nominal 5 μm section could in reality be anywhere between 4 and 6 μm thick). Section thickness is very difficult to measure and fluctuation in section thickness is a common finding. Sections will be alternately thicker and thinner than the set thickness. This may be due to a slightly loose knife or block or due to wear in the microtome producing looseness.

5. Diffusional losses of reaction product. If the final product is even slightly soluble in any of the reagents used in processing the section, then some of the colour will be lost.

Suggested further reading

Bancroft, J. and Cook, H. (1994) *Manual of Histological Techniques and their Diagnostic Application.* Churchill Livingstone.

Bancroft, J., Layton, C. and Suvarna, S. (2012) Chapter 10: Haematoxylins. In: *Bancroft's Theory and Practice of Histological Technique*, 7th edn. Elsevier Health Sciences.

Culling, C., Allison, R. and Barr, W. (1985) *Cellular Pathology Technique*, 4th edn.: Butterworths.

Kiernan, J.A. (2008) *Histological and Histochemical Methods: theory and practice*, 4th edn. Scion Publishing.

Sheehan, D.C. and Hrapchak, B. (1980) *Theory and Practice of Histotechnology.* Mosby.

Self-assessment questions

1. How can the basophilic staining of nuclei be controlled to ensure that it is due to DNA and not to other basophilic materials?
2. What is meant by 'intercalation'? Name one dye that stains by intercalation.
3. Outline the principles of the Feulgen technique.
4. Give two different ways that proteins can be stained in tissue sections and indicate how specific they are.

5. Name two ways in which enzyme activity may be lost in tissue sections.
6. What is the reason for adding potassium cyanide to substrate solutions being used to demonstrate dehydrogenase enzymes?
7. How is the naphthol azo-coupling method modified to demonstrate hydrolases other than phosphatases?
8. Why is it better to measure enzyme activity biochemically rather than histochemically?

10 Histochemistry of pigments, neuroendocrine amines and minerals

Learning objectives

After studying this chapter you should confidently be able to:

- **Describe the classification of pigments found in sections**
  Pigments are classified as artefact (caused by tissue processing), exogenous (foreign materials from the environment that enter the body whilst it is still alive), or endogenous (generated within the body by metabolism). Endogenous pigments may be subdivided into haematogenous (derived from blood) or autogenous (not directly connected to blood).

- **Describe the formalin-induced fluorescence reaction of neuroendocrine amines**
  Treatment with formaldehyde converts many neurotransmitter amines into fluorescent compounds that can be visualized directly with a fluorescent microscope. Most amines require freeze-drying and treatment with formaldehyde vapour as they are small water-soluble molecules and are lost during aqueous formalin fixation.

- **List the main minerals found in tissue sections and describe how they can be identified**
  Calcium, iron and magnesium are the main minerals in normal tissues. Smaller amounts of copper, aluminium and trace elements may be present but are normally present in too small a quantity to be demonstrated easily. Calcium can be demonstrated by the von Kossa technique or by reaction with dyes such as alizarin. Magnesium is usually found associated with calcium and is rarely identified separately. Iron in the ferric form is usually identified using the Perls technique.

- **Outline the use of ion-selective dyes**
  Ion-selective dyes identify soluble metallic salts but are only suitable for use in intact cells and not in histological sections. They are used in research and are not currently used in routine diagnosis.

10.1 Pigments

Most materials in tissue sections are uncoloured until a stain is applied, but a few materials are coloured and can be seen in unstained sections. These self-coloured materials are referred to as pigments. They include a wide variety of materials from different sources

---

Box 10.1 // Other pigments

The pigments discussed in the main text are the common ones encountered in tissue sections. Other pigments do occasionally occur but are rarely demonstrated.

The slight yellow colour of fat is due to fat-soluble plant pigments such as carotenoids. These pigments are of no significance in the usual amounts and I have only heard of one exceptional case of a person who was adversely affected by an unusually high amount. He died of a vitamin A overdose. He was taking vitamin A tablets and drinking a lot of carrot juice and turned a bright orange colour. Most people do not suffer from the small amounts present in their bodies.

---

and have differing significance. The presence of some pigments is normal but others can be related to disease, so it is often important to be able to recognize and identify them.

When identifying pigments, two features need to be borne in mind about the appearance and reactions of the pigments:

1. Pigments generally have a limited range of colour with most being yellow to brown or brown to black. The colour varies depending on how much is present. A brown pigment present as a very heavy deposit may appear almost black, whilst the same pigment present in only small amounts may appear yellowish. Thus, descriptions of the pigment colour are only a guide and not a precise description.

2. The results of a test are affected by the colour of the pigment. Schmorl's test, for example, gives a blue colour as a positive reaction, but if it is applied to a yellow pigment, then a positive result will be greenish (blue + yellow) rather than pure blue. This also means that it may be difficult to see whether a reaction has occurred. How can you tell what has happened if a positive reaction gives a black colour but the pigment is already black?

Pigments can be conveniently grouped into three major types: **artefact pigments**, **exogenous pigments** and **endogenous pigments** (which can be haematogenous or autogenous). These groups are convenient labels, although some pigments can be put into more than one category.

Artefact pigments

These were not originally part of the tissue but have been added by the processes used to prepare the tissue. The commonest are associated with fixatives.

Formalin (formaldehyde) pigment is most easily seen around old or degenerating blood in tissues fixed in acid solutions of formaldehyde. The formation of this type of pigment can mostly be prevented by using buffered formaldehyde for fixation. The pigment appears as a fine brownish deposit around the old blood and is usually birefringent. Formaldehyde pigment can be identified and removed by treatment with picric acid solutions.

A similar, sometimes identical, pigment is formed in malaria and is therefore called **malarial pigment**. The mechanism of its production is different and it may be found within red cells infected with the parasite, in the area around infected red cells and even within the cytoplasm of macrophages that have ingested affected red blood cells. The colour is similar to formaldehyde pigment and it will dissolve in picric acid solutions similarly to formaldehyde pigment. The presence of the parasite within the red cells helps to identify the source of the pigment, although large deposits of the pigment may obscure the parasite.

Mercury pigment is a dense black granular (or occasionally globular) pigment, occurring irregularly throughout the tissue. It cannot be prevented if the fixative contains mercuric chloride, but it is easily removed by treating the section with iodine solutions (this also acts as a test to recognize mercury, although all heavy metal-derived pigments will also be dissolved). Mercury pigment is less common in modern histological practice where mercury-containing fixatives are used less frequently, but it is found in tissues stored in archives, as mercuric chloride was once a very common fixative.

Chrome pigment may be found as a brownish pigment in tissues fixed in chrome fixatives and not washed in water before processing. It occurs when the chrome salt reacts with the alcohols used in tissue processing and produces a brown oxide, although the reaction is not easily produced even when it is done deliberately as a demonstration. Chrome pigment is not readily removed once formed, so it is better to avoid it by always washing chromium-fixed tissues in water.

Osmium pigment occurs as a very dense black pigment spread throughout the tissue following fixation in osmium tetroxide (commonly used for electron microscopy specimens). It is not usually a problem since in electron microscopy it acts as a stain and when it is used in light microscopy the blackening is often wanted to identify double bonds. However, it can be removed from sections by oxidation with hydrogen peroxide if required for some stains.

Exogenous pigments

Exogenous pigments were present in the original tissue but are not a normal part of the body and are contaminants that entered the body during life. The pigments were ingested, inhaled or otherwise introduced in life. Their significance varies, with some being pathological and some being indicators of the person's employment (e.g. silica) or lifestyle (e.g. tattoos).

Carbon usually occurs as a dense black pigment. Carbon may be intracellular in macrophages, when it is often seen as a fine powder, or can occur in angular clumps in the extracellular connective tissue. Carbon is very common in the lungs of city dwellers and in particular in the lungs of smokers. It also occurs in regional lymph nodes around the lungs where it has been carried by macrophages (see *Fig. 10.1*). It can sometimes be seen in skin from tattoos (see *Box 10.2*) and needle tracks. The carbon itself is not pathological, although the association with tobacco smoke does have links to smoking-

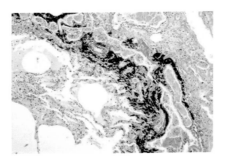

Figure 10.1
Cross-section of the lung of a cigarette smoker stained with the van Gieson method showing carbon deposited in the connective tissues.

> **Box 10.2 // Tattoo pigments**
>
> A variety of pigments are sometimes used in tattoos. Carbon is one that sometimes occurs, especially in 'amateur tattoos'. In professional tattoos the pigments used can be much more elaborate and of many different colours. The pigments do not usually need to be identified but if you wish to see the colours, then use an incident light microscope rather than the usual transmitted light microscope.

related disease. It may also act as a carrier for more dangerous materials as fine carbon particles adsorb many chemicals on to their surface (this is why 'activated charcoal' is used in many filters). The carbon produced during the burning of tobacco or in badly maintained diesel engines may adsorb carcinogens produced by incomplete combustion and transport them into the lungs.

Carbon is difficult to identify accurately as it is very inert. Usually it can be reliably identified by its location (lungs) and shape. It can be destroyed by micro-incineration but this is not commonly done. It has no pathological significance in itself but it sometimes needs to be distinguished from other black pigments that may be important. This can be done by eliminating the other pigments by testing for each one in turn.

Silica may be brown to black in colour depending on the source of the silica and is found as angular masses in the lungs and lymph nodes of miners, quarry workers, stonemasons and other people exposed to stone dust, especially from power grinding equipment.

Silica is very inert as it is the same material as glass, and it is difficult to identify with specific tests. There are no positive tests and it is unaffected by all histochemical reagents and micro-incineration. Silica may be birefringent (rotates the plane of polarized light) and shows up brightly using a polarizing microscope; this is often a useful indicator.

Silica causes the disease silicosis (see *Box 10.3*), which is a serious and often fatal fibrosis of the lungs. Silicosis is a recognized industrial disease, so that assessment of the amount of damage caused by silica may be important in cases of industrial compensation.

Asbestos occurs as a brown pigment found in the lungs but usually appears as fibres rather than the angular masses associated with silica. Inside the body, it becomes coated with an iron-containing protein that forms distinctive beaded asbestos bodies. These asbestos bodies are easily recognized and give a positive Perls reaction for iron (see *Figs 10.2* and *10.3*).

> **Box 10.3 // Silicosis**
>
> Silica causes damage to cells by its ability to form hydrogen bonds with membranes. If very fine particles get into the lungs, they will be ingested by macrophages and end up in the lysosomes. Here, the silica particles can hydrogen bond to the lysosomal membrane and cause it to rupture. Lysosomal enzymes are released into the cytoplasm and kill the cell. The enzymes break down both the cell and surrounding tissue. The silica is inert and remains unaffected. The silica will be ingested by another macrophage and the cycle will be repeated. This results in chronic inflammation and scar tissue forms where the tissue is damaged. The cycle will continue to repeat until the silica becomes coated with enough protein or fibrous tissue for it to be ignored by the macrophages. If the silica load in the lungs is large, then large areas of the lungs will be replaced by fibrous scar tissue and this is the cause of silicosis.

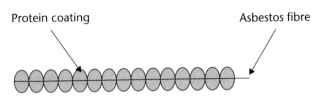

Figure 10.2
An asbestos body

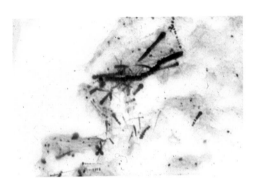

Figure 10.3
Asbestosis in a lung section stained using the Perls technique. The asbestos fibres have become coated with iron-containing pigments, which stain blue with Perls.

The fibres are inert and birefringent (different types of asbestos can be identified by their birefringence). As with silica, asbestos is not destroyed by micro-incineration. Asbestos is often associated with disease and causes asbestosis, which can be very debilitating and is a recognized industrial disease. Asbestos increases the risk of developing lung cancer and is the only known cause of the malignant mesothelioma tumour.

Endogenous pigments

Endogenous pigments are the products of metabolism in the body. They occur naturally within the tissues and so occur in normal tissues. However, they are also associated with disease states when they occur in excessive amounts or in unusual situations. The pigments can come from blood or one of its breakdown products (**haematogenous pigments**) or can be derived from other metabolic processes (**autogenous pigments**).

Haematogenous pigments

These pigments come from the blood or one of its breakdown products.

Haemoglobin is always present in the tissues inside the red blood cells and in this form it is usually ignored as a pigment. It does, however, sometimes occur outside the red blood cells, for example in haemolytic diseases, and then it becomes significant. Haemoglobin has a molecular size that is close to the limit of the glomerular filtration system. Once haemoglobin is released into the plasma, it can pass through the glomerular filter into the renal tubules and this can lead to renal problems (see *Fig. 10.4*).

In formalin-fixed sections, haemoglobin is not the bright red colour usually associated with blood but is a brownish colour. Free haemoglobin is an amorphous or globular material

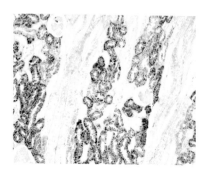

Figure 10.4
Perls stain of kidney from a case of haemolytic anaemia showing deposits of iron-containing pigment in the convoluted tubules. The haemolytic disease has resulted in haemoglobin passing through the glomerulus and being reabsorbed by the cells of the tubules.

that is strongly acidophilic. Haemoglobin can be identified by this strong acidophilia. In a haematoxylin and eosin stain, the eosin gives a more orange/pink colour to haemoglobin than to other materials in the section and this is often sufficiently distinctive to allow haemoglobin to be identified. More specialized techniques have been developed that allow an even better distinction, e.g. the kiton red/almond green stain. Histochemically, haemoglobin can be identified by the peroxidase activity of the haemoglobin molecule (see **Chapter 9** for the peroxidase reaction). This peroxidase activity is unusual in that it resists normal processing and often can be demonstrated in paraffin wax sections.

Haemosiderin and ferritin. A whole range of iron-binding proteins occurs in cells and tissues to scavenge and recycle the iron released by the degradation of haemoglobin. Many are present in such small quantities that they cannot be easily visualized by light microscopy. Larger deposits occur as brown granular or powdery deposits, particularly in macrophages of the spleen and liver (where destruction of old red blood cells occurs) and around sites of haemorrhage. These larger accumulations are called haemosiderin by histologists. Haemosiderin is easily identified by the Perls reaction, which detects the iron in the haemosiderin. The iron is detected by potassium ferrocyanide (potassium hexacyanoferrate (III)). This produces a dense blue precipitate, which is the pigment Prussian blue, hence the other names for the test: Prussian blue reaction (PBR) or Perls' Prussian blue reaction (PPBR) (see **Fig. 10.5**).

The iron in haemosiderin will not react directly with the ferrocyanide but must be released from the protein by acid hydrolysis. Thus, the Perls reagent is a mixture of 2% potassium ferrocyanide and 2% hydrochloric acid. The iron in haemoglobin is too strongly held within the porphyrin ring to be released in this way and so haemoglobin and red blood cells are negative for the Perls stain.

Other tests are available for haemosiderin (Turnbull's blue and blackening of the free iron with hydrogen sulphide) but are rarely used.

Diseases associated with excess haemosiderin include haemochromatosis and haemosiderosis. It is also an important way of identifying sites of tissue damage (remnants of bleeding), and identification of haemosiderin is useful in identifying some types of cells.

Bile pigments. These are derived from the porphyrin ring of haemoglobin. They vary in colour from red/brown (bilirubin, the commonest form) to green (biliverdin). The range

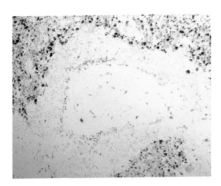

Figure 10.5
Spleen stained using the Perls technique to show macrophages containing haemosiderin (a storage form of iron).

of colours seen in a bruise as it heals is mainly due to the breakdown of these pigments. Bile pigments are slightly soluble in organic solvents, so small amounts are usually lost. Thus, paraffin sections of normal liver cells (the main site of pigment production) are usually free from pigment. Only large deposits can survive processing and appear in wax sections.

Bile pigments are difficult to identify. They can be identified by oxidation, which alters the colour. The Fouchet technique uses ferric chloride as an oxidant in the presence of trichloracetic acid. This gives a bright green colour, which is fairly stable, but not all forms of bile pigment will react.

Gmelin's test uses concentrated nitric acid as the oxidizing agent and reacts with more types of bile pigment but the colour change is fleeting. Sections are examined whilst the test is being applied, but even so the colour changes can easily be missed and the slides cannot be kept or viewed later. A positive result is usually reliable but a negative result with the normal tests should not be considered conclusive. The test involves oxidation of the pigments by irrigation. This means running the reagent under the coverslip whilst viewing the section. The reagents are quite corrosive and the difficulties of looking down the microscope and focusing on the section at the same time as squeezing nitric acid out of a Pasteur pipette mean that it is a cumbersome technique and is best carried out using three arms and two heads. For most people this means getting someone else to help while you do the test.

Excessive amounts of bile pigment are found in some liver diseases or in haemolytic disease. The main confusion likely to occur is between lipofuscin (see Autogenous pigments below) and bilirubin, both of which can occur in the liver. It is often easier to eliminate lipofuscin than to identify bile definitively.

Autogenous pigments

These bile pigments are in many ways quite diverse and are derived from metabolism but not associated directly with blood. Despite their diverse origins, the pigments described here do have one characteristic in common – they are all reducing pigments. This makes them a neat group in practice since a single reducing test allows them to be identified as belonging to this group.

Two reducing tests are in use:

- Masson–Fontana silver reduction (black result);
- Schmorl's ferricyanide reduction (blue result).

The first is more sensitive but is slow (24 h or more) and the black result may be difficult to interpret if the pigment itself is dark brown or black. The Schmorl's test is quicker (5–15 min) but it is often difficult to interpret because the background tissue takes on a greenish tinge.

Melanin. This is a black or brown pigment that is found in the eye (especially the retina), skin (especially negroid skin, tanned skin and around the nipples) and in parts of the brain (substantia nigra) (also see *Box 10.4*). Melanin can be identified by its reducing ability with the Schmorl's test, appearance and the fact that it is bleached by oxidizing agents (hydrogen peroxide, potassium permanganate or potassium chlorate/HCl).

The main pathological significance of melanin is its association with malignant melanoma (see *Fig. 10.6*).

Box 10.4 // Melanosis coli

In melanosis coli, the gut becomes almost black due to the accumulation of a pigment in the macrophages similar to lipofuscin. This condition is caused by excessive use of purgative laxatives. Despite the name, the pigment is not related to melanin and it is just the sheer amount of pigment that causes the dark colour. The pigment is sometimes called pseudomelanin.

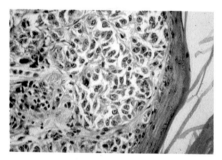

Figure 10.6
Section of malignant melanoma with areas of brown melanin pigment visible on the left.

Lipofuscin is a pigment derived from oxidized lipid. It appears as droplets in the lysosomes of cells. It accumulates with age and is sometimes called 'wear and tear' pigment. Levels increase after the use of some poisons (e.g. carbon tetrachloride) and in some chronic diseases. It can be found in any tissue but is often prominent in heart muscle (see *Fig. 10.7*), liver and brain.

Lipofuscin is the most easily identified of the autogenous pigments as there are several good tests, which, whilst individually not conclusive, become highly selective when two or three are used in combination:

1. It is basophilic: methylene blue and aldehyde fuchsin both give good strong staining (see *Fig. 10.7*).

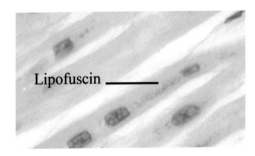

Figure 10.7
Lipofuscin in heart muscle. The brown pigment indicating lipofuscin is situated at the poles of the nucleus.

2. It is periodic acid–Schiff (PAS)-positive, but this is less useful as there are many other substances in the tissues that are also PAS-positive. It is useful to remember that this may be the cause of an unidentified PAS-positive reaction.
3. It stains as a lipid with some fat-soluble dyes, especially Sudan black.
4. It is peracetic acid–Schiff-positive and performic acid–Schiff-positive (typical of unsaturated lipids). However, as these are dangerous methods, it is always better to use a simpler and safer technique.
5. It stains positive by the Ziehl–Neelsen technique. This is usually used for the acid-fast mycobacteria associated with tuberculosis and leprosy but is a useful method for lipofuscin.

Chromaffin pigment is a brownish intracellular pigment that is found in adrenaline-producing cells (adrenal medulla and sympathetic ganglia). The pigment is not entirely natural as it only occurs after an oxidizing fixative such as chrome fixatives (hence the name, from 'chrome affinity') and is not seen in formaldehyde-fixed tissues. Chromaffin pigment represents the oxidized precursor of adrenaline.

It is often considered not to be a pigment as it is a part of normal metabolism. It can be considered along with the other neurotransmitter amines and demonstrated by the formalin-induced fluorescent technique on freeze-dried tissue (see below). It can also be considered as an artefact pigment since it only occurs after fixation in chrome fixatives and is not a natural pigment. However, it has traditionally been included as a pigment and has been retained in this section as it is often easier to demonstrate it by fixing a piece of tissue in a chrome fixative than to carry out the full formalin-induced fluorescent technique on freeze-dried tissue. It is a basophilic pigment and produces green colour when stained with Giemsa.

Excessive chromaffin is found in phaeochromocytomas, which are tumours of the chromaffin- or adrenaline-secreting tissues.

Argentaffin. This pigment is the strongest reducing pigment and will react strongly with silver reagents (hence its name). It is found in the endocrine-secreting cells of the gut (argentaffin cells) (see *Fig. 10.8*). Again, this pigment only occurs after the use of certain fixatives, particularly formaldehyde, and is strongly fluorescent (formalin-induced fluorescence; see below). It also reacts with diazo salts such as fast red B giving a red deposit. Argentaffin pigment also gives a chromaffin reaction if fixed in chrome fixatives and can be called enterochromaffin pigment (hence, the alternative name for argentaffin cells is enterochromaffin or EC cells).

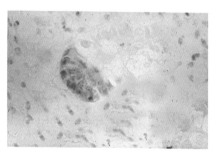

Figure 10.8
Argentaffin cells in the base of the crypt of Lieberkühn.

Other pigments also occur but are less commonly encountered (e.g. ochronosis pigment) and will not be considered here.

10.2 Neuroendocrine granules

The neuroendocrine cells are a scattered group of cells that secrete biologically active amines as hormones. These amines are more commonly thought of as neurotransmitters, but in these cells they are secreted into the bloodstream rather than into the synapses between nerve cells. They also secrete characteristic neuropeptides that act as hormones. This set of cells is also known as the 'diffuse endocrine system' and the amine precursor uptake and decarboxylation (APUD) system from their metabolic activity (see *Box 10.5*). Amines are formed by the decarboxylation of precursor materials such as tryptophan to form serotonin (5-hydroxytryptamine).

Box 10.5 // APUD cells

The neuroendocrine cells form a useful system but their embryonic origin is disputed. They may not even all come from the same original embryonic layer, but the difficulty of tracing cells as they migrate in embryonic development means that no one is certain of their origin. They form a coherent set of cells in functional terms and also have a characteristic set of tumours, apudomas, which continue to secrete their peptides and can result in clinical syndromes from this secretion.

Carcinoid tumours are tumours which arise from the gut or respiratory tract APUD cells but do not secrete any identifiable peptide hormone.

Staining techniques for neuropeptides

Neuropeptide secretions are reasonably well preserved in fixed sections and can usually be stained using dyeing techniques. There are many reasonably selective methods that will stain different cells within the same neuroendocrine organ in different colours. Methods for the pituitary and pancreas are available to distinguish the main cell types on the basis of their acidophilic and basophilic characteristics. These stains are usually quite attractive with clear bright colours but require a fair degree of skill to get good results. These methods are largely empirical and so do not have any particular theoretical interest. The peptides can also be specifically demonstrated by using immunotechniques. The main interest in neuropeptide identification is in their role in the diagnosis of tumours of the diffuse endocrine system.

Demonstrating neuroendocrine amines

The other secretory components of the neuroendocrine cells are much more difficult to demonstrate because of their low molecular weight and solubility. Substances such as adrenaline and histamine are usually lost during processing as they readily diffuse and are quickly destroyed by enzymes. The only efficient way to preserve amines is to freeze-dry the tissues. The snap freezing prevents diffusion and sublimation of the water prevents their loss. Some amines can be detected in routine paraffin sections if they are bound to a larger tissue component. This seems to occur with serotonin (5-hydroxytryptamine) in argentaffin cells (and rat mast cells) and histamine in mast cell granules (see *Box 10.6*).

Box 10.6 // Mast cells and amines

Although amines are difficult to demonstrate in tissues, they are very important in disease. They play a crucial role in the inflammatory response. Mast cells are the main cells that produce these amines, but the amines that are present vary with species. Human mast cells, like all mammalian mast cells, contain histamine, but rat mast cells also contain serotonin and ruminant mast cells contain dopamine.

Individual methods for some amines

A number of demonstration methods have been developed to detect individual amines. The specificity of some of these methods is not very high but they can be useful for tumours to demonstrate that a substance is present. An example of this is silver reduction by argentaffin cells. The method is not specific for serotonin in these cells but is a useful histological method and can be helpful in the recognition of carcinoid tumours. Similarly, the chromaffin method can be used for catecholamines in the adrenal medulla using an oxidizing fixative to preserve the adrenaline and noradrenaline. Serotonin can also be demonstrated using the azo-coupling method, which reacts with the phenolic group of serotonin.

Formaldehyde-induced fluorescence

Amines will react with formaldehyde to give a fluorescent product. This is known as the formalin-induced fluorescence (FIF) technique. The reaction does not occur readily in solution (the amines will rapidly diffuse away) and instead it must be done on freeze-dried sections using formaldehyde vapour. The tissues are freeze-dried and then transferred to a closed container along with some paraformaldehyde and heated to generate formaldehyde vapour. The temperature and humidity are critical factors for maximum fluorescence. The temperature and humidity may need to be altered slightly to adjust for differences in tissues and preparation. Generally a temperature of between 60 and 80°C for 1 h is used at a relative humidity of 50–75%.

The tissue can then be embedded in paraffin wax. This does not need any processing since the tissue is freeze-dried; simply put the tissue into molten wax and gently reduce the pressure to remove air bubbles. The tissue can be blocked out and sectioned, but water must be avoided so they are not floated on a waterbath to flatten them. The tissue is dewaxed, but not rehydrated, and mounted. Fluorescence can be seen if the specimens are irradiated with blue-violet light of 410 nm wavelength. Dopamine and noradrenaline fluoresce green, whilst serotonin gives a yellower colour.

Glyoxylic acid can replace formaldehyde

Instead of using formaldehyde, a similar reaction can be produced using glyoxylic acid. Glyoxylic acid can be used to perfuse the animal, thus generating a weakly fluorescent compound before removing the tissue from the animal. Glyoxylic acid will react with cryostat sections and this is more convenient than the block method used for FIF and can give a brighter result. The fluorescence is intensified by heating or treating with formaldehyde vapour. The sensitivity of this method is very high and it has been estimated that 10^{-7} picomoles of adrenaline can be detected in cells.

Other fluorescent methods

Although formaldehyde and glyoxylic acid are the most commonly used forms of induced fluorescence, it is possible to get condensation reactions with other substances such as o-phthalaldehyde (histamine), hydrochloric acid (dopamine), acetaldehyde, glutaraldehyde and acetic acid (similar to FIF).

FIF and related methods have been used in research mainly to identify and locate nerves using the amine neurotransmitters. FIF has been replaced to some extent by antibody techniques using antibodies against the enzymes producing the amines. Immunocytochemistry is more convenient as routine paraffin sections can be used, but these methods only show which cells are producing amines and not how much is present. In some cases, it is useful to detect which nerves are depleted of their transmitters, as well as which ones are capable of producing the transmitters. The depletion can only be seen using FIF or a related method.

10.3 Metals and their compounds

The histochemistry of metals and their salts is partly dependent on their solubility. In most cases of sectioned material, the only metallic substances that are likely to be present are those that are insoluble. Metal-containing compounds can be divided roughly into four different types:

- elemental metals;
- soluble metal salts;
- insoluble metal salts;
- organically bound metals.

Elemental metals

Elemental metals are found as the solid element and not combined with any other substance. Such metals are uncommon in sections except as contaminants or exogenous implants. Metals are used in medicine in the form of dental fillings, radiotherapy implants and metal prostheses such as replacement joints, pins and plates in bones, and metal implants in arteries used to open out the narrowing lumens ('stents'). The identification of the metal is rarely a problem in these cases. Preparing sections of these substances, however, can be extremely difficult. Radiotherapy implants may be left in place permanently and the tissues may not be removed until much later. They can ruin a knife-edge if they are unwittingly left in the paraffin wax block and may rip through the tissues at the same time, damaging the tissue itself. Occasionally sections are required, usually

> **Box 10.7 // Many trace elements and poisons are difficult to demonstrate histochemically**
>
> Metals are extremely important in metabolism and there are a huge number of 'trace elements' that are needed by the body. However, these are often present in such minute quantities that there is no method sensitive enough to detect them in single sections. These include elements such as cobalt (found in vitamin B12) and zinc (found in the zinc-finger structures of proteins).
>
> Many metals are also poisonous but even in toxic amounts they may still be difficult to demonstrate histochemically. Some metals are always toxic, e.g. cadmium, mercury and lead. The role of other materials is less certain, for example arsenic, which is certainly highly toxic, yet some evidence exists that it may be used in metabolism, although only in minute amounts.

for research, with the implant in place and the tissue still adhering to it. Depending on the metal used, sectioning can be near to impossible but histochemical identification is not a problem.

Soluble metal salts

The soluble metal salts are ionic and are not able to be demonstrated or localized in sections produced by any method since they will rapidly either be lost or relocated by any technique. Metal salts are ionic and very mobile. Simply cutting the tissue will allow them to diffuse. Frozen sections are not necessarily any better as the section often thaws briefly during cutting and then refreezes and relocation can occur in that brief instant. Even freeze-drying may not prevent movement since the salts may be soluble in wax or plastic monomers. It is sometimes possible to precipitate the metal by some means, but even then relocation is more likely than accurate localization. The only method of localizing these materials is to examine intact living cells without sectioning using ion-sensitive dyes (see below); however, this is not always a reasonable method to apply.

Insoluble metal salts

Insoluble metal salts are found mainly in the minerals of bone, but other insoluble calcium salts such as calcium oxalate do occur. As with the first group, it is also possible to have contamination with metal salts. The metals may even be introduced by the processing. Rust in tap water is not unknown and may contaminate tissues during a washing step in the preparation.

Organically bound metals and their unmasking

The final group, where the minerals are bound into organic form, is the most important group and includes the iron in haemoglobin and the cytochromes.

There are several methods available for releasing such organically bound metals. Minerals need to be soluble and ionic before they can be demonstrated since it is only the ions that react. Many of the difficulties with the techniques are related to releasing or unmasking the minerals without losing or relocating them. Releasing the minerals can sometimes be done by chemical treatments such as acid hydrolysis with hydrochloric acid or oxidation by hydrogen peroxide. If these can be shown to release all of the metal, then they are certainly the simplest solution.

Micro-incineration can unmask many metals

As an alternative to chemical unmasking, the release of metals can be achieved by burning off all of the organic matter using **micro-incineration**. This changes all of the organic minerals into salts and so the total content of metal can be seen. Incineration is done in a laboratory furnace at 500–600°C. Glass slides cannot withstand this high temperature, so sections need to be mounted on quartz slides. For the best results, oxygen needs to be excluded whilst the temperature is low (<500°C) and only then is air admitted. Oxidation then occurs rapidly and completely. The exclusion of air helps to remove the organic matter by preventing incomplete combustion products forming that then resist further oxidation. The slide is then carefully removed and cooled. The pattern of ash is quite extensive and usually the main tissue structures can be seen in outline. The pattern of ash is sometimes known by the somewhat grand name of a **spodogram**. It is very fragile and must be protected with a coverslip and then ringed. An alternative is to coat it gently by gently dipping into a solution of celloidin. The celloidin then hardens and protects it, whilst still allowing reagents to penetrate and react with the released salts.

Micro-incineration can also be done more crudely with a standard Bunsen burner. The tissue section on an ordinary slide is placed on a heat-proof mat. A Bunsen burner is then picked up and the hot blue spot of the flame is played over the section so that the section passes through the bright blue edge of the flame where it is hottest. The section will turn brown initially and then turn from brown to white as the organic material burns off. There is no need to dewax paraffin sections as the wax will also burn off. Once the section has turned white, it consists of only the mineral salts. The slide should then be covered to slow down cooling as the slide is likely to shatter if cooling is rapid. The final ash can then be protected as before.

Micro-incineration is very rarely done. It is not suitable for all minerals; for example, mercury will evaporate. It is a simple way of distinguishing pure carbon deposits from silica as carbon will be burnt off in the process but silica will remain. It is also very difficult to locate minerals precisely within the tissue as by the time you have incinerated the section there is no tissue left and accurate location is not possible.

Spot test sensitivity and contamination

One further problem with mineral identification is contamination during preparation of the sections. Many of the histochemical spot tests are so sensitive that they will pick up contamination at very low concentrations. For the best and most reliable results, ALL reagents should be completely free of the ions that are going to be demonstrated. This can be surprisingly difficult since some reagents will pick up ions very easily. Copper contamination from wax baths is not unknown and sections can be contaminated from metal forceps, scalpels and even microtome knives. Contamination should always be considered if results seem at all unexpected. If the wax block is prepared before you know that a histochemical test for minerals is needed, it may be too late and contamination may already have occurred. The opposite problem is the loss of minute deposits during processing, especially in acidic reagents. More materials are soluble at acid pH than in alkaline solutions but ideally neutral pH should be used.

Calcium

Calcium deposits in the body come in a few main forms. Hydroxyapatite is the main mineral in teeth and bone. Mineralization in bone is critical for health, whilst for some diseases the measurement of calcification is an important diagnostic aid (see **Box 10.8**).

> **Box 10.8 // Calcium deficiency in bone**
>
> Diseases in which calcium is deficient include rickets and osteomalacia, which are caused by lack of vitamin D. Although dietary deficiency is now quite rare in developed countries as vitamin D is added to many foodstuffs, these diseases can still occur due to malabsorption in Crohn's disease.
>
> Vitamin D can be synthesized in the skin by the action of UV light but the darker skins of some races prevent this formation of vitamin D in the skin. There is an occasional case of vitamin D deficiency in some Asian communities in Britain. Some chapatti flours also seem to inhibit the absorption of calcium and this exaggerates the disease. Bone biopsies may be used to confirm the diagnosis.
>
> Osteoporosis is a very important disease in post-menopausal women and causes skeletal deformities, bone pain and fractures. In the years immediately following the menopause, bone mass may reduce by as much as 3% per year. Osteoporosis is usually diagnosed from the radiographic appearance rather than by bone biopsy.

Strictly, calcification should be called mineralization as it includes metals other than calcium.

The von Kossa technique is useful for calcification but is not specific for calcium

The von Kossa technique using silver to demonstrate mineralization has been explained in *Chapter 7* and, although important, this technique is not specific for calcium itself (see *Fig. 10.9*). Other forms of calcium often do not react using the von Kossa technique, for example crystalline calcium carbonate and the calcified materials in invertebrates, so it is best regarded as a measure of bone mineral and not as a method for calcium detection.

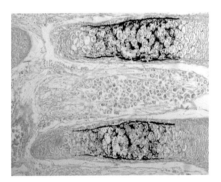

Figure 10.9
Calcium in bone demonstrated by the von Kossa method of silver impregnation. The bones at the top and bottom show calcium deposits, whilst the middle bone shows no mineralization.

Some dyes will complex with calcium

Some dyes can bind selectively to calcium and are more specific methods for calcium than the von Kossa method. They are also equally good at determining mineralization. Simple haematoxylin staining as with haematoxylin and eosin gives a distinctive blue colour with mineralized bone compared with the pink colour of osteoid. This is seen even more clearly with an unmordanted haematoxylin when only the calcified material will stain.

Alizarin red sulphonate (alizarin red S) is probably the most used of the simple dyes. Alizarin is not absolutely specific for calcium and also reacts with aluminium, barium and strontium. Aluminium and barium, however, give a slightly different colour but strontium is usually mixed with the calcium and is indistinguishable from it.

Alizarin can be used to show bone development in embryos

One use for alizarin is in the Dawson technique for examining bones in embryos. This is a method of treating whole embryos so that the bony structures can be seen through the body tissues. The soft tissues of the embryo are macerated by treatment with strong hydroxide. This renders them softer and less likely to take up the stain. The bones are stained with alizarin and the soft tissues are then made transparent by soaking in glycerol, which has a high refractive index and, like the organic clearing agents, makes soft tissues almost transparent. The stained bones can then be seen clearly through the translucent soft tissues (see *Fig. 10.10*).

Figure 10.10
Embryonic bones in a rat pup stained by Dawson alizarin red method.

Drug and dye incorporation into bone during normal bone growth

Xylenol orange is another dye that can be incorporated into bone but is usually used as a vital stain when it is incorporated into new bone as it mineralizes. This is a useful way of locating where new bone is being formed but is of course limited to animal studies. An alternative to the xylenol orange dye is the use of tetracycline antibiotics. These work in the same way and fluoresce quite strongly when bound to calcium in bone. Different tetracyclines fluoresce in different colours, so, depending on the tetracyclines that have been used, it is possible to identify when particular parts of calcified structures have formed. This can be useful even in humans – tetracyclines are used as antibiotics for the treatment of infections, so some samples may include tetracyclines from their clinical use. This technique can be used in a pulsed delivery with a short dose given to label newly formed bone and then a gap before a second dose. The doses rapidly localize in the newly forming bone as discrete lines and the distance between them represents the rate of bone growth in the period between the two pulses. I have even seen the same effect in the tooth of a child who had had three courses of tetracycline – the tooth showed distinct lines corresponding to the growth between the tetracycline treatments.

Magnesium

Magnesium is often found along with calcium in mineralized tissues but it can be identified as an independent ion by chelation with dyes. The dyes are thiazol yellow G and 4-(*p*-nitrophenylazo)-resorcinol (Magneson). Thiazol yellow gives a red colour to magnesium deposits, whilst Magneson gives a blue colour.

Iron

Iron occurs in haemoglobin, myoglobin, cytochromes, ferritin and haemosiderin, in all of which it is organically bound. Iron is constantly recycled by the body, with only small amounts being lost or gained under normal circumstances. Iron is poorly absorbed from the diet, particularly from vegetable sources, and women can become iron deficient due to the monthly losses of iron in the menstrual blood flow. Many women are prescribed iron tablets to supplement their dietary sources. If large amounts are absorbed, it can prove toxic and iron poisoning has been a problem with children taking their parents' iron pills (see *Box 10.9*). This toxicity may be due to the iron generating free radicals, which then damage the tissue.

Box 10.9 // Iron and toxicity in children

Iron poisoning is one of the commonest causes of poisoning in children. In the USA, there were 110,000 cases between 1986 and 1992 of whom 33 died. The immediate effects of iron overdose are vomiting, diarrhoea and gastrointestinal bleeding. If the child recovers from these problems, there may be liver damage, heart failure and coma within 1–2 days. If the child survives and recovers from the initial toxicity, there may still be problems including liver damage up to 6 weeks later.

Haemochromatosis is caused by excessive iron uptake

In the disease primary haemochromatosis, there is an excessive absorption of iron and the iron deposits as haemosiderin in many organs. In the liver, the deposits may lead to cirrhosis following hepatocyte death. In the heart, it may cause cardiac myopathy leading to heart failure, and in the pancreas, it can cause diabetes mellitus. The disease is inherited as an autosomal recessive disorder, so it will appear suddenly in families rather than showing a strong familial inheritance.

Haemosiderosis occurs following repeated blood transfusions in thalassaemia

In secondary haemochromatosis or haemosiderosis, there is a similar accumulation of iron but this is not due to a genetic defect and instead is the result of other problems such as alcoholism or thalassaemia. Thalassaemia patients need regular blood transfusions as their own red cells are destroyed rapidly. The constant influx of blood overloads their iron excretion system, so iron will build up unless it is removed constantly using drugs.

Iron is difficult to unmask when bound in a porphyrin ring

Iron is usually only detected as the protein-bound forms of haemosiderin and ferritin, both of which give positive Perls reactions. The use of the Perls reagent for detecting haemosiderin has been covered in the section on pigments earlier in this chapter. In haemoglobin, myoglobin and the cytochromes, iron is not simply attached to protein but also complexed into a porphyrin ring to form the haem group. This stronger link makes it resistant to removal and the simple use of low pH in the Perls reaction will not allow it to be released; thus, any compound with the iron bound in this way will usually be Perls-negative. The iron can be detached from the haem using an alkaline solution of peroxide, but this is less reliable as a method than the usual Perls reaction and is not often used.

Ferrous iron can be detected with Turnbull's blue reaction

The Perls method only detects ferric iron, but since this is the main form in the body this is not a major limitation. If ferrous iron needs to be detected specifically, then a modified Perls reaction should be used. This uses potassium ferricyanide instead of the potassium ferrocyanide used in the standard Perls reaction. The final pigment produced by the reaction is the same deep blue colour and is usually called Turnbull's blue, although it is probably exactly the same material as the pigment in the Perls reaction. Usually Turnbull's blue reaction does not detect much ferrous iron in tissues unless there is a deposit of exogenous ferrous material.

If all of the iron present in a piece of tissue is required, regardless of the ionic form, then the section can be treated with hydrogen sulphide or ammonium sulphide. All of the iron will be converted to ferrous sulphide, which is black, and this can then be stained using Turnbull's blue reaction. The result is usually little different to a simple Perls reaction. Iron, which is resistant to most methods of unmasking, can usually be detected by micro-incineration, followed by a sulphide/Turnbull's blue reaction.

Other (rarely used) methods for detecting iron

There are also various methods of detecting iron using chelating agents. Unoxidized haematoxylin will give a blue/black colour but is not very specific. A more specific chelating agent is 4,7-diphenyl-1,10-phenanthroline, which gives a red colour with iron and a yellow colour with cobalt. The copper chelate is uncoloured. Ferric iron chelates only slowly, so if the reaction is being used specifically for ferrous iron the treatment should be kept short. If both ferrous and ferric iron are required to be demonstrated, then adding some thioglycolic acid will reduce the ferric ions in haemosiderin and unmask it from the protein without the need to add hydrochloric acid as in the Perls reaction.

The detection of non-haem iron is very important and frequently done, but despite the variety of methods available, it is the oldest method, Perls, that has proved to be the simplest, most reliable and most sensitive.

Copper

Copper is an important ion in metabolism but is present in only small amounts in a normal person and these amounts are below the detection limit of histochemical tests. Copper is normally handled efficiently by the liver, with copper from the diet being linked to caeruloplasmin for transport. Copper caeruloplasmin is recycled by the liver and free copper can be excreted into the bile. Copper accumulates in the liver in Wilson's disease, which is a rare autosomal recessive genetic disorder. In Wilson's disease, the copper and caeruloplasmin are not secreted into the plasma and there is an accumulation in the liver cells. Free copper may then appear in the blood and can deposit in the brain and cornea. The excessive levels of copper in Wilson's disease can be detected histochemically, although it may need unmasking if it is associated with protein. It can be unmasked by treating the section with hydrogen peroxide or hydrochloric acid.

Copper can be detected with rubeanic acid

The best detecting agents for copper are rubeanic acid (dithio-oxamide) and *p*-dimethylaminobenzylidine rhodanine (DMABR). Copper can be detected using an unmordanted haematoxylin solution but this is less sensitive and is non-specific.

Lead

Lead was at one time a widespread material used in plumbing (hence the name, *plumbum*, which is Latin for lead), paints and as a petrol additive. The average total body load is estimated at around 120 mg. Environmental exposure to lead compounds can result in significant problems with anaemia, motor neuropathy and deposition of lead in bone but it is rarely diagnosed on tissues (see *Box 10.10*). Lead poisoning can be recognized in blood samples by the stippling of red cells or can be measured chemically in serum samples. It is rare that there is sufficient lead in a normal person to be detectable by histochemical methods. The most sensitive method of detection is the sodium rhodizonate method, which produces a red chelate with lead ions.

Box 10.10 // Toxic metals in diet

Both copper and lead have been used in water pipes but only lead seems to dissolve significantly in the water and cause poisoning. Lead will only dissolve significantly in soft water areas. In hard water areas, the pipes rapidly become lined with a deposit of calcium salts. This furring causes problems of blockages and narrowing of the pipes but protects humans. Lead pipes are being gradually replaced but the vast numbers of these pipes in older houses means that it will be many years before they are all completely replaced.

Soft water is generally slightly acid due to dissolved carbon dioxide and it is this acidity that allows metals to dissolve more rapidly. Acid solutions will generally dissolve more metals than neutral or alkaline solutions. In cooking it is not wise to allow any acidic material (such as fruits and fruit juices) to stay in contact with anything containing potentially dangerous metals. This includes pans (copper, iron and aluminium are all toxic), glazed pots (many coloured glazes contain dangerous metals including uranium) and enamel containers (again it is the coloured pigments that are the problem).

Aluminium

Aluminium accumulates in the bodies of patients on haemodialysis and can cause an osteomalacia-like condition that resists normal treatment. In the brain, aluminium causes encephalopathy, which again is found in some long-term dialysis patients. Aluminium is commonly used in normal life (see *Box 10.11*) in the form of cooking utensils and aluminium foil. Aluminium hydroxide is also used in some antacid remedies. Aluminium has also been reported in the neural plaques found in the brains of patients with Alzheimer's disease but it is not thought to be the cause of Alzheimer's.

Box 10.11 // Aluminium and tap water

Aluminium can occur in tap water naturally and the World Health Organization recommends that aluminium concentrations should not exceed 0.1 parts per million. Some parts of the world have much higher levels; for example, a survey in Malawi found levels up to 100 times higher.

Aluminium is also used as a flocculating agent to clarify drinking water. Normally the levels of aluminium are very low but in the town of Camelford there was an accident in which a delivery of aluminium salts was put into the wrong manhole and ended up being pumped directly into drinking water. The local inhabitants were severely affected by this release. The long-term effects of this release are not known and the affected people will no doubt provide more information about the effects of aluminium on humans.

Aluminium can be detected in sections using solochrome azurine, which forms a chelate complex with a deep blue colour. This reaction is also given by **beryllium**. Beryllium is used in fluorescent lights and may get into the body by inhalation or through cuts. Even quite small amounts produce accumulations of macrophages forming a **granuloma**.

Other heavy metals

Other heavy metals will also sometimes be seen but are not common. Silver and mercury are seen in the form of dental amalgam, which can become impacted into soft tissues during dental treatment. Silver is associated with silver earrings in badly pierced ears (or other pierced organs). Both metals are removed by iodine and hypo treatments as for the mercury artefact pigment.

Another method of identifying minerals is the use of X-ray analysis. This topic will be dealt with in *Chapter 19*.

10.4   Ion-selective dyes

These are a group of specially designed reagents that alter their fluorescent characteristics depending on their immediate local environment (see *Box 10.12*). The fluorescence change is related quantitatively to the concentration of particular ions in the fluid surrounding them. The change in concentrations of the ion can be followed dynamically in living cells. These dyes are quite different to most of the histochemical methods in the rest of this chapter in that they do not produce a coloured precipitate in slices of dead tissue but instead measure minute changes in the cytoplasmic concentration of selected ions. The dyes are only suitable for use inside living cells; they are not conventional stains and cannot be used on fixed sections. They are a research tool for examining metabolism, particularly in cultured cells.

Box 10.12 // Aequorin

The ion-selective dyes are a masterpiece of synthetic chemistry since many of these compounds have been designed specifically to identify one particular ion. The chemists start with a relatively crude ion-sensitive dye and alter it to make it more selective and more sensitive. This designer approach requires great knowledge of the interaction of molecules and ions and relies on the use of computer modelling to determine the best configuration. The improvements are still continuing.

There is also a range of ion-selective compounds that have been produced by nature that are also useful probes. Aequorin (a bioluminescent tracer from a jellyfish) and luciferase (an enzyme used to detect ATP) can both be used as tracers but such biological molecules are difficult to load into cells and are often more useful in cell-free extracts.

One advantage of biological molecules is that the genes for these molecules can be incorporated into cells and animals, so that all of the cells will exhibit fluorescence without the need to introduce the probe molecules. This has already been done with mice that glow green in the right light.

Ion-selective dyes can be used to monitor metabolism in living cells

By introducing these dyes into cells, it is possible to see and measure directly how particular ions alter with time or following specific events. The cytoplasmic changes in calcium concentration are critical in the control of many activities. It is the release of calcium ions into the cytoplasm that triggers muscle contractions, changes in the ovum following fertilization and the response of cells to many other stimuli. The calcium-selective dyes fura and quin-2 allow direct observation and measurement of these calcium fluxes. Other dyes of the same type can be used to measure a variety of changes, but each is more or less selective for a particular type of ionic change. These ion-selective dyes have enabled visualization of many internal processes within the cell.

A list of some of the changes that can be measured and examples of the probes used are given in *Table 10.1*.

Introducing ion-selective dyes into cells

Ion-selective dyes need to penetrate the living cells before they can be used to measure the internal concentrations of the ions. This is termed 'cell loading' and can be done in two ways:

- **Bulk loading,** where a suspension of living cells is treated so that many cells are labelled at once. One of the more popular ways of achieving this is to use the acetomethoxy ester (AM derivative) of the dye, which is able to penetrate through the cell membrane. Once inside the cell, these esters are hydrolysed by the cell's esterase enzymes releasing the active dye. The active dye is polar and cannot move out of the cell, so it is trapped in the cytoplasm. Other bulk loading methods include the use of osmotic shock or liposome delivery. Liposomes are membrane vesicles containing solutions of materials. They can be made in the laboratory and fused with cell membranes to deliver their contents into the cytoplasm.
- **Single-cell loading** involves directly injecting the dye through the cell membrane using a micro-pipette or micro-electrode.

Measuring the change in fluorescence

The change in fluorescence needs to be measured and calibrated. The method varies with the dye being used but two methods are commonly used:

- **The ratiometric method.** The dye has two emission peaks and the ratio between the heights of the two peaks measures the concentration of the selected ion. This method is used to measure calcium concentrations using fura-2. The advantage of this technique is that it is self-calibrating for the amount of dye since both peaks are measured and if the dye concentration changes, then both peaks change in parallel.

Table 10.1   Examples of ion-selective dyes

Ion or process measured	Example of a selective probe	Concentration range
Calcium	Fura	14–1450 nM
Magnesium	Mag-fura	2–200 nM
pH	SNAFL (seminaphthofluorescein)	pH 7.2–8.2
pH	BCECF (bis-carboxyethylcarboxyfluorescein)	pH 6.6–7.5
Sodium	Sodium green	0.6–600 mM
Chloride	MQAE (N-(ethoxycarbonylmethyl)-6-methoxyquinolinium bromide)	1–100 mM

- **Single-wavelength change.** Here only a single wavelength is measured and the system needs to be calibrated using a buffer and an ionophore. At the end of the experiment, the cell is immersed in a calcium buffer with a constant and known concentration of calcium ions. The cell membrane is made fully permeable to calcium using an ionophore. The calcium concentration inside the cell will be in equilibrium with the buffer, so the system can be calibrated from the readings of the equilibrated cell.

The measurements are often made using a confocal microscope, which allows accurate measurements not only within a single cell but also within compartments of the cell. This system of measuring inside living cells in a dynamic manner is extremely useful in research but as yet has no real use in diagnosis. The only drawback to the technique is the capital cost of the equipment needed to carry out these techniques and the cost of the ion-selective dyes.

Suggested further reading

Bancroft, J., Layton, C. and Suvarna, S. (2012) Chapter 13: Pigments and minerals. In: *Bancroft's Theory and Practice of Histological Techniques*, 7th edn. Elsevier Health Sciences.
Culling, C., Allison, R. and Barr, W. (1985) *Cellular Pathology Technique*, 4th edn. Butterworths.
Kiernan, J.A. (2008) *Histological and Histochemical Methods: theory and practice*, 4th edn. Scion Publishing.
Lyon, H. (1991) *Theory and Strategy in Histochemistry*. Springer.
Sheehan, D.C. and Hrapchak B. (1980) *Theory and Practice of Histotechnology*. Mosby.

Self-assessment questions

1. What are the three most common exogenous pigments in sections?
2. How does the occurrence of lipofuscin vary with age?
3. Which pigment(s) are positive with the Perls test?
4. Which pigments act as reducing agents and will give a positive Schmorl's test?
5. What are the two main types of secretion from APUD cells?
6. An excess of iron in the tissues causes which disease(s)?
7. How can calcium salts be demonstrated in sections?
8. Which metal accumulates in Wilson's disease and how can the metal be demonstrated?
9. What is the difference between ratiometric and single-wavelength ion-selective dyes?

11 | Infective agents and amyloid

Learning objectives

After studying this chapter you should confidently be able to:

- **Outline the limitations of histological identification of bacteria**
  The fixation and processing of tissues means that bacteria will be killed, so many important microbiological methods are not applicable.

- **List the types of organisms found in sections and name examples of each type of organism**
  Organisms may be bacteria, fungi, viruses or parasites.

- **Outline methods for the detection of micro-organisms in tissue sections**
  Only staining and immunotechniques can be used. Staining techniques are limited in scope and selectivity, making definitive identification difficult or impossible. Gram staining using methyl violet and iodine can subdivide the bacteria into Gram-positive and Gram-negative, whilst some methods such as Ziehl–Neelsen staining can identify single groups of organisms. Immunotechniques using antibodies raised against a single species of pathogen can be precise but are limited to confirmation of a diagnosis rather than identifying an unknown organism. Molecular biological techniques have similar advantages and limitations to immunotechniques and can also detect viral DNA in cell nuclei.

- **Describe the nature of prions and associated diseases**
  Prions are abnormally folded proteins which can act as catalysts to cause normal proteins to fold in the same abnormal manner. They can be infective and are classed as transmissible spongiform encephalopathies. Prions cause diseases such as Creutzfeldt–Jakob disease, scrapie and BSE.

- **Describe the nature of amyloid**
  Amyloid is a pathological protein formed by a normal protein being converted into a β-pleated sheet form. The altered protein is not easily degraded in the body and accumulates as amyloid deposits. The proteins altered in this way include immunoglobulins, peptide hormones, membrane glycoproteins and keratin.

- **Describe the major staining methods used to detect amyloid**
  These are Congo red staining combined with polarizing microscopy and staining with the fluorescent dye thioflavine. Other methods are much less useful, although IHC is an option.

Infection of tissue is a common finding in cellular pathology and the ability to recognize the causative agent can be invaluable in diagnosis. Identification from tissue sections is not the best method available and whenever possible identification should be done by microbiological techniques, which allow greater sensitivity and accuracy. Most microbiological identification relies on growing the organism and detecting its growth requirements, sensitivity to inhibitors and reactions to biochemical tests. Identification from histological sections suffers in comparison with these culturing methods since the organisms are killed by fixation. Formaldehyde fixation in particular is a very effective method of killing infective organisms. All of the useful microbiological culture methods are lost once the tissue is fixed. Identification of the organism may still be possible but is less certain or more difficult, or both.

In infected tissues the causative organism may be relatively uncommon and unevenly spread. This will mean that in an individual section there may be only a few organisms. Unless the staining technique makes these pathogens stand out strongly, it is easy to miss them altogether. The use of fluorescent techniques (see **Chapter 17**) where the organism is the only material staining against a dark background can help to identify their localization in these cases.

Organisms can also be altered by processing and care may be needed. The simple staining techniques of microbiologists often need adapting to cope with formalin-fixed sections, e.g. Gram-positive bacteria may lose their positive reaction in sections and appear Gram-negative.

Finally there may be the possibility of harmless bacteria proliferating in the tissues unconnected with the disease (see **Box 11.1**).

Box 11.1 // Commensals

Accurate identification is often crucial since bacteria are common in many sites. Non-disease-causing organisms are called **commensals** and these are common and may even be beneficial. Innocuous commensals need to be eliminated as only the pathogenic organisms are of concern in disease. Rapid fixation of the tissues is needed since many organisms will continue to grow after death and in a few hours or days they can completely destroy the tissue making diagnosis impossible.

Commensal organisms can also mask pathogens by their sheer numbers. One slightly different pathogenic organism in a group of commensals may easily be missed. This can be a problem in post-mortem material where the growth of a harmless commensal may overwhelm the pathogenic organism.

11.1   Identification of causative organisms

Although simple histological staining may be limited, there is often other information available or other tests that can be tried if a diagnosis is needed. The attempt to diagnose an agent in fixed tissue can use the strategies detailed below. The following approach is particularly informative when investigating a tissue sample, say a skin biopsy, from an immunocompromised patient which may have several different and unusual infectious agents present.

- **Clinical diagnosis and symptoms** may be helpful. Often a clinical diagnosis may already have been made and all that is needed to identify the organism is to confirm the clinical assessment. The patient's history may also be helpful with regard to recent travel abroad, ethnic origin, medication and immune status.
- **Appearance of the section.** The body may respond in a characteristic way to certain organisms. Langhans' giant cells (see **Figs 11.1** and **11.2**) are very strongly linked to tuberculosis, even if the organism itself is not easily seen in the standard haematoxylin and eosin stain. The appearance of the tissue may not be specific but may limit the possibilities and thus simplify identification.

Figure 11.1
Giant cells. These are large cells, often more than 200 μm in diameter, with a large number of peripheral nuclei

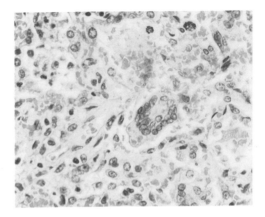

Figure 11.2
A multinucleated giant cell is shown in the centre of the image

- **Special stains or histochemical tests.** The stains may be fairly general (e.g. Gram stain) and may simply identify whether the organism belongs to one particular group or they may be more specific indicating a limited range of organisms.
- **Immunofluorescent or immunochemical detection.** This method will identify any organism very selectively and can be absolutely specific. The only difficulty is that this can only be used as a confirmatory method since you need to know which antisera to try. Once you have narrowed the options down to a very limited range of possible organisms, this is probably the best way to identify and confirm an individual organism. The use of immunotechniques is also a good way to spot organisms that are too small to be visualized with the light microscope (e.g. viruses).
- **Nucleic acid hybridization techniques.** In many ways, these resemble the antibody techniques in their applicability. They are excellent confirmatory methods provided you know what you are looking for. They do have the advantage over immunotechniques for virus detection in that they will detect the nucleic acid incorporated into the cell even when no virus particles are present. Some viruses can insert their own nucleic acid into the genome of the host cell resulting in a latent infection. Herpes simplex virus does this and can regularly reactivate, causing repeated bouts of cold sores. These latent infections can be quite important and the use of DNA hybridization allows the detection of the virus even when the infection is latent and not active. Kits are available that will distinguish between types of a particular virus, e.g. HPV 16 and 18 in colposcopic biopsies of cervix.

11.2   Types of micro-organism

Infective agents can be of a variety of types and these vary in their size and life history. Only the major types of organism are dealt with in this chapter although others exist and can be important. Mycoplasmas, for example, are considered to be sufficiently different from the classical bacteria to have their own classification and are important in causing up to 10% of community-acquired atypical pneumonia (caused by *Mycoplasma pneumoniae*).

The main groups of organisms are:

- **Viruses.** These are very small, obligate intracellular parasites. Individual virus particles can usually only be seen with the electron microscope.
- **Bacteria.** These are larger organisms with most being between 1 and 10 μm in size. Bacteria are visible with the light microscope and a few types can be stained selectively, although not usually identified positively, using simple staining methods. They are probably the most important group in histological identification although not necessarily the most important in terms of disease.
- **Fungi.** These are found in two forms: **yeasts**, which are single-celled forms, and **hyphae**, which are groups of cells forming threads. They are often larger than bacteria and vary from about 2 to 200 μm. They have a distinctively different cell wall to bacteria.
- **Protozoa.** Tropical diseases such as malaria and sleeping sickness are caused by protozoa. Although once uncommon in Britain, these diseases are becoming increasingly important as a result of immigration and international travel. Some previously minor diseases have also become important in immunosuppressed patients.
- **Helminthic infections.** These are caused by parasitic worms and are more common in countries where hygiene is poor. Helminths are usually quite large and can be recognized by their morphology alone.

Box 11.2 // Smallpox

Viral diseases are often more difficult to treat than many other forms of infection, so identification may only be of academic interest rather than clinically useful. Viruses can be prevented from causing disease by immunization. This was extremely effective in the case of smallpox, which has been completely eradicated by a World Health Organization programme that started in 1967 and ended in 1969 after 2 years with no reported cases. The disease was once prevalent in many countries with 15 million cases and up to 2 million deaths. The organism caused a distinctive inclusion body called a Guarnieri body in infected cells.

By contrast, there has been a rise in the incidence of human immunodeficiency virus infections but this virus is best seen in cultured cells and cannot be identified easily in tissue sections.

Viral infections

There are many viruses that cause disease (see *Box 11.3*) and so can occur in tissues. Recognition of viral infections is often possible on haematoxylin and eosin sections since they can have distinctive effects on the cells. Common signs are syncitial (grape-like) accumulations of nuclei following the fusion of cell membranes initiated by viral coat

> **Box 11.3 // Pathological effects of viruses**
>
> Viruses can cause damage to the host in three ways. Firstly, the virus may directly damage or kill the cell in which it replicates. This **cytopathic effect** is found with hepatitis A virus, which kills liver cells.
>
> Secondly, the virus may trigger an immune response in the host. The infection results in viral antigens appearing on the surface of the cell and these initiate an immune response. The result is that the cell is killed but it is the host's response that causes the damage. Since without a host response the virus causes very little damage, it allows the development of carriers who have a weak or non-existent immune response to infection. The host remains healthy but infective, i.e. is a carrier of the disease. This can occur with hepatitis B virus.
>
> Finally, the virus may transform the cell into a tumour cell. This can occur if the virus carries an **oncogene** or if it interferes with gene regulation. The Epstein–Barr virus and some human papilloma viruses can affect cells in this way.

proteins, and large nuclei with pale areas (halos) at their periphery. Although individual viruses are very small, they may occur as aggregates where the viruses clump together within the tissue forming **viral inclusion bodies**. The inclusion bodies are rich in both proteins and nucleic acids so they stain very strongly. These inclusion bodies can often be better seen with trichrome stains (e.g. Lendrum's phloxine tartrazine) where the differences in acidophilia and basophilia of the inclusion body can be exaggerated. Inclusion bodies often have individual names after the person who first described them. Inclusion bodies can be in the cytoplasm (e.g. Negri bodies in rabies) or in the nucleus (e.g. herpes virus) or in both (e.g. cytomegalovirus). In some cases, such as with the surface antigen of the hepatitis B virus (HBsAg, previously also called Australia antigen), there are more selective methods available. HBsAg can be demonstrated by using an orcein stain after permanganate oxidation, which converts sulphur-containing protein into sulphonate residues that then react with the orcein.

Electron microscopy is one of the most useful methods for detection of viruses as many viruses have distinctive shapes, although this will often only indicate the type of virus involved rather than the individual species. The use of antisera against viral proteins is probably currently the best method in widespread use but the use of *in situ* hybridization techniques may become more important as they can be even more specific and have the advantage of detecting viral DNA hidden inside the nuclei of cells. This is particularly useful because a range of viral infections can progress beyond the simple inflammatory and immune responses when there is a failure to fully clear the embedded viral genome from the DNA of its host. It is estimated that at least 20% of malignancies are associated with an earlier viral infection and many autoimmune reactions are also believed to be triggered by viral infections that generate an abnormal immune response.

The **rickettsiae** are intermediate between the viruses and bacteria in that they are more bacteria-like in structure but are obligatory intracellular parasites. Like viruses, they produce inclusion bodies in the cell rather than extracellular colonies and so their demonstration in tissues is more related to viruses than bacteria.

Important virus types causing human diseases

Herpes viruses. These are DNA viruses that are often acquired during childhood and then become latent when they integrate their DNA into the host's DNA. They may re-erupt into active disease in later life. All are similar in structure with a roughly spherical shape of about 120 nm and are covered with a lipid membrane. Individual diseases include varicella-zoster virus, which initially causes chickenpox but can re-emerge from latency in dorsal root ganglia as shingles; herpes simplex virus 1, which causes cold sores; herpes simplex virus 2, which causes genital herpes; Epstein–Barr virus, which causes glandular fever; and cytomegalovirus, which often occurs as an opportunistic infection in AIDS.

Retroviruses. These viruses contain RNA rather than DNA and include human immunodeficiency virus, which causes AIDS, and human T-cell leukaemia virus.

Paramyxoviruses. These are RNA viruses and include the viruses that cause measles and mumps.

Rhinoviruses. These are RNA viruses and include the common cold virus.

Togaviruses. These are RNA viruses and include rubella virus, which causes German measles.

Rhabdoviruses. These are RNA viruses and include rabies virus.

Enteroviruses. These are RNA viruses and include poliovirus. Although best known for its paralysing effects from infecting the anterior horn cells of the spinal cord, poliovirus is an enteric virus that first infects the gut; only in some cases does it get into the blood stream and affect the nerve cells.

Papillomaviruses. These are DNA viruses and include human papilloma viruses, which cause warts. Some types have also been linked to cervical cancer.

Bacterial infections

The principal method used to demonstrate bacteria is the Gram stain or one of its modifications. The Gram stain detects a difference in the cells of the two types of bacteria although there are several possible explanations for the differential staining. The permeability of the cell wall differs (Gram-positive cells have thicker walls); this is the simplest reason and explains the staining well. The dye (usually methyl violet or crystal violet) penetrates the cell wall and is then aggregated by adding iodine. The large dye aggregates are still easily removed from the more permeable Gram-negative bacteria by a decolorizing agent such as alcohol or acetone but are retained by the more impermeable Gram-positive bacteria (see *Fig. 11.3*). The difference is only in the rate of removal and prolonged washing in acetone will decolorize all bacteria. The Gram-positive nature also alters with fixation and processing. Usually another modified Gram stain is used rather than the simple technique but the method is similar.

Some bacteria can be stained more selectively

Mycobacteria can be stained using the Ziehl–Neelsen (ZN) stain for acid-fast bacilli. The original method used hot carbol fuchsin to stain the mycobacteria followed by differentiation in acid/alcohol. The presence of a hydrophobic material (mycolic acid) in the wall of these organisms is the usual explanation for their retention of the dye. The ability of mycobacteria to resist decolorization is called acid fastness and the organisms can be referred to as AFBs (acid-fast bacilli) or AAFBs (alcohol- and acid-fast bacilli). The acid fastness varies with the species. *Mycobacterium tuberculosis* (which causes tuberculosis) is more robust than *Mycobacterium leprae* (which causes leprosy) and will show stronger acid fastness. The ZN technique is often modified to retain the acid fastness

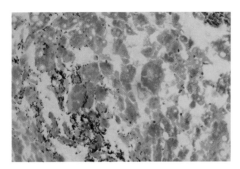

Figure 11.3
Infected lung stained with Gram stain. Gram-positive organisms can be clearly seen as dark blue rods (bacilli).

by avoiding strong hydrophobic solvents that would extract the mycolic acid. It has also become popular to use variations of the staining technique that do not involve the use of hot carbol fuchsin and are often referred to as a 'cold ZN'. The best known of these is the Kinyouin's modification which uses a 4× strength carbol fuchsin solution.

The demonstration of AFB can also be performed using the fluorescent dyes auramine O and rhodamine. This makes it easier to spot isolated bacteria, which fluoresce bright yellow against a dark background (see *Fig. 11.4*).

Selective methods have been published for *Helicobacter pylori* (which can occur in the stomach and is linked to ulceration, gastritis and stomach cancer) and the use of silver staining for spirochetes. Details of these methods are available in many books on histological techniques. Antibodies specific for *H. pylori* are available commercially and are preferred by some workers when the number of organisms are small and may be difficult to spot using, for instance, a Giemsa stain.

Important organisms seen in tissue sections

Gram-positive organisms. *Staphylococcus aureus* is a common pathogen and can cause boils, infections in wounds, abscesses and septicaemia. It is a great problem in hospitals where some strains are becoming resistant to several antibiotics including methicillin (methicillin-resistant *Staphylococcus aureus* or MRSA). The clumps of Gram-positive cocci are fairly distinctive. *Streptococcus* species are also Gram-positive but form chains rather than clumps and include *Streptococcus pneumoniae*.

Box 11.4 // Bacterial morphology

The shape of bacteria and their method of associating together can also be useful indicators of the type of organism present. **Cocci** are rounded, whilst **bacilli** are rod-shaped. These two forms of bacteria are the commonest. Other shapes such as comma-shaped **(vibrio)** and spiral forms are less common, so if seen they can be more informative. If the organisms form chains they can be called streptococci (or streptobacilli), whilst clustering indicates staphylococci.

The shape needs to be examined carefully since sections will often slice through the organism so that rods can appear as cocci. If there are many organisms present, then some will lie entirely within the section and the shape and grouping will be identifiable, but if there are only a few bacteria in a section, it becomes more difficult.

Figure 11.4
Lung tissue stained with the auramine/rhodamine technique showing fluorescent bacilli from a case of tuberculosis.

Lactobacillus acidophilus is a commensal found in the vagina ('Döderlein's bacillus'). *Corynebacterium vaginale* may cause cervicitis and is fairly common in cervical tissues (5–7% of women). *Corynebacterium diphtheriae* was once a common disease causing a severe form of laryngitis but effective vaccination has made it a rare disease in modern Britain.

Clostridia species are associated with a variety of diseases including gas gangrene (*Clostridium perfringens*), botulism (*Clostridium botulinum*), tetanus (*Clostridium tetani*) and pseudomembranous colitis (*Clostridium difficile*).

Mycobacteria are also Gram-positive but only weakly so. Although leprosy and tuberculosis are the main mycobacterial diseases in humans, other mycobacterial species may become opportunistic infections in AIDS patients.

Gram-negative organisms. The Gram-negative 'gonococcus' *Neisseria gonorrhoeae* is the causative organism in gonorrhoea. The organisms are difficult to find in sections of infected tissues although easier to find in smears. *Legionella pneumophila* is a small coccobacillus that causes the pneumonia associated with droplets of water from air-conditioning units. It has a high mortality and is also known as legionnaires' disease. This is again difficult to see in sections. It is easier to stain with silver techniques than with dyes.

The spirochetes are an unusual group of long, slender, spiral, rod-shaped organisms. They cause the diseases syphilis (*Treponema pallidum*) and Weil's disease (*Leptospira interrogans*). The slender rods are difficult to see in sections and need to be stained with a silver technique to see them easily. Suitable silver techniques include the Dieterle's, Warthin–Starry, and Steiner and Steiner methods.

Fungal infections

Fungi are extremely common organisms in the environment but only a few are pathogenic in humans (see *Box 11.5*). The commonest site of infection is on a surface such as skin or in the mouth. The general term for fungal diseases is **mycoses**. Superficial mycoses such as athlete's foot and ringworm are usually fairly mild diseases but systemic mycoses, where the disease is more widespread inside the body, can be dangerous.

Fungi can occur as a yeast or hyphal form. The hyphae are often distinctive, with features such as the presence or absence of cross-walls (septae) and the degree of branching being useful indicators. Yeast forms are less easy to differentiate on morphological criteria.

Fungi can be stained and demonstrated by the periodic acid–Schiff (PAS) technique, which detects their polysaccharide cell walls. The use of chromic acid as an oxidizing agent (Gridley's technique) or methanamine silver (Grocott's technique) are essentially similar to the PAS method but are still used instead of a simple PAS.

Box 11.5 // Fungal infections

Fungal infections were once either comparatively mild and usually superficial diseases (e.g. athlete's foot and thrush) or if they were life-threatening they were unusual and often limited to certain occupations. The increase in immunosuppression and immunodeficiency has changed the situation and several once minor or unusual diseases have become common causes of death. In a normal immunocompetent person, these infections would be countered effectively by the immune system and either eliminated or kept in check as a low-level infection. The loss of immune capability in AIDS demonstrates the importance of the immune system since these diseases have now become killers in AIDS patients. AIDS patients are said to be **compromised** by the immunosuppression; they are much more susceptible to infection and often succumb to minor diseases. These infections are referred to as **opportunistic infections** since they seize the opportunity of growing and spreading in the compromised host.

Important fungi seen in tissue sections

Aspergillus fumigatus is a very common organism in soil and can be a commensal in the upper respiratory tract. If it invades the lung, as happens in some AIDS patients (see ***Box 11.5***), it may cause pneumonia.

Candida albicans is found in small numbers as a common commensal on wet mucous membranes but can become pathogenic following antibiotic therapy that kills the normal suppressive bacterial flora. It then causes the condition known as **thrush**. In immunosuppressed patients, it can also become systemic.

Cryptococcus neoformans and *Histoplasma capsulatum* can both be acquired from bird droppings and cause systemic disease but are only significant in immunosuppressed patients.

Pneumocystis carinii again is an organism that has only become important in immunosuppressed patients and is a major cause of death in AIDS patients where it can cause pneumocystis pneumonia or PCP. Originally classified as a protozoan because of an amoeboid trophozoite stage, and also found in rats, this organism has since been reclassified. It is now described as a yeast-like fungus and only *Pneumocystis jirovecii* preferentially colonises human lung, with *P. carinii* colonizing the rat lung. PCP pneumonia and identification of the causative agent is one of four key signs of an HIV +ve status.

Parasitic infections

Parasites are large enough and distinctive enough for most diagnoses to be made on the shape and structure of the organism and the nature of the host's response. A haematoxylin and eosin stain may be supplemented with the Romanowsky method (such as Giemsa) or stronger haematoxylins (iron or phosphotungstic acid haematoxylin) or even using PAS to show glycogen, but there are no staining techniques specific for the organisms.

Entamoeba histolytica causes amoebic dysentery. The organism is up to 50 μm in diameter and may contain the remains of ingested blood cells.

Toxoplasma gondii is common in cat faeces and can cause an acute inflammatory condition in lymph nodes (lymphadenopathy) but the infection is often unnoticed (a subclinical infection). In immunosuppressed patients, it can be more dangerous causing brain infections and inflammation of the meninges.

Plasmodium species cause the various forms of malaria. Malaria is common in tropical countries and is regularly seen in immigrants and travellers from these countries. The parasites are usually diagnosed from blood films but they can also be seen in tissues (see *Fig. 11.5*). The associated malarial pigment is similar to formalin pigment and can be removed by alcoholic picric acid. Removal of the pigment allows the malarial parasites to be seen more easily.

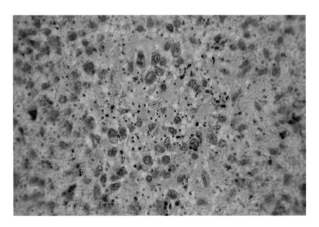

Figure 11.5
Malarial spleen. The spleen filters out damaged red blood cells and other debris from the blood. The malarial parasite becomes concentrated within the splenic pulp and leads to enlargement (splenomegaly). The parasites can be seen as the dark dots in the centre.

Leishmania donovani causes the systemic disease kala-azar, whilst *Leishmania tropica* causes the skin disease 'oriental sore'. The infection is spread by the bite of the sandfly and the organisms can be seen in infected areas in the cytoplasm of enlarged macrophages.

Trichomonas vaginalis is found in the urinogenital tract of both sexes. It is commonly seen in smears but less often in sections.

Schistosomiasis mansoni is found in the lower digestive tract, has a characteristic spike and a ZN positive wall. *Schistosomiasis haematobium* is found in the urinary tract and is ZN negative with no spike.

Trypanosome species is found in Africa where it is transmitted by bites from infected Tsetse flies and causes sleeping sickness. It is also endemic in South America where it is spread by reduviid bugs causing Chagas' disease. It is found in the blood and can be seen in blood films and in tissues such as brain (see *Fig. 11.6*).

Larger parasites include tapeworms and nematode worms living in the gut lumen (see *Fig. 11.7*).

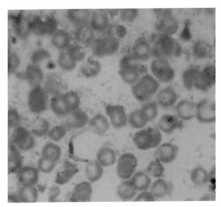

Figure 11.6
Trypanosome in blood smear.

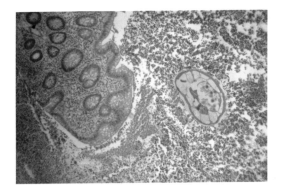

Figure 11.7
Appendix containing a nematode pin worm. This section of appendix shows the presence of a pin worm (or thread worm) in the lumen. The appendix shows appendicitis with pus, hence its surgical removal, but pin worms are very common although only very rarely are they the direct cause of appendicitis. This worm is a chance finding rather than a cause.

Prion diseases

Prion diseases represent a comparatively recently recognised pathology. The name was coined by Stanley Prusiner in the early 1980s from the words *protein* and *infection* to describe the causative agent for a group of transmissible diseases causing brain degeneration, with the brain tissue becoming spongy (transmissible spongiform encephalopathy, TSE). The sheep disease scrapie has been recognised for a long time but its mode of transmission and the nature of the transmissible agent were unknown. No bacteria, fungi or virus could reliably be demonstrated in all cases of the disease. The agent could be transmitted by injection of infected brain tissue, but even when the infected brain tissue was 'sterilised' by the usual biocidal agents (disinfectants, formaldehyde, boiling, autoclaving or UV irradiation) the tissue retained its infectivity. This suggested that no known type of infective agent (bacteria, fungi or virus) was involved because none of these could resist all of these agents.

It was suggested in the 1960s by Alper and Griffith that the particle consisted exclusively of a novel self-replicating protein. This was a radical suggestion because nucleic acids were believed to be essential in any biological replicative activity such as infection and so it was strongly opposed at first, with suggestions of 'virino' and 'atypical slow viruses' being involved, but the 'protein only' concept is now widely accepted. In the 1980s Prusiner isolated and purified the infectious protein with no trace of nucleic acid and so confirmed the hypothesis.

It is now believed that the mechanism involves a change in the folding pattern of particular proteins that results in the degeneration and transmissibility. The protein involved has been called PrP (from prion protein) but it is a natural cell component with the gene (the PRNP gene) being found on chromosome 20. The cellular form (PrP^C) is a membrane protein and is probably a receptor. The protein from scrapie (PrP^{Sc}) differs from PrP^C in having a higher content of beta-pleated sheets and less alpha-helix which represents a change in protein conformation rather than a mutation or change in amino acid sequence. The scrapie protein seems to be able to bind to the normal PrP^C protein and act as a catalytic template to reconfigure the PrP^C folding to form more PrP^{Sc} which can detach and then both the original and new PrP^{Sc} molecules can recruit and convert more normal protein in a chain reaction. The scrapie protein is resistant to the usual proteases and so accumulates and can form a type of amyloid deposit.

There are many forms of TSE in both animals and humans. Human forms of TSE caused by prions are Creutzfeldt–Jakob disease (CJD), fatal insomnia, Kuru, and Gerstmann–Sträussler–Schenker disease. Animal forms include scrapie and bovine spongiform encephalopathy (BSE).

CJD is the most important human form (see *Fig. 11.8*) and has several subgroups.

• It can be sporadic with no obvious predisposing cause for the disease.
• It can be familial, where the PrP^C has a mutation that predisposes to the pathological refolding.
• It can be iatrogenic where the prion is transmitted by treatment (from infected instruments, tissues such as corneal transplants and blood transfusions, pituitary hormones).
• Variant CJD (vCJD) is related to the bovine form of scrapie, BSE (also known as 'mad cow disease').

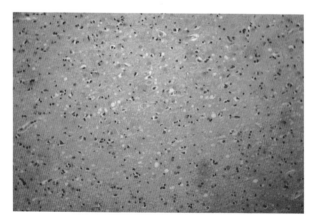

Figure 11.8
Spongiform encephalopathy. This is brain tissue from a case of the prion disease, CJD.

> **Box 11.6 // Prions and safety**
>
> In the histological laboratory specimens believed to be infected with prions need a great many extra precautions because the causative prion is not inactivated by routine formalin or glutaraldehyde fixation, as is the case with the usual bacterial, viral and fungal infections. Routine formalin fixed TSE material remains infective for long periods (possibly indefinitely). The prions can be inactivated by initial formalin fixation followed by treatment with formic acid (95%) and then given a further formalin fixation. However, preparation of infected material is probably best left to specialist centres rather than being handled in routine histology laboratories.
>
> There are no reports of laboratory acquired infections.

11.3 Amyloid

Amyloid is an abnormal protein that accumulates between cells in many organs and can be associated with different pathological conditions. Fresh amyloid has a waxy appearance and absorbs acid dyes although only very weakly. This name is inaccurate as there is no chemical resemblance to starch and it is mainly protein (see *Box 11.7*). The cut surface of affected organs when tested at autopsy will, however, give a positive reaction to the iodine starch test.

Composition and structure of amyloid

Component	*Percentage by weight*
Water and salts	75
Protein (including plasma protein)	25
Polysaccharides (mainly mucins)	1–2
Lipid	0–3

Amyloid protein is a β-pleated sheet

Despite always having a very similar appearance and sharing staining properties, amyloid protein is very variable in its amino acid composition. All of the different forms of the protein have a β-pleated sheet structure and this seems to be the common link between these different materials. Although amyloid has little structure when viewed with the light microscope (i.e. it is **amorphous**), it does form fibrils that are visible with the electron

> **Box 11.7 // Origin of the term amyloid**
>
> The name amyloid is misleading since it suggests a starch-like material. The eminent pathologist Rudolf Virchow found that amyloid reacted with iodine/sulphuric acid and produced a blue colour and since this is reminiscent of the blue colour given by starch when treated with iodine he introduced the name **amyloid** (*amylo* = starch).
>
> The later finding that it is in fact protein has led to people suggesting other names but in an attempt to be more descriptive these names have often been clumsy. One suggestion was **idiopathic fibrillar glycoprotein**. This name is certainly more accurate than amyloid but such names are unlikely to replace the succinct, if inaccurate, amyloid, which is much easier to say and remember.

microscope. These fibrils are 7.5–10 nm in diameter but are variable in length and are striated at 10 nm intervals. The β-pleated sheet structure is unusual in human proteins (although it is found in silk). This structure is very resistant to enzyme degradation and this probably accounts for its accumulation and deposition in tissues since the enzymes cannot remove the deposits quickly enough.

The different amyloid proteins are related to normal proteins

There are several types of amyloid that have been identified but not all of them are considered to be pathological. ASc-type amyloid is common in old people but does not seem to be linked directly to any disease. Only two of the pathological types are common (AA and AL) and the others are included for completeness. The two common pathological amyloid types occur mainly as a secondary effect to other diseases and amyloidosis may be the first clinical finding that indicates an underlying disease. The different types of amyloid have amino acid sequences that are similar or identical to normal body proteins and each amyloid type has a set of diseases or conditions with which it is often associated. *Table 11.1* lists several different types of amyloid.

Amyloid P is associated with all amyloids

In addition to the main fibrillar amyloid, there is a minor protein component that is found in all amyloid deposits except for those in the brain. This component is called amyloid P. It is derived from blood proteins (serum amyloid P, SAP) and is an acute-phase reactive protein. The normal blood form of the protein (SAP) appears to bind to

Table 11.1

Amyloid type	Normal protein the amyloid resembles	Diseases with which the amyloid is associated
AL Amyloid light chain immune-associated amyloid	Immunoglobulin light chain	Multiple myeloma, Waldenström's disease
AA Reactive amyloid	Serum amyloid A (acute phase reactant, apolipoprotein of high-density lipoprotein)	Rheumatoid arthritis, tuberculosis, Hodgkin's disease, familial Mediterranean fever
AE Endocrine-related amyloid	Peptide hormones (calcitonin, insulin)	Insulinomas, medullary carcinoma of the thyroid
β-Amyloid	Membrane glycoprotein	Alzheimer's disease
AS Senile amyloid		Old age (80+)
ASc ASc1	Transthyretin (pre-albumin) Atrial naturetic peptide	
AF Familial amyloid	Transthyretin (pre-albumin)	Familial amyloidosis
AD Dermal amyloid	Keratin	Lichen amyloidosis
Haemodialysis-associated amyloid	β2 microglobulin	Renal failure with haemodialysis
AScr	Scrapie protein	Transmissible spongiform encephalopathies (CJD, GSS, etc.)

amyloid deposits, hence its accumulation, but instead of forming fibrils it forms a 9 nm ring structure made up of five subunits of SAP with a central hole of about 4 nm. Amyloid P is immunogenic and has been used to make amyloid-specific antibodies which have some use for demonstrating the β-amyloid accumulations in the senile plaques seen with Alzheimer's disease.

Distribution of amyloid

Amyloid distribution varies with the disease

Amyloid can occur in many organs and the exact distribution varies with the associated disease. Three types of distribution have been identified:

1. **Systemic primary.** Amyloid is mainly found in the heart, gastrointestinal tract, tongue, skin and nerves. This is seen in cases of primary amyloidosis and neoplasms of B lymphocytes. The amyloid type is usually AL but this primary distribution pattern can occur with AA amyloid in rheumatoid arthritis.
2. **Systemic secondary.** In this case the amyloid is found in the liver, spleen, kidney, adrenals, gastrointestinal tract and skin. This distribution is associated with cases of chronic inflammatory diseases. The amyloid type that is found is usually the AA variety.
3. **Localized.** Localized tumour-like nodules occur in tongue, bladder, skin or lung and are associated with localized endocrine neoplasms. The β-amyloid in Alzheimer's disease is also localized but occurs as plaques rather than nodules.

Amyloid can be found in many organs but the frequency with which a particular organ is involved varies and there is no single site that is always affected. Taking a single biopsy sample is therefore never fully reliable as the organ may not be affected in that patient. Common sites include the liver (48%), kidney (87.5%), gingiva (19%) and rectum (75%). Rectal biopsies have become a popular method of investigating possible cases of amyloidosis because of their less-invasive nature (often an endoscopic biopsy) and frequent involvement.

Amyloid accumulates between cells and causes their death

The first accumulation of amyloid occurs between the cells and often close to the basement membranes of a blood vessel. As the deposit increases, it gradually infiltrates into tissues and can trap and destroy the cells. There is still doubt about the mechanism of cell death. It may be due to simple mechanical 'strangling' by cutting off the cells from their blood supply or it may involve a toxic effect of the amyloid. Amyloid has been found to be toxic to nerve cells in cultures but whether it is also toxic *in vivo* is not known.

Effects on organs

Kidney. Amyloid is first deposited in the glomerular mesangium and the basement membrane of blood vessels. As the accumulation continues, it eventually obliterates the capillary lumens and destroys the glomerular cells. The amyloid deposit can eventually completely replace the glomerulus. Renal arterioles can also be affected leading to ischaemia and tubular atrophy. The destruction of the renal tissue produces proteinuria and eventually the **nephrotic syndrome**. Renal involvement is usually fatal (see *Box 11.8*).

Spleen. The spleen becomes enlarged (splenomegaly) and the deposits may be of two types: (i) generalized deposits in both the red and white pulp ('lardaceous spleen'), and

---

Box 11.8 // Treatment of patients with amyloid

Amyloid deposition has no direct cure but effective treatment of any underlying disease will often slow down or even stop the deposition. Identification of the type of amyloid can therefore help by indicating which underlying cause is implicated and so suggest possible therapeutic measures. Amyloid diagnosis before death has become much more common than in the past due to better biopsy techniques and a greater awareness of the deposition of amyloid. Most patients are now detected in the early stages rather than late in the disease. Whilst this is often beneficial to the patient, it has made life more difficult for the laboratories. The large deposits in organs that are easy to find and easy to stain are being replaced with small biopsies containing less amyloid, which is more difficult to detect.

(ii) deposits only in the white pulp ('sago spleen'). If only the spleen is infiltrated and other organs are not affected, it is not usually fatal.

Heart. Usually deposits in the heart are smaller than in other organs. The deposits occur in the subendocardium, with more generalized deposits in the myocardium. The deposits may cause pressure atrophy of the heart cells and cardiac amyloid often leads to cardiac arrhythmia particularly if localized within the SA or AV nodes or conducting pathways between.

Liver. The amyloid deposits cause massive enlargement (hepatomegaly) and the liver becomes pale and waxy. Amyloid appears first in the space of Disse and progressively squeezes the adjacent hepatocytes to death. If only the liver is involved, then it is rarely the cause of death.

Detection of amyloid

Targeted sampling is important when attempting to demonstrate the systemic accumulation of amyloid. It is common to take a small rectal biopsy in the first instance, rather than attempting to sample a major organ. It is often easier to demonstrate amyloid deposits in the walls of small blood vessels in the sub-mucosa. Skin and gingival biopsies are easily accessible and so are also favoured for the same reason.

Staining

Since amyloid deposits vary in their composition, they will not all stain in the same way. A technique that works well with one sample of amyloid may stain a different

---

Box 11.9 // Origin of deposits

The deposition of amyloid in the tissues probably reflects a generalized problem with the destruction and removal of large amounts of unusual proteins. The wide range of proteins involved and the relationship with excess protein secretion suggests that during the degradation of the extracellular protein, the protein alters its conformation and forms the β-pleated sheet form. The β-pleated sheet form is then more difficult for enzymes to destroy. If other proteins can similarly change their form during degradation, then it is likely that other forms of amyloid will continue to be discovered.

---

case of amyloidosis only very weakly or not at all. A negative staining reaction should be considered inconclusive rather than definitely negative. If an amyloid deposit fails to stain with one technique, then it may be worthwhile trying a different one, which may give a much better result. It is also believed that the reactivity of amyloid decreases with the length of time between cutting and staining a section which is another source of variability.

There are a variety of methods that have been used to demonstrate amyloid and some of these are outlined below.

Iodine/sulphuric acid

This method is mainly of historical interest although some pathologists still use it as a macroscopic method on whole organs whilst performing a post mortem. The mechanism is still unclear, with the original idea of a carbohydrate in the amyloid acting in a similar way to starch being largely replaced by the concept that it is the β-pleated sheet arrangement that traps the iodine.

Methyl violet metachromasia

Methyl violet has been used for a long time but how it produces metachromasia is not apparent. Amyloid deposits often contain mucinous materials that might explain the metachromasia but methyl violet does not produce metachromasia with any carbohydrates or mucins. Amyloid is not metachromatic with other dyes such as toluidine blue that give strong metachromasia with mucins. The results with methyl violet are inconsistent and often difficult to observe. This has led to the suggestion that the methyl violet contains an impurity that is simply more selective for amyloid. This would be analogous to the trichrome methods where differences in staining are relative rather than absolutely selective. It is no longer as popular a method as it once was.

Congo red

This is the preferred method in most laboratories. Most amyloids react and give some staining. The red colour (see *Fig. 11.9*) may be weak and some other materials may occasionally stain slightly, but it is still probably the best of the current methods. It is even more specific if the section is examined using crossed polarizers in a polarizing microscope (see *Chapter 17*). The amyloid fibrils will show up as a dichroic green/yellow colour (see *Fig. 11.10*). The Congo red attaches using hydrogen bonding and inserts between the fibrils in the β-pleated structure and this results in a regular lattice or crystalline-like arrangement that gives rise to the birefringence. Chitin and cellulose show a similar birefringence but these look quite different to amyloid in sections so there should be no confusion. The section thickness and orientation can be crucial to this colour – too thin and the colour is red, too thick and it becomes yellowish. The fibres need to be oriented correctly relative to the plane of polarization of the light to see the full effect so only a proportion of the amyloid will be seen at any one time (see *Fig. 11.10*). The remaining amyloid deposits can be seen by rotating the section. A Congo red technique from 1962 uses an alcoholic solution, high salt concentration and high pH to minimise ionic staining and so enhance hydrogen bond staining; this does not require differentiation and is better than simple congo red staining (e.g. Highman's) and is recommended by UK NEQAS. Congo red is also fluorescent so the extent of the amyloid can be seen using this property. (Sirius red is an alternative to Congo red for staining amyloid but is not fluorescent.)

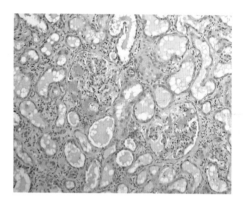

Figure 11.9
Kidney stained with Congo red for amyloid. The amyloid can be seen staining red in the two central glomeruli.

Figure 11.10
Amyloid stained with Congo red viewed with crossed polarizers. The bright dichroic colour indicates the Congo red-stained amyloid.

Pre-treatment of the section with potassium permanganate destroys the congophilia of AA amyloid but AL amyloid is resistant so this property can be used to help to distinguish the two types.

Thioflavine

This is possibly more sensitive than Congo red because of its intense fluorescence but it can be non-specific. It still remains popular as an alternative or complementary method to Congo red and is excellent for detecting very small deposits that are easily missed with Congo red staining.

Immunohistochemistry

Because amyloid is proteinaceous it is possible to raise antibodies against it for use in immunological techniques. A sensitive IHC detecting most types of amyloid utilizes antibodies to Amyloid P protein and IHC for the individual proteins found in amyloid can indicate the type of amyloid.

Proteomics

Proteomics is very sensitive and uses mass spectrometry to analyse protein fragments from amyloid. Congo red-stained amyloid is first isolated from sections by laser micro-dissection and then digested to generate a mixture of protein fragments that can be separated by HPLC and then analysed by mass spectrometry. This can identify the type of proteins present in the amyloid and is also capable of recognising if a new amyloidogenic protein is involved. This is a highly specialised and expensive technique and is likely to be restricted to a few specialist research centres.

Other methods

Other methods include immunotechniques, staining for tryptophan and X-ray crystallography but these tend to be limited to research into amyloid rather than reliable and easy identification techniques.

Infectivity of amyloidosis

The role of prions in the transmissible spongiform encephalopathies that produce amyloid deposits in the brain is widely accepted, but the infectivity or transmissibility (see ***Box 11.10***) of other forms of amyloid is not so well documented or investigated. There is no significant evidence of amyloidosis being transmitted in humans, but in animals some reports of transmissibility have been published. These suggest that injection of amyloid fibrils can accelerate the development of amyloidosis which in turn suggests a catalytic action increasing abnormal protein folding analogous to that of prions in TSEs. Mice are the main model for studying amyloidosis, but whether the results found in mice are relevant to humans has not yet been established.

Box 11.10 // 'Catching' a disease

The possibility of a disease passing from one person to another has generated a lot of descriptive words which are often confusing because their precise meaning is unclear. This lack of clarity applies not only to the general public but even to the medical literature. The words include infective and infectious. In the medical literature these generally mean that an outside agent invades the body and replicates causing the disease. This would include bacterial, fungal, viral, parasitic and prion diseases. The bacteria, virus, etc. would be called infective whilst the disease in an individual would be an infection, and the general term infectious disease would be used for all occurrences of the disease. Infections can spread from one infected individual to another by direct contact, inhalation of infected droplets, ingestion of contaminated foodstuff and a variety of indirect routes.

The term contagious and contagion can be synonymous with infectious, but they have been used to specify that the disease is spread by direct physical contact. Communicable disease is used to cover diseases that can be passed directly from one person to another by any mechanism.

Transmissible is a more widespread description of diseases that can be acquired directly from another person but also diseases that can only be acquired by injection into the body and cannot be simply passed to an uninfected person without the intervention of an infecting (or injecting) vector. This would include diseases such as malaria and Lyme disease in which the route of infection is from an insect bite (i.e. injection) and direct human to human transfer does not occur.

Suggested further reading

Bancroft, J. and Cook, H. (1994) *Manual of Histological Techniques and their Diagnostic Application.* Churchill Livingstone.

Bancroft, J., Layton, C. and Suvarna, S. (2012) Chapter 14: Amyloid, and Chapter 15: Microorganisms. In: *Bancroft's Theory and Practice of Histological Techniques,* 7th edn. Elsevier Health Sciences.

Culling, C., Allison, R. and Barr, W. (1985) *Cellular Pathology Technique,* 4th edn.: Butterworths.

Kiernan, J.A. (2008) *Histological and Histochemical Methods: theory and practice,* 4th edn. Scion Publishing.

Sheehan, D.C. and Hrapchak, B. (1980) *Theory and Practice of Histotechnology.* Mosby.

Swisher, B.L. (2002) Microorganisms. In: *Theory and Practice of Histological Techniques* (eds Bancroft J.D. and Gamble M.), 5th edn. Churchill Livingstone.

Vowles, G.H. and Francis, R.J. (2002) Amyloid. In: *Theory and Practice of Histological Techniques* (eds Bancroft, J.D. and Gamble, M.), 5th edn. Churchill Livingstone.

Self-assessment questions

1. Explain why it is better to identify bacterial infections from microbiological specimens rather than using paraffin wax sections.
2. Outline the types of method that can be used to help identify micro-organisms in tissue sections.
3. What are the major types of infective organism that can occur in sections? Name one example in each group.
4. What is amyloid? Why is amyloid not destroyed in the body?
5. List the different types of amyloid that are usually recognized and give an example of an associated disease or condition for each.

12 Immunological techniques

Learning objectives

After studying this chapter you should confidently be able to:

- **Describe how antibodies can be used as histological reagents**
  Antibodies are specific and can be used to identify individual proteins in tissues. The antibodies can be labelled in a variety of ways including the use of fluorescent dyes and biotin or can be used in bridging techniques, which do not need the antibodies to be modified.

- **Outline the principles of the direct and indirect immunofluorescent techniques**
  In the direct technique, only one antibody is needed and this antibody is labelled with fluorescent dye. The antibody can be applied directly to the section and will bind to its specific antigen, which can then be localized by its fluorescence. In the indirect technique, the primary antibody is not labelled fluorescently. A second antibody capable of binding to the primary antibody is needed and this needs to be labelled with fluorescent dye.

- **Outline the use of enzymes as labels and the advantages of immunohistochemistry**
  Enzymes can be used instead of fluorescent dyes as labels. They can be detected using enzyme histochemical techniques. They can give greater amplification of the signal than simple fluorescence, they can be viewed without the need for a fluorescent microscope and it is possible to use unlabelled bridging techniques. All of these make immunohistochemistry more popular than immunofluorescence for histological sections.

- **Describe the reasons for antigen retrieval techniques and name the common methods of antigen retrieval**
  Antigen retrieval is needed to allow antibodies to bind to antigens in fixed tissues as the fixation effectively masks the antigen. Antigen retrieval can use enzymes, microwave heating or pressurized heating in an autoclave or pressure cooker.

- **Outline the use of lectins in histochemistry**
  Lectins can be used in a similar way to antibodies but bind to specific sugars rather than antigens. Lectins are non-immune proteins and do not need to be elicited by immunization.

- **Give examples of the use of immunohistochemistry (IHC)**
  IHC is useful in research to identify and localize specific proteins of interest. In diagnostic pathology IHC can be used to identify autoimmune diseases and to identify tumours and their subtypes.

12.1   Use of antibodies as histological reagents

Animals are constantly threatened by pathological bacteria, fungi and viruses that cause infective diseases. As a defence against these organisms, evolution has produced the vertebrate immune system. The immune system of animals has two major parts, one of which is the cellular immune system and is concerned with the killing of invading cells by direct cellular action. The cell types involved are the T lymphocytes and macrophages. The second type of immunity is humoral immunity, which depends on antibodies produced by B lymphocytes and plasma cells. Antibodies are glycoproteins that react with the invading organisms and bind strongly to them. Antibodies are members of the immunoglobulin family and can be extracted from the blood of an immune reactive animal or human being. The binding of the antibody to the foreign cells either inactivates the invading cells or coats the cells and makes them more likely to be attacked by the cellular immune system.

As antibodies are soluble proteins, they can be used as biochemical and histochemical reagents. Immunological methods have become very important in histology because of their high degree of specificity, sensitivity and reliability.

Antibody production

Antibodies are specific proteins produced in response to antigens. The trigger for the immune system to produce antibodies is the presence of a foreign material called an **antigen.** Antibodies are produced by B lymphocytes and each B lymphocyte is specific for one antigen and has on its surface a receptor that is specific for the antibody it produces, indicating a particular **affinity**. When this receptor is stimulated by the specific antigen, it causes the cell to divide rapidly and produce a clone of identical cells, all producing exactly the same antibody. This proliferation produces a large number of cells and hence increases the production of antibody.

An antigen is any material that can be recognized by the receptors on the immune cells. Antigens can be solid particles such as cells or bacteria, but they can also be soluble materials such as foreign proteins or certain other complex molecules. It requires a fairly large molecule to trigger the response and small molecules are not usually antigenic. An antibody that has been triggered in this way is very selective and will react only with the antigen that stimulated its production. Other antigens will not be bound by the antibody, i.e. it is highly **specific**. Antibodies are also extremely active in binding to their antigen and are not easily dislodged by washing or by minor changes in the fluid environment i.e. they have a high **avidity**. It is this specificity and the high binding strength that make antibodies so popular as reagents.

Antibodies used in histology

There are a number of different types of antibodies (see *Box 12.1*) but the main ones used in immunotechniques in histology are immunoglobulin G (IgG) and immunoglobulin M (IgM). They differ in the number of binding sites they have for antigens. IgG has two identical binding sites and consists of a Y-shaped molecule with a binding site at the end of each arm of the Y. IgM is a larger molecule and resembles five IgG-like molecules bound together to form a pentamer. IgM is usually produced early in the immune reaction, with IgG being produced later. There are some differences in these two proteins and this can sometimes be relevant, but for most purposes in histology the differences between the major types are minor and either type will work as an immunohistochemical reagent.

> **Box 12.1 // Immunoglobulins**
>
> Antibodies are found in the blood and form part of the **globulin** fraction: they form the γ-globulins, which can be identified by electrophoresis. Since they are a part of the immune system, they are also termed **immunoglobulins**, which is abbreviated to Ig. There are several subtypes of immunoglobulins. IgG and IgM are important in the normal immune reaction in blood. IgA is found in secretions such as milk and saliva and IgD is found as membrane receptors on B lymphocytes. IgE is important as it is the cause of allergies but its normal function is not understood. For histochemistry, only IgG and IgM are used as reagents but detection of other immunoglobulins is sometimes needed.
>
> The antibodies form the **humoral** response of inflammation. The term humour, in this context, comes from the ancient theory of health in which there were four body fluids or **humours** (blood, phlegm, green bile and black bile) that needed to be kept in balance for health. Any material found in body fluids was described as humoral. The second form of immune response is the **cellular immune system**; this is less important in histochemistry although vital for health.

Animals do not usually produce antibodies against their own proteins

Antibodies are not usually produced against the body's own proteins as this would cause the body to attack itself. If the method of controlling the production of antibodies fails to prevent this happening, then this results in autoimmune disease. The diagnosis of autoimmune disease is one of the areas in which immunotechnique is important. The fact that an animal does not react to its own proteins does not mean these proteins are not antigenic. If the protein is injected into a different species of animal, then it will act as an antigen. Some proteins are very similar in all species of animal and will not be recognized as being different and will not produce an immune response and antibodies will not build up in the blood. It is for this reason that insulin from other species such as pigs or cattle can be used to treat human diabetes. The insulin molecule is classed as a small protein and is not sufficiently different in the different species to trigger antibody production (at least in most people), even with many years of therapeutic use. This low antigenicity of some proteins sometimes causes immunization problems when scientists try to produce antibodies for use in immunotechniques.

Raising antibodies in animals can be difficult

Often it requires ingenuity and persistence to produce some of the antibodies that are commercially available. Once an animal is immunized against an antigen, it is then kept immune or hyperimmune by regularly challenging the animal with fresh antigen injections. This keeps the concentration of antibody high in the blood and helps to make an efficient reagent. The complexities of immunizing the animal (see ***Box 12.2***) and keeping it immune mean that it is unusual for most scientists to produce their own antibodies and instead they rely on commercial firms to produce and test them.

The immune system often produces many antibodies against the same protein

The actual binding site between an antibody and an antigen is quite small compared with the size of the complete protein. If a large antigen is injected into an animal, then it is common for more than one antibody to be produced. Antibodies, each derived from a different B cell, will be specific for their respective binding site but a range of antibodies will be produced that react with different parts of the antigen molecule. Some of the antibodies will react only with the ends of the antigen molecule, whilst others will only

---

Box 12.2 // Problems with immunization

Raising antibodies often requires adding **adjuvants** to improve the reaction. Adjuvants include a variety of nasty materials such as killed bacteria, oil droplets or irritant chemicals. The method of preparing these adjuvant cocktails is often little more than guesswork or trial and error, but their common goal is to boost the inflammatory response.

The raising of antibodies is also an 'experiment' using a live animal and in the UK this requires a special licence from the Home Office. The actual interference with the animal is usually comparatively minor (one or two simple injections and taking samples of blood) but it is still strictly controlled. These two difficulties are the reason why most people prefer to buy antibodies and so save themselves considerable trouble.

---

react with the centre. Each antibody will come from a different clone of lymphocytes so the antibody is referred to as a **polyclonal antibody** and will all recognize the same antigenic molecule but from varying perspectives. Each separate antigenic section of the overall molecule is known as an epitope.

Polyclonal antibodies are not specific

Antibodies may react with proteins other than the specific antigen used to trigger their production. Usually these will be closely related proteins that share a sequence of amino acids but this cross-reactivity with other proteins can be a problem. The antibodies prepared in an animal may also contain small amounts of natural antibodies other than the specific hyperimmune antibody. This impurity is usually only a theoretical problem as the antibody solutions are used at high dilutions. The hyperimmune antibody will still be active because of its high concentration, but most other antibodies will be too weakly represented to cause difficulties. There is, however, always the possibility of a contaminant antibody giving a false positive reaction.

Monoclonal antibodies are more specific

A second type of antibody is now available and these are produced by growing cells in tissue culture. Lymphocytes generally do not grow well or for long periods in culture, so the antibody-forming cells are fused with a lymphocytic tumour cell. The hybrid cell, usually described as a hybridoma, can both grow well and produce antibody. Since the cells are all derived from a single original cell the antibody that they produce is called a **monoclonal antibody**. Monoclonal antibodies have several advantages:

- They are much more specific and react with only one small part of an antigen.
- They have less likelihood of cross-reacting with other proteins.
- They are not contaminated with any other antibodies and can be purified more easily.
- They will always be identical, whereas polyclonal antibodies will vary depending on the source animal.
- They can be produced relatively simply by culturing the hybridoma cells as a factory scale production line.

Monoclonal antibodies are so precise that they are usually able to react with just one part of the antigen and will be designated as reacting with a particular part of the molecule (e.g. the Fc portion of the immunoglobulin heavy chain). Many monoclonal antibodies are designated by simple codes that make them almost unintelligible unless you are familiar with them.

Monoclonal antibodies can sometimes be too precise

The binding of monoclonal antibodies is more precise and consistent since they can be produced under highly controlled conditions. However, they do suffer from the problem that they are absolutely monospecific and will only react with a single binding site. A polyclonal antibody, on the other hand, will have several different component antibodies, each of which reacts with a different binding site on the target molecule. Although polyclonal antibodies are less specific, they are sometimes more effective for detecting particular proteins since their specific sites of binding differ. For example, in an autoimmune reaction, it may not be known which type of immunoglobulin is acting. It might be IgG, IgM, or even one of the less common antibody types. With monoclonal antibodies, it might be necessary to use several monoclonal antibodies to detect whether there was any antibody. A negative reaction with any single reagent would simply indicate that the specific immunoglobulin was not present rather than that immunoglobulins were absent. A less-specific polyclonal antiserum prepared against immunoglobulins in general rather than a specific type could detect the binding with a single reagent. For this and other similar reasons, polyclonal antibodies are still used for some tests. When more detailed information is needed, however, it is usually better to use monoclonal antibodies to provide precise information about the nature of the sample being stained.

Fluorescent antibody techniques

Although antibodies are very precise reagents, they are not visible even with high-power microscopes and need to be labelled in some way to allow their binding to be localized. Historically, the most widely used method of labelling has been with fluorescein, a fluorescent dye, and the technique is therefore called **immunofluorescence** (IMF). Other labelling techniques have become popular since their introduction in the 1960s and are now probably more common in histology. These methods usually involve the use of enzymes and are collectively called **immunohistochemistry** (IHC) or **immunocytochemistry** (ICC; discussed later in the chapter) as a histochemical or cytochemical test is used to visualize the enzyme.

Use and storage of antibodies

The reactivity of all antibodies can be destroyed surprisingly easily and they need to be handled with care. The most important thing to be aware of is how to dilute with an appropriate buffer. Many commercial companies supplying antibodies include detail of antibody strength as mg or pg per ml, both for the stock reagent and also its recommended working dilution. The recommended working dilution can vary enormously depending on the application (e.g. IHC, IMF, Western blot) or by antibody type; monoclonal antibodies are often used at higher concentrations than polyclonal by a factor of ×5–×50. Equally important is the storage temperature, and the decision to freeze or not is a major consideration. Generally it is best to aliquot small amounts (10–20 μl) for long-term frozen storage, thawing these only once when making them up to working strength.

Most suppliers will have added the preservative sodium azide to their products and this will to a certain extent inhibit bacterial growth. Being protein by nature, antibody solutions are a target for microorganisms seeking nutrients and a home in which to multiply; they should not therefore be left at room temperature for any longer than is necessary.

Labelling with fluorescein

Fluorescein is a highly fluorescent dye related to eosin and is used as a fluorescent label. It can be detected even in minute amounts using fluorescent techniques. The dye must be attached firmly to the antibody. Simple ionic staining (see **Chapter 7**) of the protein is not satisfactory since the fluorescein could dislodge readily during simple washing and mounting. By using the isothiocyanate derivative of fluorescein (FITC), it is possible to attach the dye permanently. FITC forms covalent links with amino groups in the antibody protein forming a strong bond that does not break easily and resists removal during staining and processing (see **Box 12.3**). Attachment of the dye by this method is termed conjugation to distinguish it from simple staining.

The steps involved in producing a fluorescently labelled antibody are:

1. Prepare a pure antibody solution against a known antigen (e.g. antibody against actin is prepared by injecting purified actin into an animal).
2. Treat the purified antibody with FITC to conjugate the dye and label the antibody.
3. Wash the labelled antibody solution to remove unbound FITC. This can be done by dialysis or gel filtration.

Box 12.3 // Antibody inactivation

Chemical labelling could potentially affect the ability of the antibody to bind to its antigen. In practice, this is not a major problem. The term **epitope** or **antigenic determinant** is used to describe the part of an antigen that is directly bound by the antibody. The binding site is quite small and the binding region of the antibody is believed to be only 20–30 amino acids in size. Provided that these critical amino acids are not altered by the labelling process, then the binding ability of the antibody will remain unchanged. In normal labelling conditions, only a few labelling molecules, perhaps four to six, are added to the antibody. The binding sites are only a very small part of the antibody (perhaps 5%), so the chances of the label attaching to the binding sites are small and the majority of the binding sites are unaffected.

Labelled antibody will react with its specific antigen in the tissues

Labelled antibody can then be used at a suitable dilution to stain sections. Careful washing is needed but the technique is quite simple. The section can then be viewed with a fluorescent microscope, when any bound antibody will fluoresce showing the sites of antibody binding and this in turn identifies the sites of the antigen. Using the example of actin, the binding sites for anti-actin antibody would indicate the location of actin. This simple technique involving just one antibody conjugated to FITC is called the **direct immunofluorescent technique**. As with all antibody labelling techniques, it is important to minimize non-specific binding by thoroughly washing the section with buffer after the completion of the antibody incubation stage.

Detection of anti-nuclear antibodies in rheumatoid arthritis

In some diseases (e.g. rheumatoid disease), the patients produce an abnormal antibody that reacts with nuclear material (see **Box 12.4**). This anti-nuclear antibody is important for the diagnosis and prognosis of the disease. The antibody is present in the patient's serum and can be identified using tissue sections or cell smears and staining by immunofluorescent methods. The antibody will react with the nuclear proteins of several species and is not limited to human nuclei. Frozen sections of rat tissue or cells cultured

---

Box 12.4 // Autoimmunity

Anti-nuclear antibodies are found in a variety of disorders including rheumatoid disease, lupus erythematosus, mixed connective tissue disease and some liver disease. The antibodies are classed as autoimmune antibodies since they will react with antigens present in the patient. Autoimmune diseases can be diagnosed by detecting the free antibody in the patient's blood or by detecting bound antibody on the patient's cells.

Other autoimmune antibodies are also found and many seem to react with endocrine tissues, although the significance of this is not known.

In some cases, autoimmune antibodies are directed against a single organ (e.g. Hashimoto's thyroiditis) or against a whole number of different organs (e.g. systemic lupus erythematosus). The damage to the tissues may be caused directly by antibody or indirectly by cellular immunity but can also occur by the production of immune complexes in the tissues. An immune complex is a combination of antibody and antigen, which can become deposited in delicate tissues such as the kidney. The precipitated complex then causes the damage.

---

in vitro can be used as test slides to detect whether there is any antibody present in the patient's serum.

Human antibodies are antigenic in other animals

Detection of immunoglobulins requires an antibody directed against immunoglobulin. This is an antibody against an antibody. This production of antibodies against antibodies is a little difficult to understand and follow the first time you come across it but it is a very useful technique. Human antibodies are proteins and can therefore act as antigens in a species other than humans. The animal produces, as usual, an antibody that will bind with its specific antigen, which in this case is human antibody. Antibodies are immunoglobulins, so the reagent is called anti-human immunoglobulin; if the animal it was prepared in was a rabbit, it would be called rabbit anti-human immunoglobulin.

Human antibodies can be detected using anti-immunoglobulins

To detect whether there are any anti-nuclear antibodies in a patient's serum, the following steps are needed:

1. The first step is to apply a volume of the patient's serum to a frozen section. If there is any anti-nuclear antibody present, it will bind to the nuclei.
2. The patient's serum is then washed away.
3. The section is treated with FITC-conjugated anti-human immunoglobulin. The anti-human antibody will only bind to human antibody. In a rat section, the only way that human antibody could be present would be if there were anti-nuclear antibodies in the test serum that had bound to the rat nuclei.
4. Finally, it is necessary to wash away any unbound FITC-labelled antibody and examine the section for fluorescence. If the nuclei fluoresce, then the patient's serum contained anti-nuclear antibodies (see ***Fig. 12.1***).

Measurement of antibody concentration

The immunofluorescent technique can be used to estimate the concentration of antibody by preparing a series of dilutions of the patient's serum and testing each one for binding. The dilution that just gives a detectable positive reaction is called the **titre** of the serum. A

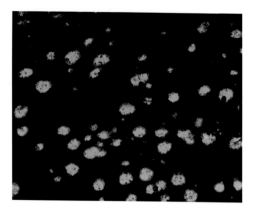

Figure 12.1
Anti-nuclear antibody shown by fluorescein-labelled anti-human immunoglobulin.

serum that has a titre of 1/512 means that if the serum is diluted to 1 part of serum in 511 parts of buffer (×9 doubling dilutions) it will still give a detectable reaction, but if diluted once more to 1/1024 (twice as dilute) then no reaction can be detected.

Direct and indirect immunotechniques
The anti-nuclear antibody technique uses two antibodies; the first or primary antibody is in the patient's serum and the second is prepared as an anti-antibody. Only the second antibody is conjugated. This is an example of an **indirect technique** (see *Fig. 12.2*) and can be used to detect any antibody, not just anti-nuclear antibodies in disease states. When a primary antibody is prepared by immunizing an animal, as for direct immunofluorescence, there is no need to conjugate this primary antibody. Instead a second antibody is prepared in a different species against the immunoglobulins of the first species and this secondary antibody is conjugated with FITC or another label (see *Box 12.5*). The indirect technique tends to give less background staining as the primary antibody can be used at a higher dilution than for the direct method.

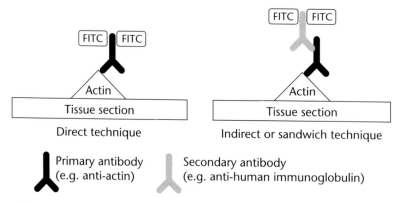

Figure 12.2
Diagram illustrating direct and indirect antibody techniques

> **Box 12.5 // Other fluorescent labels**
>
> Fluorescein labelling is the most common method of fluorescently labelling antibodies but other fluorescent dyes are also used. Rhodamine and Texas red both have a red fluorescence and can be used as a second label when two different antigens are to be demonstrated in the same preparation. This double staining is useful for seeing how two materials are distributed and the relationship between them. The plant pigment phycoerythrin has also been used as it has a high fluorescent capability.
>
> As well as labelling using the isothiocyanate derivatives, it is also possible to use succinimidyl esters and sulphonyl halides as reactive forms of the dyes and these will also react with amine groups. These are more stable but are more difficult to produce and are used less often.

Indirect techniques can be used for any antigen

The idea of an indirect technique is perhaps easier to understand if the individual species are named. Anti-actin antibody can be prepared from a rabbit by immunizing it with pure actin. This antibody is not conjugated with fluorescein. Anti-rabbit antibody can be prepared by immunizing a goat with purified rabbit antibody. This is called **goat anti-rabbit immunoglobulin**. The antibody used to immunize the goat does not need to be the specific anti-actin antibody – any rabbit antibody will produce the same effect. The goat anti-rabbit antibody will react with any antibody from any rabbit. This antibody is conjugated and will be a general reagent for the detection of any rabbit antibody. Thus, if we need to identify keratin, all that is needed is to prepare a new first antibody. This would mean raising anti-keratin antibody in a rabbit and this could then be detected using the same goat anti-rabbit immunoglobulin that was used with the anti-actin antibody.

The indirect technique has many advantages

The indirect technique is a better method than the direct technique for the following reasons:

- As just explained, it is not necessary to conjugate every antibody. This makes it cheaper since conjugation adds to the cost of an antibody.
- It is more sensitive since there can be more than one conjugated antibody bound to a single antigen site. In the direct technique, only one conjugated antibody can be bound to an antigen site. In the indirect technique, although there can still only be one primary antibody bound to the antigen site, there can be several secondary antibodies bound to the primary antibody. Each of the secondary antibodies will have fluorescein attached to it. This increases the intensity of the fluorescence, so smaller amounts of antigen can be detected. This increase in intensity is termed **amplification**.
- The increased sensitivity means that the antibodies can be used in a more dilute solution, which means that the same amount of serum will be able to treat more sections. This reduces the cost of preparing a large number of slides.

Disadvantages of fluorescent labelling

Fluorescent labelling is widely used, not only in histology but also in microbiology and haematology. However, it is less suited to staining sections than other types of specimen. The main disadvantages are:

1. Observing fluorescence requires a special fluorescence microscope with a high-power lamp and this is usually significantly more expensive than a simple microscope using visible light. This cost is both in the initial cost and in the cost of replacement lamps.

2. Fluorescence is always quite dim and in order to observe the faint glow a darkened room is required and the observer needs to allow their eyes to become adjusted to the dark before looking at the slides.

3. It is difficult to counterstain the section with a dye since any fluorescence of the background might mask a dim positive reaction. In the absence of a counterstain, it is therefore more difficult to locate the position of the fluorescent material accurately in relation to other structures. Many workers use a simple intercalating fluorescent dye (DAPI) to highlight nuclei as a characteristic bright blue colour. The DAPI is often incorporated into the mountant that is applied before coverslipping, and nuclear staining occurs simultaneously.

4. The fluorescence is not permanent. The dye will start to fade or bleach in a few days and will fade more quickly if it is examined under the microscope. Fading under the microscope illumination can be sufficiently fast that it will become lost in minutes. The process can be slowed down by adding sacrificial quenching agents to the mounting medium, e.g. *p*-phenylenediamine.

5. In the past it was difficult to photograph the fluorescence. The image is dim, so a long exposure was needed but the dye fades with long exposures, which made photography difficult. However, the recent introduction of sensitive digital cameras into microscopy now facilitates easy image capture, manipulation and storage.

Immunohistochemistry

The problems with fluorescein labels led to a search for other antibody labels including the use of enzymes and the use of these histochemical labels is termed **immunohistochemistry**. Immunohistochemistry uses enzymes to label the antibodies and histochemical tests to identify the sites of binding.

As explained in **Chapter 9**, enzymes can be demonstrated in tissues and can produce highly coloured insoluble reaction products. An enzyme can also produce large amounts of coloured product if the incubation time with the substrate is long enough. This allows visualization of very small amounts of enzyme. This makes enzymes an excellent labelling material provided some way can be found to link the enzyme and the antibody together. Although there are many enzymes that could be used, the ones that have proved most popular are horseradish peroxidase (HRP), alkaline phosphatase (AP) and glucose oxidase (GO). The use of just these three enzymes provides the ability to triple label a single section, highlighting three different structures or products. Classically the colours would be brown, red and blue, respectively, to provide sufficient contrast between the target features.

Initially chemicals were used to link the enzyme to the antibody

Having chosen an enzyme, the next step is to link the enzyme and the antibody together. One of the first techniques used glutaraldehyde, which is a bifunctional aldehyde. One aldehyde group can react with the protein of the antibody, whilst the other reacts with the protein of the enzyme:

Thus, glutaraldehyde cross-links the antibody and enzyme in the same way that it cross-links tissue proteins when it is used as a fixative. This technique worked quite well but there was a lot of waste of antibody since simply mixing the glutaraldehyde with a mixture of antibody and enzyme resulted in a variety of combinations (see *Fig. 12.3*).

- The antibody can be cross-linked to another antibody. This combination could bind to the antigen but would not be able to be visualized since there is no enzyme present.
- The enzyme could be cross-linked to another enzyme. This combination could not bind to the antigen.
- Larger aggregations of several cross-linked proteins could form. These would either precipitate out of solution or would simply be too large to approach the antigenic site and would not bind because of steric hindrance.
- Finally, there would be a mixture of an antibody linked to an enzyme in small enough aggregations to be useful. This group is the only type of reactant that could be used. The previous three types of combined proteins were not wanted and had to be removed from the solution.

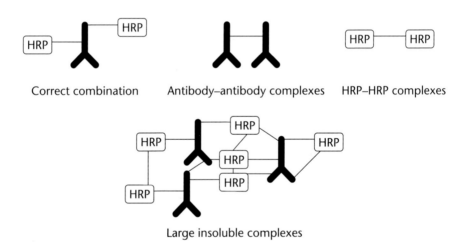

Figure 12.3
Unwanted reactions between enzymes, glutaraldehyde and HRP

The various forms could be separated effectively using gel filtration techniques so that the antibody could be prepared in a pure form. This was very wasteful of reagent since significant amounts of the antibody were being discarded and lost.

The purified antibody–enzyme combination could be used in either the direct or the indirect method, in the same way as for fluorescein labelling. In this case, indirect techniques are even more favoured since the labelling technique is so wasteful.

The major steps in the indirect enzyme-labelled antibody method using HRP are:

1. Apply the primary antiserum (see *Box 12.6*) to the sections. If there is any antigen present, then the antibody will bind to it.
2. The primary antiserum is then washed away with buffer.
3. The section is then treated with enzyme-linked anti-immunoglobulin. This will bind to any primary antibody held in the section.

Box 12.6 // Terminology of antibodies

The terminology of antibodies often hinders understanding. The terms antibody and antigen are fairly well defined. An antigen is a foreign material that, when introduced into the body, stimulates the production of an antibody. The terminology for naming antibodies is to prefix their antigen with 'anti-'. Thus, anti-actin is an antibody against actin. In some cases (as in the accompanying text), it is important to know the animal in which the antibody was prepared. Thus, if an anti-actin antibody was prepared in a rabbit it is termed rabbit anti-actin. If the antibody is prepared against an immunoglobulin (i.e. anti-antibody), then two species are needed. The one in which the antibody was made and the species from which the antigen (an immunoglobulin) came. Injecting rabbit immunoglobulin into a goat will produce goat anti-rabbit immunoglobulin. The first species is the species used to make the reagent, the second is the name of the antigen (in this case rabbit immunoglobulin).

The term serum is also in common use and can cause confusion. Serum is the protein-rich fluid left after plasma has been allowed to clot and the clot has been removed. Serum contains all the antibodies but has lost the ability to clot. Serum has been the traditional material for immunology and the subject is often referred to as **serology** from the widespread use of serum. Unfortunately a serum containing antibodies is referred to as an **antiserum**. This does not mean that it necessarily has antibodies against serum, just that it contains antibodies. The term antiserum is also loosely applied to monoclonal antibodies, which are not serum at all but antibody solutions from cell culture media.

---

4. Wash away the excess labelled antibody with buffer.
5. Treat the section with the substrate appropriate to the corresponding enzyme demonstration technique to develop the final colour. For HRP, this could be a mixture of hydrogen peroxide and diaminobenzidene in buffer.
6. Counterstain the tissue with a contrasting nuclear stain to show the general structure.

The sites of the enzyme will show as deposits of brown chromogen and these sites will be the position of the antigen.

N.B. Before step 1 it is advisable to i) treat with dilute hydrogen peroxide to saturate any endogenous tissue peroxidase enzyme, and ii) treat with non-immune serum to suppress the non-immune charge binding of antibody. Both steps are called **blocking** steps and are designed to reduce non-specific binding / false negative staining and also background staining.

Unlabelled antibody techniques

Since both the enzyme (e.g. HRP) and the antibody are proteins, they can both act as antigens. It was realized that it was possible to link these components together by using a third antibody. This requires two different species of animal and three antibodies.

The three antibodies needed are:

1. The primary antibody against the antigen prepared in a stated species (e.g. rabbit anti-actin).
2. An anti-HRP antibody prepared in the same species (e.g. rabbit anti-HRP).
3. A bridging antibody against the immunoglobulin of the species used to prepare steps 1 and 2 (e.g. goat anti-rabbit immunoglobulin).

These are applied in sequence to the section (see *Fig. 12.4*).

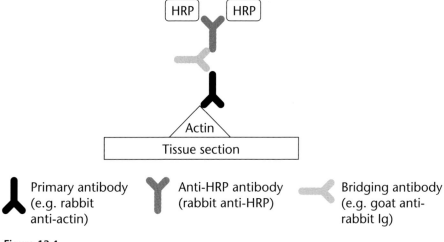

Figure 12.4
Diagram of the unlabelled antibody method

The primary antibody (rabbit anti-actin) is applied first and binds to the antigen. After washing away the excess antibody with buffer, the goat anti-rabbit immunoglobulin is applied and will bind to the rabbit anti-actin antibody. As there is a large amount of the goat anti-rabbit immunoglobulin, the binding will only use one of the binding sites available on the secondary antibody. When the final rabbit anti-HRP antibody is applied, it will be trapped by the unused binding sites of the secondary antibody. The secondary antibody thus binds the primary antibody and the tertiary (HRP) antibody by acting as a bridge between them.

The next step would be to add HRP, which would be trapped by the anti-HRP antibody. The enzyme can then be demonstrated as before and a brown chromogen deposit will indicate the sites of actin within the tissue. The process can be repeated to increase the amount of enzyme (see *Box 12.7*).

The peroxidase–anti-peroxidase technique (developed by Sternberger)
Variations on this technique became the most popular technique of immunohistochemistry. The HRP enzyme and the anti-HRP antibody can be prepared as a single ready-prepared complex (peroxidase–anti-peroxidase, or PAP) rather than having to add the final antibody and the enzyme as separate steps. The PAP complex consists of two antibodies and three peroxidase molecules in an almost cyclical structure. The complex is soluble and, even though the enzyme is bound by the antibody, its activity is not inhibited.

Box 12.7 // Increasing the amount of bound enzyme

With the unlabelled antibody technique, it is also possible to repeat the processes involving the anti-HRP and HRP. By adding extra anti-HRP after the HRP has been added, the anti-HRP will bind to the HRP. By using a lot of anti-HRP, it is possible to ensure that only one of the two binding sites is used. The other binding site can now bind more HRP. Thus, by alternating anti-HRP and HRP solutions, the amount of HRP can be greatly increased. In practice, the process is limited since there is also the risk of an increase in the background staining.

The PAP reagent can easily be purified

The PAP complex has another advantage. Since the PAP complex is prepared separately, it can be purified more easily and this gives greater sensitivity and less background staining. The anti-peroxidase antibody needs to be very pure. If it is prepared in an animal, it is likely to be contaminated with non-specific antibodies. These antibodies will compete with the anti-peroxidase antibody for binding on to the bridging antibody and this will reduce the final colour. It is almost impossible to separate the specific and non-specific antibodies whilst they exist as individual molecules but the PAP complex is easier to purify.

If an impure serum from a rabbit immunized with peroxidase is mixed with peroxidase it will form the complex. The complex is now much larger than the other antibodies and can be separated by gel filtration to give a much purer preparation. The same technique can be used with alkaline phosphatase and is referred to as the APAAP (alkaline phosphatase–anti-alkaline phosphatase) method. By 1990, PAP and APAAP techniques had virtually replaced the simple unlabelled bridge technique from which they were originally developed. Subsequent advances have led to more advanced methodologies, e.g. the avidin–biotin complex (ABC) and enzyme-linked polymer system (EPOS).

The PAP technique is very sensitive but is longer and more complex to perform than the simple direct immunofluorescence technique (see *Fig. 12.5*).

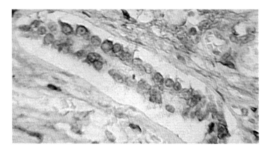

Figure 12.5
Autoantibodies in the liver shown with peroxidase-labelled antibody and diaminobenzidine chromogen. The site of the autoantibody is shown by the brown deposit.

Endogenous enzymes may cause background reactions

One possible disadvantage of using enzymes as labelling reagents is that the tissue may contain similar enzymes. Alkaline phosphatases and peroxidases are common in tissues. The enzymes within the tissue are **endogenous** and if they are allowed to react with the substrate they will produce the same chromogen as the exogenous enzyme used to label the antibody.

The presence of endogenous enzyme can be seen if a section is treated with the same technique as the test section but omitting the enzyme-labelled antibody step. Any colour produced can only be due to the tissue enzymes. This background is a nuisance and needs to be suppressed. One method is to inactivate the endogenous enzyme before performing the immunocytochemistry reaction and this is probably the best way if it can be done conveniently.

Blocking may be specific for the enzyme, e.g. using methanol containing 0.3% hydrogen peroxide will inhibit peroxidase enzymes. Alternatively, the enzyme may be blocked non-specifically, e.g. using formaldehyde or other fixatives. This works quite well for the

APAAP method and since the technique is often carried out on paraffin wax-processed tissue there is no need to deliberately inactivate the enzyme. The peroxidases in tissues are less easily inactivated in this way and even paraffin wax-processed tissue may still retain significant peroxidase activity, especially in lymphocytes, and may need to be inactivated specifically using methanol/hydrogen peroxide.

Antibodies may be adsorbed non-specifically on to the tissues

The high sensitivity of the PAP method may itself cause difficulties if the bridging antibody is absorbed or adsorbed by charge binding on to the tissue at sites where there is no antigen. The bridging antibody will then bind the PAP complex and give a non-specific background colour. This non-specific background can be eliminated by treating the section with a normal serum from the same species as the bridging antibody. Thus, if the bridging antibody was goat anti-rabbit immunoglobulin, then the section would first be treated with diluted normal goat serum. Any binding sites capable of non-specific reaction with goat serum will bind the non-immune serum and be blocked. The immune serum will not be bound and as a result the background will be cleaner.

The PAP technique is not reliably quantitative

Enzyme labelling techniques, and particularly the bridging techniques, are not quantitative (see *Box 12.8*). The method produces such high amplification that even small amounts of antigen can give strong results for very little antigen. If the antigen is highly concentrated in the tissues, then it may even give a weaker response.

High concentrations of antigen can block the adsorption of the PAP

This incongruous result is due to the bridging antibody effectively being blocked by the high concentration of primary antibody that is bound when there is a lot of antigen in the tissue.

Box 12.8 // The PAP method has some limitations

The PAP and APAAP methods are better for most purposes in histology but fluorescent methods still have their advantages. Firstly, the fluorescence can be easier to locate if there is not a great deal of antigen present, for example trying to locate a few bacteria in a large section. Secondly, the method can be quantitative by measuring the fluorescence. Finally, the fluorescence shows where the antibody is located and can move with the antibody. Enzyme labelling only shows where the antibody localized if the antibody does not move. If the antibody can move, then it can only demonstrate where the antibody was at the time the chromogen was being produced. This can be important in some applications. For example, the membrane of a cell is constantly being recycled and the recycling can be demonstrated by 'capping'. An antibody against a cell membrane protein is applied to living cells. The membrane will be evenly labelled all around the membrane. As the membrane is recycled by being drawn into the cell, the membrane proteins form a cap just above where the membrane is being drawn back into the cytoplasm.

Initially even labelling   'Capping' of the label

Following the previous example, the goat anti-rabbit IgG antibody is divalent and can bind two rabbit antibodies. Provided the rabbit antibodies are well separated within the tissues and there is a large excess of goat antibodies, then each goat anti-rabbit IgG molecule will only have one of its binding sites attached to the primary rabbit antiserum. The second binding site will then attach to the PAP complex. However, if there is a lot of antigen, then the antibodies will be closer together and the concentration may be higher locally than the concentration of the goat anti-rabbit immunoglobulin. This results in many, or indeed most, of the bridging antibodies binding with *both* of its binding sites to the primary antiserum (see *Fig. 12.6*). There will be few or no free binding sites to bind the PAP, so the final colour will be less. This mechanism also explains why using a more dilute reagent will give a stronger result when there is a lot of antigen in the specimen. The PAP technique should therefore be used to detect the presence of antigen but not used to attempt to quantify the amount, even approximately.

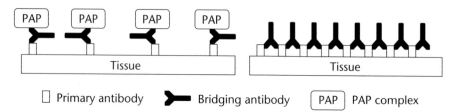

Figure 12.6
Effect of high antigen density on staining. On the left, the primary antibodies are widely spaced, so the bridging antibody is not saturated and has free binding sites to bind the PAP complex. On the right, the primary antibodies are close enough to allow the antibody to bind to the antigen using both binding sites, resulting in no free binding sites for the PAP complex

Controls are needed to ensure specificity and correct technique

To ensure the technique is working correctly, a minimum of three controls is needed. These must be done at the same time and in the same way as the test section:

1. A known positive control (weakly reacting is best). This control must show a positive reaction. A negative shows a failure of the whole complex system to react, indicating that a negative result in the test section is unreliable.
2. A known negative control. This should be completely negative. Any positive result shows whether there is a non-specific result. Any positive reactions in the test section would be unreliable.
3. A control using the test section but omitting the first (primary) antibody. This control should be negative. If this is positive, it may indicate an endogenous enzyme in the test section, or non-specific adsorption of one of the other reagents.

The above three controls should detect any discrepancies. If any unexpected results are seen, they must be investigated further by using extra control steps to determine the nature of the discrepancy (e.g. blocking controls, no application of any antibody, pre-treatment with HRP).

Avidin–biotin methods

Avidin is a glycoprotein found in egg white (see *Box 12.9*) that binds strongly to the vitamin biotin. This binding is much stronger than the usual antibody binding and is very

> **Box 12.9 // Avidin**
>
> Avidin was originally discovered when rats that were fed a diet rich in raw egg whites began to suffer from biotin deficiency. The avidin was holding the biotin so effectively in the rat gut that the animals became deficient. The investigation of this 'toxicity' of egg protein led to the discovery of avidin. It is unlikely that eggs will have the same effect in humans as the eggs are usually cooked and the number of eggs eaten is usually quite small. Biotin is produced in quite large quantities by the normal gut bacteria and deficiency in humans is extremely rare.

difficult to disrupt; it is the strongest known non-covalent bonding interaction. By linking biotin to the antibody, a complex can be formed that is very stable.

Proteins can be labelled with biotin

Biotin can be conjugated to most proteins, carbohydrates and nucleic acids by using an active derivative. For proteins, this is usually sulpho-*N*-succinimidobiotin, which can add the biotin on to free amino groups. Biotin is a small molecule and does not usually interfere with the activity of the protein. Biotinylation of antibodies is therefore an alternative method of labelling.

Avidin is used to bind on to the biotinylated protein

The biotin is then detected by using avidin, which binds to the biotin. The avidin itself has a reporter label attached, which can be a fluorescent molecule or an enzyme. The most popular of these techniques uses an avidin–biotin complex ('ABC'), which combines several avidin molecules and several peroxidase molecules. The enzymes are biotinylated so that they will bind to the avidin molecule (see *Fig. 12.7*). This complex is much larger

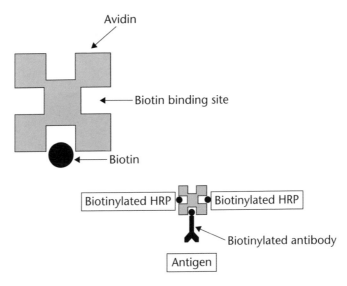

Figure 12.7
Avidin–biotin method. The avidin acts as a bridge between the biotinylated antibody and the biotinylated HRP. The avidin in the lower diagram shows a free biotin binding site, which can be used to bind the 'ABC' reagent

than the PAP complex and contains more peroxidase, but despite this the ABC method is about comparable in sensitivity to the PAP method.

The ratio of biotin and avidin can be varied

A biotin-rich complex, known colloquially as 'CBA', can be used to boost the activity. The CBA is used after the ABC reagent and is bound to the ABC complex by its extra biotin, thus adding an extra layer. The CBA can then be followed by an extra treatment with the ABC reagent, which can bind to the excess biotin in the CBA. This in turn can be followed by ABC and so on. This alternation of biotin-rich and avidin-rich layers could be compounded to give a number of repeats but this is pointless as beyond two or three extra steps the added sensitivity will be counterbalanced by a loss of specificity.

The avidin in all of the above techniques can be replaced by the protein **streptavidin,** which is extracted from the bacterium *Streptomyces* and has similar binding properties. It is preferred by many workers because it gives less background staining.

The use of biotin as a label also has its own problems as many tissues, including liver and kidney, have significant amounts of endogenous biotin and can give false-positive results. If this is suspected the tissue biotin can be blocked from reacting by treating first with an avidin solution and then biotin. Although it sounds like cookery, some methods recommend the use of an egg white solution, followed by a powdered milk solution, to achieve this aim.

Immunogold labelling

Immunogold labelling uses particles of colloidal gold to label the antibody. The colloidal gold is adsorbed on to the protein and acts as a visible label. This is very popular with electron microscopists as the spherical gold particles are easily seen in the electron microscope. The colloidal gold can be made in differing particle sizes so that it is possible to use two different-sized labels to identify two antigens in the same preparation. Although mainly used for electron microscopy, colloidal gold can be made visible for light microscopy using silver solutions. The untreated bound gold has a very pale pink coloration but silver reagents react with the gold to produce an enhanced blackening easily seen by light microscopy. This signal can be amplified further simply by viewing with standard polarizing microscopy, which produces a characteristic intense pale blue birefringence. This is generated by rotation of polarized light as it interacts with the regular arrays of gold spheres that are bound to the antigenic sites.

Polymer-conjugated systems (EPOS and EnVision patented systems)

This system utilizes a polymer spine (dextran) with both multiple (c. 10) antibodies and reporter enzymes (up to 70) attached (see *Fig. 12.8*). The construct binds and reacts in a way similar to the standard ABC complex but with the higher density of enzyme molecules localized at antigen binding sites producing a much stronger signal. To what extent the spine interferes with penetration is still uncertain.

Polymer-conjugated systems have several advantages. The inert dextran backbone has no great affinity for tissue protein or other components, so there is little endogenous binding (unlike the avidin technique, which binds to biotin in the tissues). There is possibly some increased sensitivity compared with basic streptavidin methods. It is much quicker as there are fewer steps than alternative techniques. There are some disadvantages including the possibility that the size of the polymer may impede penetration into the tissues and thus reduce binding. The reagents are also more expensive than the reagents used in simpler techniques.

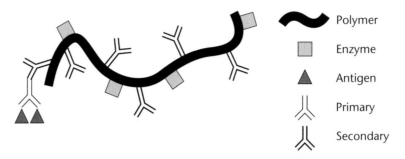

Figure 12.8
Arrangement of components in polymer-conjugated IHC reagents

Antigen retrieval methods

The fluorescent antibody technique usually only works well on fresh frozen tissue but the high sensitivity of the PAP and ABC methods allows them to be used on paraffin-embedded sections. This has probably been one of their major advantages as the method can be applied to routine FFPE blocks with no need to plan in advance for the use of immunostaining and old blocks from the archives can be used quite effectively. Indeed, it is even possible to perform the techniques on stained sections from which the coverslips have been removed. The results from paraffin sections can, however, sometimes be disappointing as the fixation and processing can alter the antigens and may vary significantly between laboratories. To restore the full immunoreactivity of the sections, various techniques have been tried. These techniques are referred to as unmasking of the antigen or **antigen retrieval methods** (see *Fig. 12.9*).

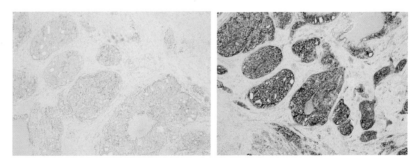

Figure 12.9
The left-hand image shows the result from IHC on untreated paraffin wax sections and the right shows sections from the same block with the same IHC technique following antigen retrieval with citrate buffer (pH 9) and microwave heating.

Enzyme retrieval

The first method involved the use of enzymes at 37°C to digest away some of the protein to expose the masked antigenic sites. Usually it is a proteolytic enzyme, such as trypsin, pepsin or streptokinase, that is used to remove protein. The enzymes need to be prepared freshly as they lose activity once in solution. This works well in most instances, but can be very sensitive to the conditions of incubation. Too long an incubation may remove excess protein and associated antigen, whilst too short an incubation may not unmask the antigen fully. The timing is also very dependent on fixation. This dependency is on both the type of fixative used, with formaldehyde generally being preferred, and also on the length of fixation, with long fixation times needing longer subsequent digestion. Because of these variables, it is usually necessary to test the timing and the technique for each batch of enzyme and sometimes for each individual specimen to get optimal staining.

This variability has led to some workers questioning the validity of the technique and arguing that some of the results may represent an artefact of the digestion process rather than the site of preformed antigen. However, the technique does seem to give reliable and useful results if well controlled and has been widely adopted for paraffin sections; it is less useful for fresh tissue.

Heat retrieval

An alternative to using enzymes is to heat the section. Simple heating is insufficient (see *Box 12.10*) and it requires prolonged exposure (5–10 min) to boiling or superheated fluid to achieve the retrieval. Usually a sodium citrate solution buffered to a slightly acid pH is used and the heating can be achieved using a microwave oven, a pressure cooker or a combination of both. The sections do have a much greater tendency to detach from the slide, so the use of adhesives such as silane for coating the slides is needed (see **Chapter 4**).

Box 12.10 // Antigen retrieval needs superheated water

Dry heating of sections does not seem to unmask antigens in the same way as wet heat. It seems likely that there is a hydrolysis effect and the energy needed to activate this is quite high. Simple boiling of the section is less efficient. Microwaves probably produce temperatures above the normal boiling point within the solution and pressure cookers also have a higher temperature than the standard boiling point of water.

Microwave retrieval. In the microwave method the dewaxed sections are placed in buffer in a microwave oven and heated at full power for 10–30 min depending on the antigen to be unmasked, the power of the microwave and the fixation of the tissue. The fluid rapidly boils and if this is being done in a Coplin jar, which holds about 50 ml of fluid, it results in boiling of the fluid and often the buffer spills out, so the fluid needs to be constantly replenished. By using larger containers with an excess of buffer, it may be possible to have sufficient fluid to avoid the need to replenish. The sections are allowed to cool in the buffer and then transferred to distilled water. If the sections are removed whilst still hot, this may result in loss of reactivity to the antibodies. This method is rapidly gaining popularity as it does not need to be optimized for different batches of enzyme, as happens with the digestion method.

Pressure cooker retrieval. Pressure cookers are an alternative way of heating the sections and utilize higher temperatures since water boils at a higher temperature at higher pressures. An ordinary stainless steel domestic pressure cooker is used. It is partially filled with citrate buffer and brought to the boil at normal pressure (the cooker is not sealed) before putting in the slides. The pressure cooker is then sealed and brought back to the boil at raised pressure. Most pressure cookers work at 1 atmosphere above normal pressure and this raises the boiling point of water to about 120°C. The increased temperatures reached mean that antigen retrieval times are much shorter, typically 2–4 min once the full pressure is reached. Pressure cookers can also treat many more slides at once than a microwave oven.

It is also possible to use two methods of retrieval by microwaving the sections first, cooling them to 37°C and then treating them with a protease enzyme.

The choice of retrieval method is largely a matter of personal choice, the method of tissue preparation and the antigen being sought. Some antigens will be demonstrated by heat retrieval but not by enzyme digestion and vice versa. Often there are several different techniques being employed within the same laboratory for different purposes. One beneficial aspect of the use of retrieval methods is that the antibodies can often be used in more dilute solutions with dilution factors in excess of 1/10,000 being cited in some papers.

12.2 Labelling of other proteins

Proteins other than antibodies can also be labelled using the same general methods as explained above.

Lectin histochemistry

Lectins are plant, or occasionally animal, proteins that bind to sugar molecules rather than antigens but otherwise the reaction has much in common with antibody binding. They are not immune proteins but occur endogenously in the tissues of the plant without needing to 'immunize' it. Like antibodies, they can bind more than one sugar (they are polyvalent). Unlike antibodies, many lectins need specific cations to bind effectively and these need to be added, e.g. concanavalin A needs calcium ions to bind effectively. Lectins may be highly toxic (see *Box 12.11*) as they can interfere with sugar interactions.

Box 12.11 // Lectins

Lectins were originally identified by their ability to agglutinate red blood cells. The agglutination can be specific for certain blood groups since the blood group antigens contain sugars and the different blood groups have different terminal sugars. Lectins can be toxic in some cases since, if they are absorbed intact without being digested, they will agglutinate the red cells in blood and cause a haemolytic disorder. Some types of bean contain sufficient lectin concentrations to cause disease in this way if they are eaten uncooked. Not all lectins are dangerous poisons, however, and they are currently being investigated for possible use in drug delivery to specific sites.

Only terminal sugars react

The sugars to which the lectins bind can be part of proteins (glycoproteins) or can be complex carbohydrates or in the form of mucins, but in all cases only the terminal sugars can be accessed. Sugars buried inside protein or bound inside polysaccharides are not usually available to the lectins and will not be detected. Some methods utilize a neuraminidase unmasking step to reveal sugars at a deeper level in the complex carbohydrate sequence. Lectins can be used as very precise reagents for locating sugars in tissues. *Table 12.1* lists some lectins and their specificities. Lectins sometimes react with other sugars but with a decreased binding strength. The use of lectins in histochemistry has greatly increased in the past few years as the importance of carbohydrates in cell recognition, cell adhesion and metabolic control has been realized.

Lectins can be located by several different methods

Lectins can be treated simply as an antigen and injected into an animal to raise antibodies against them and used in one of the labelling strategies already explained. Lectins can

Table 12.1   Examples of some lectins and their specificities

Lectin	Abbreviation	Source	Specificity
Concanavalin A	Con A	*Canavalia ensiformis*	α-Mannose, α-glucose, N-acetylglucosamine (GlcNAc)
Wheat germ agglutinin	WGA	*Triticum vulgare*	GlcNAc-β-1,4-GlcNAc
Horse gram lectin	DBA	*Dolichos biflorus*	GalNAc-α-1,3-GalNAc
Gorse lectin	UAE	*Ulex europaeus*	α-L-Fucose
Horseshoe crab lectin	LPA	*Limulus polyphemus*	N-Acetyl neuraminic acid

be applied to tissue sections and, after washing away the excess, any bound lectin can be identified by a standard immunotechnique. Lectins are also available as fluorescent conjugates, biotinylated lectins and enzyme-labelled lectins. These can be employed in a similar fashion to antibodies labelled in these ways.

Finally, lectins can occasionally be demonstrated using carbohydrate-containing enzymes. Glucose oxidase, for example, contains about 18% carbohydrate. The lectin can bind to the sugars present in the enzyme. The glucose oxidase can then be demonstrated simply using an enzyme technique. This greatly simplifies the technique compared with most other labelling methods. This works well for concanavalin A but is less successful with other lectins as the specific sugars they bind to are not always available in the enzyme. HRP can also be used in the same way but suffers slightly from the presence of endogenous peroxidase, whilst human tissue does not contain any glucose oxidase activity so there is no background staining.

Protein A

Protein A is a staphylococcal protein with a highly specific affinity for the Fc part of mammalian IgG. It can be used in the same way as an anti-immunoglobulin. Protein A can be obtained commercially labelled with fluorescent dyes, biotin and enzymes. One benefit of using protein A is that it is not species specific and will detect most mammalian IgGs. This is sometimes useful as it reduces the need to have several different anti-immunoglobulins when using primary antisera from different animal species. Its disadvantage is that it cannot be used in tissues containing endogenous IgG. Anti-human IgG can be used on rat lymph nodes as they will not contain human IgG but protein A cannot be used on any lymph nodes as it will react with all mammalian IgGs.

Specific receptors

Hormones will be bound by their specific receptors within the tissues and this can be used to detect the location of such receptors. This is fairly straightforward if the hormone is a protein and can be labelled in the same way as other proteins.

Some receptors can be detected using exogenous proteins that will bind to the receptor. These techniques using specific binding materials are often unique and, although useful and important, do not have much in common beyond the facts that they bind tightly with a receptor within the tissue and that they can be labelled or conjugated by one of the methods already outlined without altering their binding ability. For example, α-bungarotoxin is found in the venom of a Taiwanese snake and binds to the acetylcholine receptor of the neuromuscular junction and this can be used to locate and identify the receptors. α-Bungarotoxin is available commercially labelled with biotin and fluorescein.

The general concept of labelling a protein that binds to the tissue in some specific way applies to all such exogenous binding materials. The source of the binding agent, however, can be quite varied.

12.3 Applications of immunohistochemistry

The use of antibodies is an extremely powerful tool in biomedical science and as a technique is ubiquitous in both the research and diagnostic arenas.

Research

Antibody techniques are often used in research to supplement and make sense of the findings from strictly molecular approaches. They are a way of linking form and function. Their use in research has been vital in understanding the functional mechanisms of detailed molecular structures at a protein level, e.g. the contractile apparatus in muscle, ultimate filtration unit in kidney or the synaptic junction. This has been achieved by the use of antibodies against each of their functional components, thus effecting their 'dissection' protein by protein. In combination with an approach that involves the capture and combination of images from IHC-stained serial sections through complex cellular structures, it has been possible to generate a 3-D colour-coded model of functional structures that can then be digitally rotated or deconstructed. This approach greatly enhances the understanding of how these structures are put together and therefore how they work.

Diagnostics

Immunohistochemistry may be the only way to confirm the nature of a malignancy and is often the only way to resolve a difficult question; for instance, what is the nature and origin of a poorly differentiated metastatic deposit?

The satisfactory resolution of this question would better inform a more directed clinical approach to treat the primary malignancy. This treatment may be as diverse as chemotherapy, radiotherapy or even a bone marrow transplant. For some poorly differentiated malignancies there may be very few concrete clues to the cell of origin if cellular morphology has become significantly altered. Occasionally even the use of specialized staining techniques to identify normal or abnormal cellular products may be inconclusive. For these 'difficult to diagnose' cases, the use of an antibody panel will often help to shed more light on the problem (see *Table 12.2*).

The panel of antibodies is usually constructed to help indicate the broad classification of a cell type, with possibly some that are more definitive if there is a better indication of what the cell type might be. It should be noted that antibodies and IHC is an expensive approach to take and it is advisable to include up to a maximum of just five in a panel. Since new and 'better' antibodies are constantly being made commercially available, it is difficult to construct a definitive table that will retain currency for more than a few months. This is particularly evident for CD markers and haematological malignancies.

The antibodies listed in the table are by no means exhaustive but represent those most likely to be chosen when considering the construction of an antibody panel to identify the origin of an unknown malignancy.

In the case of a very poorly differentiated cell type the first step would be to select the full range of **intermediate filament** antibodies plus LCA to inform the confirmation of a main tissue type.

Table 12.2 Antigens which can be targeted by specific antibodies to assess cell types

Antibody	Target	Diagnostic significance
Possibility of lymphoma?		**Relevance to lymphoma**
CD45 (LCA) (leucocyte common antigen)	White blood cells	Indicates a lymphoma
CD3	T cell surface	T cell lymphoma
CD20	B cell surface	B cell lymphoma
K / λ light chains	Immunoglobulin light chains	Refines B cell lymphoma decision
CD68	Macrophage surface glycoprotein	Histiocytic tumour
Possibility of carcinoma?		**Relevance to carcinoma**
CK7	**Intermediate filament** Basic type II cytokeratin – low mwt	Most epithelial carcinomas including breast and cervix
CK20	**Intermediate filament** Acidic type I cytokeratin – high mwt	Colorectal carcinomas, Merkel cell tumour of skin
HMWCK [34βE12]	High molecular weight cytokeratin of basal cells (prostate)	Adenocarcinoma of prostate, loss of basal cells demonstrated
EMA (epithelial membrane antigen)	Surface glycoprotein of secretory epithelium	Carcinoma of eccrine epithelium, extra-mammary Paget's disease
Chromogranin	Peptide found in neurosecretory granules	Carcinoid tumours, also adenomas of hormone-producing cells
Possibility of sarcoma?		**Relevance to sarcoma**
Desmin	**Intermediate filaments** located at intercalary discs and z lines	Malignancy derived from muscle muscle: skeletal, cardiac or smooth
Smooth muscle actin	Actin filaments	Malignancy derived from smooth muscle: leiomyomas, leiomyosarcomas
Myoglobin	Oxygen transfer protein	Indicates muscle origin
Possibility of nerve tumour?		**Relevance to nerve tumour**
Neurofilament	**Intermediate filaments** in axons	Confirms neuronal origin
Neurone specific enolase (NSE)	Transmitter synthesis enzyme	Confirms neuronal origin
Synaptophysin	Structural protein	Confirms neurosecretory cell of origin
GFAP (glial fibrillary acidic protein)	**Intermediate filaments** in glial cells	Confirms glial cell origin
Possibility of melanoma?		**Relevance to melanoma**
S-100	Calcium channel binding protein found in melanocytes	Highly expressed in melanoma cells
HMB 45	Pre-melanosome vesicles	Highly expressed in melanoma cells
MART-1	Melanoma antigen	Unique to melanoma cells

Once known, this may then be followed by a more refined panel, especially when investigating the origin of an epithelial malignancy that has been sampled as a secondary deposit elsewhere in the body.

There is one final labelling technique that can be applied to many materials and that is autoradiography. This is covered in **Chapter 16**.

Suggested further reading

Brooks, S.A., Leatham, A.J.C. and Schumacher, U. (1997) *Lectin Histochemistry.* Royal Microscopical Society.

Jackson, P. and Blythe, D. (2012) Chapter 18: Immunohistochemical techniques. In: *Bancroft's Theory and Practice of Histological Techniques*, 7th edn. (eds K.S. Suvarna, C. Layton and J.D. Bancroft). Churchill Livingstone.

Kiernan, J.A. (2008) *Histological and Histochemical Methods: Theory and Practice,* 4th edn. Scion Publishing.

Polak, J.M. and Van Noorden, S. (2003) *Introduction to Immunocytochemistry,* 3rd edn. BIOS Scientific Publishers.

Renshaw, S. (ed.) (2005) *Immunohistochemistry: Methods Express.* Scion Publishing.

Self-assessment questions

1. Define the terms antibody and antigen and explain briefly why antibodies are useful as histochemical reagents.
2. Outline how fluorescein is used as an antibody-labelling material.
3. What are the disadvantages of the immunofluorescent technique?
4. What enzymes are used in immunological staining?
5. Outline the steps needed in the PAP (or APAAP) technique.
6. How can the PAP technique be used in electron microscopy?
7. What is meant by antigen retrieval and how can it be achieved?
8. Describe the use of lectins in histochemistry.
9. What is the main use of immunogold labelling?
10. What is the use of avidin in the ABC technique of antibody labelling?

13 Gynaecological cytopathology

Learning objectives

After studying this chapter you should confidently be able to:

- **Outline the advantages of exfoliative cytology**
  Collection of material for exfoliative cytology is simpler, quicker and cheaper than the surgical collection of biopsy specimens.

- **Indicate why cytology can be used for screening**
  It is less invasive and can be repeated at regular intervals without undue stress to the patient.

- **Outline the methods used in liquid-based cytology**
  Specimens are collected using a cervical brush and then dislodged into a vial of liquid transport medium. Cells are concentrated by either centrifugation or filtration and attached to a slide for staining and examination.

- **Describe the normal constituents of a cervical smear**
  Normally there will be squamous epithelial cells from different layers of the cervical epithelium (superficial, intermediate and parabasal cells) and endocervical cells from the endocervix. Some blood cells may be present depending on the stage of the cycle (neutrophils, leukocytes, erythrocytes) but large amounts or their presence at certain times may not be fully normal. Some bacteria are normal (lactobacilli) and small amounts of mucin and spermatozoa are not abnormal, nor are a small number of red blood cells from the trauma of sampling.

- **Describe the changes that occur in gynaecological smears with age, menstrual cycle and pregnancy and the underlying causes of these changes**
  The appearance of the smear is controlled by the hormones oestrogen and progesterone. Oestrogen increases the number of mature, superficial cells and decreases the amount of mucin and the number of neutrophils seen. Oestrogen effects are seen in sexually mature women during the proliferative stage of the menstrual cycle. Progesterone decreases the maturation so more intermediate and parabasal cells appear, with fewer superficial cells. Progesterone also increases the glycogen content of cells and this may result in an increase in bacteria and as a consequence an increase in the number of leukocytes. Progesterone effects are seen in the secretory phase of the menstrual cycle and also during pregnancy. In the absence of progesterone and oestrogen, there is less growth of the epithelium and more parabasal cells are seen. Effects from the lack of both hormones are seen in pre-puberty and post-menopausal smears.

- **Outline the appearance of a smear showing inflammation and name some common causes of inflammation**
  Neutrophil leukocytes are more common, mucin may be increased and there may be signs of haemorrhage with increased numbers of erythrocytes. The causative organism may be visible. There may be degenerative changes in the cells and altered staining of the cells. Common organisms include *Candida albicans*, *Trichomonas vaginalis* and herpes and human papilloma viruses.

- **Describe the changes in smears in pre-malignant and malignant states**
  Loss of uniformity is common. Nuclear abnormalities include altered nuclear size and altered nucleocytoplasmic ratio, abnormal chromatin distribution, prominent nucleoli, multinucleation and abnormal staining.
- **Outline why a smear may be considered inadequate**
  This can be due to too few cells, excessive mucus, pus or erythrocytes that obscure the cells, incorrect sampling technique resulting in air-drying or inadequate fixation artefacts.

One of the major problems with histological diagnosis is the need for a piece of solid tissue to prepare sections. This means surgery. Surgery is expensive since it needs anaesthetics, sterile conditions and highly trained personnel. It would be much cheaper and simpler if there was no need to cut into the body and cells could be removed without the need for surgery. Exfoliative cytology is just such a method of obtaining cells for diagnosis without having to cut into the body.

Exfoliative cytology relies on cells being shed naturally from a surface (see **Box 13.1**). These cells can be collected without damage to the body so it is simpler and safer. Because malignant cells are less adhesive than normal cells there is a good chance that exfoliated cells will include malignant cells from any lesion that is present. The commonest form of exfoliative cytology is the use of smears from the uterine cervix to recognize the early stages of cancer (malignancy), but it is also useful for diagnosis of other diseases and for monitoring treatment. Although cervical smears are the most common form of smears, they are not the only type and specimens can be obtained from a wide variety of sources. The use of isolated cellular techniques is one of the most rapidly changing and expanding branches of cellular pathology and is becoming a subject in its own right, and most cytology departments in the NHS are now separate departments from histopathology.

Box 13.1 // The origins of exfoliative cytology

At first the idea of using shed cells was resisted as it seemed unreliable. Since only a sample of surface cells is examined, it was thought that a small or deep tumour would be missed and the accessibility of the cervix for histological biopsy made cytology unnecessary. Cytological diagnosis also relies on different criteria to histological sections since the relationship between cells is lost. It was only the persistence of Dr George Papanicolaou that showed the potential of the method, adapting his early work on the effect of hormones on the rat menstrual cycle. It still took many years before it became fully accepted. Papanicolaou published his first paper on cancer diagnosis using smears in 1928 but it was not until after the Second World War that it became a recognized clinical method. Cervical screening in the UK began in the late 1940s with an early British pioneer in the field being Stanley Way in Gateshead who set up a screening service in 1948.

13.1 Advantages and disadvantages of exfoliative cytology

Advantages

Exfoliative cytology provides a useful alternative to histological examination since it has several advantages when compared with conventional biopsy techniques:

- The collection of specimens is relatively simple and causes very little damage since there is no cutting or incisions. It is considered a non-invasive technique.
- Both specimen collection and processing are quick. Cytology can be used for rapid diagnosis as an alternative to frozen sections. It is easily possible to have a smear ready for diagnosis in a similar time scale to that needed to prepare a frozen section, especially if using a rapid Giemsa-type stain. Speed is not usually the reason for using exfoliative cytology and most smears are processed as non-urgent and routine reports may take several days, but smears can be processed and reported on within an hour if necessary. This has particular relevance to drop-in breast screening clinics.
- The technique of specimen collection is cheaper since it does not require expensive aseptic surgery or even getting the patient into hospital. In many cases, it can be done during a visit to the GP and it can be carried out by a trained nurse or equivalent trained personnel and does not need a surgeon.
- Exfoliative cytology can be used as a screening method (see **Box 13.2**). Screening involves checking large numbers of apparently normal people for disease in the hope of finding the very early stages of disease in a few of them. It is always easier to treat a disease in the early stages and the most at-risk individuals within a population should be targeted. Early detection is of great benefit to the patient and can be cost-effective by reducing the need for expensive treatment of advanced disease. Screening with cytology is only possible because of the above advantages. It would not be possible to screen patients without symptoms if specimen collection was expensive, involved significant risk to the patient or was very painful.

Box 13.2 // Screening

Whether to create a screening programme is a complex decision since it involves ethical, economic and social dimensions. The economic decision is the easiest. The cost of screening is compared with the costs of not screening. The costs of not screening include the extra cost of treating later stages of the disease and the loss of earnings from the disease. Another significant financial consideration is the cost of treating all individuals with a positive test result. The social problems relate to the effects of worry and stress from any false positives or recalls. The ethical problems relate to how much pressure is used to get people to comply with the screening and whether it is morally justifiable *not* to screen for economic reasons. The aim of the Government is to have at least 80% of women between the ages of 25 and 60 screened for cervical disease at least once every 5 years.

Disadvantages

Exfoliative cytology does have some disadvantages when compared with a full histological biopsy:

- A negative result from a smear is less certain than with a histological section since the cells may reflect a poor sampling of the diseased area and not a genuine lack of disease. For this reason, a negative result on a smear should not be allowed to override a clinician's suspicions that there is a genuine disease present. If there is doubt, then the smear should be repeated.
- Even in a situation where a tumour is detected by the smear, the actual site of the tumour may be uncertain. For example, if a fluid sample is taken, then abnormal cells could have come from anywhere that has come in contact with the fluid.

- The screening of smears can be very tedious. Most of the cells in most of the samples will be normal and screeners can find it very difficult to maintain the concentration needed to detect the occasional abnormal cell. This is especially true for screening of 'well-women' samples where less than one in a hundred smears may show abnormality and even in the positive cases the majority of cells on a smear may be quite normal.
- The exact nature of any tumour is less certain than when a biopsy is taken since only isolated cells can be seen and their relationship with adjacent cells is lost in the smear preparation process. Important characteristics such as tumours invading surrounding tissues cannot be seen directly in smears but only inferred from the appearance of the individual cells.
- If the sampling of a site is inadequate then it will result in a false negative (see **Box 13.3**) where the screening test result indicates there is no disease present but there is actually a tumour. This will give the patient a false sense of security and will delay treatment, with a consequently poorer prognosis. If the testing regime results in false positives, where the report of the screening test indicates disease but the patient is actually disease free, it will result in anxiety for the patient and unnecessary investigations or treatment.

Because of these disadvantages, many pathologists prefer to confirm a positive cytological diagnosis with a biopsy (e.g. colposcopy for cervix) before planning and applying any treatment.

Box 13.3 // False positives and false negatives

The effectiveness of any screening method is determined by the number of false positive and false negative results. The relationship between the reported result and the existence of the disease is shown in the table below.

	Positive report	Negative report
Disease present	A True positive	B False negative
No disease present	C False positive	D True negative

Two measures of the effectiveness are used.

The **sensitivity** is the proportion of people with a disease who are detected by the screening test. It is calculated by dividing the number of true positives by the total number of people with the disease, i.e. A/(A+B), and is usually converted to a percentage.

The **specificity** is the proportion of people who don't have the disease and don't test positive for it. It is calculated by dividing the number of true negatives by the total number of people in whom the disease is not present, i.e. D/(C+D), which is again usually converted to a percentage.

A high sensitivity indicates that there are few false negatives whilst a high specificity indicates that there are few false positives. Unfortunately, in some cases adjusting the testing regime to improve the specificity will result in a worsening of the sensitivity, and vice versa.

Potential infection of exfoliative cytology samples

In all smear preparation techniques it is important to take full precautions against infection risks, as the samples are usually unfixed when they arrive and so may be contaminated with pathogenic (disease-causing) organisms. This is most likely in specimens from patients who are obviously unwell. Bacteria may even be able to multiply in some samples, making the bacterial count even higher if the samples are delayed or stored inappropriately.

13.2 Gynaecological smears

The use of smears for the detection of cervical and endometrial malignancy is probably the best known of all cytological techniques and has been used for many years. When applied effectively, it can lead to significant reductions in deaths in the target population. George Papanicolaou pioneered the use of cervical smears as a diagnostic method in the 1940s. He developed the main staining technique for smears and the technique is often referred to as a 'Papanicolaou test' and is often shortened so that the term 'pap smear' is a common designation for the cervical exfoliative cytology test.

Although the actual techniques of preparing the slides may be identical, there are two distinct types of specimen:

1. **Asymptomatic patients.** Samples from asymptomatic patients are collected as part of the cervical cytology screening programme to detect disease before it becomes apparent to either the patient or the clinician. The patients generally appear and feel healthy and the screening is done on a large section of the population, significantly those most at risk. Since most smears are normal, the initial checking of these smears is not done by pathologists but by trained **cytoscreeners** who are mainly trying to separate those smears that have no abnormality from those that are possibly showing some suspicious signs. The cytoscreeners do not make a final diagnosis although they may indicate the type of abnormality that they find. The positions of all abnormalities are marked on the slide (e.g. by a dot of ink next to the suspicious-looking cell) and the smears are then passed to the pathologist or consultant cytologist for final diagnosis and reporting. All abnormal findings are reported and checked, not only malignancy and its early stages. Inflammation and infections are also significant findings and should not be ignored. To ensure that no abnormal cells are missed, the entire slide is reviewed cell by cell for each case. Each slide will have over one thousand cells making this a time-consuming and tedious task. Improvements in cell recognition systems have allowed the use of computers for the initial screening. Computers can work continuously, do not tire and are more consistent than human screeners. These are not currently fully automated but simply select the most abnormal cells on each slide and mark them, with the result that manual screening is much less labour-intensive. (More details of the system can be found in **Chapter 18**.) A large study of over 70 000 samples was carried out in the UK between 2005 and 2010 to compare manual assessment with automated reading in cytology (MAVARIC trial). This trial found that automation-assisted screening was 8% less sensitive than manual screening, with only a slight increase in specificity. The conclusion drawn from this study was that automation-assisted screening could not be recommended. However, it is likely that the systems will improve and automation in screening will eventually become accepted.
2. **Symptomatic patients.** In this case, there is a clinical reason for taking the smear (see **Box 13.4**). The patient may report symptoms or the clinician may detect

> **Box 13.4 // Symptomatic patients**
>
> As explained in *Chapter 2*, patients will only be seen if they go to see their GP with a particular problem. Patients will only visit their GP if the problem is one of the following:
>
> - painful or uncomfortable
> - worrying
> - embarrassing or socially limiting
> - disabling
>
> Cervical cancer falls into none of these categories until the disease is well advanced and it is then difficult to treat. In the early stages of the disease, the signs are quite slight and easily missed. This makes cytological screening essential to detect the early and treatable stages of the disease.

something during a physical examination. In this case, the purpose of the smear is to diagnose the condition and is not just for general screening. The selection of which cases to investigate is done by the clinician, taking into account the symptoms and case history of the patient. The smears may also be used to monitor the effectiveness of treatment. These can often involve more invasive techniques and more frequent sampling since they can be justified by the symptoms. Smears from symptomatic patients should always be seen by the pathologist and not just screened and reported as 'no abnormality seen', since in this case a negative result is not expected and would be considered a significant finding.

The main use of exfoliative cytology is for gynaecological screening

Cancer of the cervix has proved to be a good disease to tackle with such screening. One reason for this is that the earliest signs of malignant change can be seen as simple nuclear changes (dyskaryosis) and alterations in the maturation of cells (dysplasia) in the cervical epithelium. These early changes are not malignant and often subside without progressing to malignancy, but in some cases they can lead to further change resulting in a carcinoma *in situ*. This may then develop into an invasive carcinoma if not treated. *Table 13.1* shows a typical time course of the disease.

These times will vary with individual patients and there is evidence that these time scales are changing, even as an average for the population. Some of these cytological changes are now being seen earlier than was previously the case although the reasons for this altered incidence are still disputed.

The pattern, however, remains the same with a significant delay in most cases between the time when the lesion becomes malignant and the time when it begins to invade other

Table 13.1 Typical time course of malignant changes in the cervix

	Age at onset (years)	Persistence
Dysplasia	30	10–20 years
Carcinoma *in situ*	40	5–10 years
Invasive carcinoma	60	1–4 years before symptoms become apparent

tissues. If each woman is checked every 3–5 years, then at least one of the samples should fall in the period between the tumour becoming overtly malignant and the time when it spreads beyond the initial site.

The use of exfoliative cytology for detecting cervical disease has been a success story for screening. The rate of cancer deaths has fallen consistently since the introduction of screening (42% fall in the first 20 years). Mortality rates fell from 8.3 per 100,000 women in 1971 to 3.3 per 100,000 by 2003.

It is estimated that by using screening and early treatment 4500–5000 deaths are prevented in the UK each year. In total, the number of lives saved between 1987 and 2004 is estimated at 100,000.

The degree of protection or confidence that is offered by a negative smear depends on the frequency of testing and the age of the patient. The degree of protection from a single negative smear has been estimated and is shown in *Table 13.2*.

Table 13.2   **Degree of protection offered from a single negative smear**

	Age of patient (years)		
	20–39	40–54	55–69
3 yearly	41%	69%	73%
5 yearly	30%	63%	73%

The current scheme of testing in the UK is:

Age group	Frequency of screening
25	First invitation
25–49	3 yearly
50–64	5 yearly
65+	Only screened if not screened since 50 or has had a previous abnormal test

These are targets for screening but the frequency of screening is also dependent on how willing women are to be screened (see *Box 13.5*).

The logic of using a variable screening regime depending on the patient's age is that below 25 years of age abnormalities in the smear are common (but most regress) and invasive cancer is very rare so screening has little benefit. Patients over the age of 65 with three clear screens have almost no risk of developing the disease. By reducing the amount of screening the costs are reduced, the system is less overloaded and the psychological/social problems are reduced.

In the USA, screening is based on private medical care rather than a national screening programme and annual smears are carried out from the age of 28 onwards.

Types of gynaecological sample

The cells for cervical smears can be collected in a variety of ways, some of which are largely historical; others are likely to be replaced by the use of liquid-based techniques. Whatever method of sampling is used, the aim of the exercise is to take an adequate number of epithelial cells from the area of the transition zone (TZ) where the epithelial

> **Box 13.5 // Rescreening**
>
> Although cervical cytology is considered non-invasive since it does not break the surface of the body it is not a trivial method of sampling and many women find it embarrassing and undesirable. Getting women to take the test even when it is in their own interest needs good advertising and information systems, as well as the scientific ability to screen the samples. Compliance with requests for rescreening is one of the major reasons for differences in the effectiveness of screening in preventing deaths from malignancy. Scandinavia has historically benefited from a greater response to screening and has therefore had a better record of disease prevention and treatment. This includes screening for breast cancer as well as cervical cancer.

type changes from squamous to secretory/glandular. It is the externalization of the TZ onto the outer rim of the cervical os at puberty that leaves young girls particularly at risk.

Cervical scrape

This was the commonest technique when screening became widespread. The early method used an Ayres spatula (see *Fig. 13.1*). The original Ayres spatula is possibly not ideal as it does not have a sufficiently elongated tip to ensure adequate penetration into the external os and sampling of the endocervix. The cervix should be viewed whilst taking the smear to ensure correct sampling. This requires the use of a speculum but it is best to

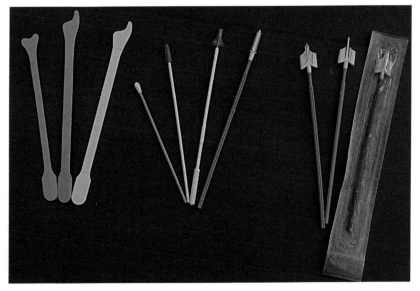

Figure 13.1
Sampling devices for cytology. In the left-hand grouping are a beechwood Ayres spatula with two Aylesbury spatulas (one made of beechwood and one plastic) with elongated 'beaks' to penetrate further into the endocervical canal. In the centre are devices for sampling the endocervical canal; the leftmost is a 'cotton bud' (which has been discredited as a reliable sampler) and the other three are endocervical brush samplers. The right-hand grouping shows Cervex brushes for sampling the cervix and endocervical canal; these brushes are used for liquid-based cytology. The brush heads are made of flexible plastic and are supplied in sterile packaging (shown far right).

avoid using lubricating materials as they can contaminate the smear. The solid spatulas have been replaced by cervical brushes (or brooms) which have an extended chevron shape. The bristles are of different lengths and are designed so that the central peak of the brush penetrates into the cervical canal. The curved shoulders are shaped to match the contours of the cervix and the bristles are flexible to give a close contact. The bristles also have a **D** cross-section with the flat of the **D** scraping the surface while the curve of the **D** gives more rigidity. The asymmetric nature means the brush is most effective if rotated clockwise. The brush is used to lightly scrape the squamo-columnar junction where the columnar epithelium lining the cervical canal changes to the stratified squamous epithelium found in the vagina. The squamo-columnar junction is a very common site of malignant change, so sampling of this precise area is essential in screening. The cells can be smeared on to a slide and fixed but more commonly the cells are rinsed off into the transport medium used for liquid-based cytology (LBC) which is considered later in the chapter. Cervical brushings are the main method of obtaining cervical smears and are usually performed by clinicians, GPs or nursing staff rather than by cytologists.

Endocervical brushings
The brush is gently inserted into the endocervical canal and rotated. The cells are then transferred to a slide and fixed. This is less common but will detect endocervical malignancy more reliably and can be used if the squamo-columnar junction is located inside the endocervical canal.

Liquid-based cytology techniques
These are replacing the more traditional techniques given above and will be examined later in the chapter.

Smear fixation
Smears must be fixed immediately whilst still wet. Air-drying results in artefacts (blurring of nuclear detail, for example), especially when the Papanicolaou staining method is to be used, although it is a common method of smear preparation for Romanowsky staining. In addition, the liquid-based cytology techniques often use their own proprietary fixatives and transport fluids. Many fixatives have been tried over the years but two generic fixatives are in common use:

Alcohol (95%, sometimes with the addition of 1–3% acetic acid to improve nuclear preservation). This is the usual fixative but the wet solution is flammable and cannot be sent through the post.

Spray fixatives. These consist of alcohols and polyethylene glycols (PEG or carbowax). The alcohols fix the specimen and then it is coated with a thin layer of PEG that protects it from drying. The PEG film is easily removed before staining. These are often bought as ready-prepared spray cans or bottles but can be easily made and used from a simple dropper bottle. The advantage of these fixatives is that the PEG acts as a protective coating and stops artefacts caused by drying. The slides can be allowed to 'dry' and this is useful for slides that have to be transported or posted.

Staining methods
The staining of smears has to bring out both nuclear and cytoplasmic detail and help to differentiate between the different cell types. Although no single method can be perfect for every situation, the Papanicolaou technique has become almost universal in

> **Box 13.6 // Routine staining**
>
> It is a little surprising that in three cellular diagnostic disciplines a single stain has become dominant, yet it is a different stain in each case. In histology, haematoxylin and eosin is dominant, in cytology it is the Papanicolaou stain and in haematology it is the Romanowsky method. The techniques do not transfer effectively across the disciplines. The use of haematoxylin and eosin is non-existent in haematology and minimal in exfoliative cytology. Papanicolaou is not used in routine histology or haematology. The Romanowsky stains do have uses in the other two disciplines but only for very limited applications. This specialization partly reflects genuinely different requirements in the three disciplines but also reflects a historical inertia. Histologists have always used haematoxylin and eosin so they will continue to use it. Papanicolaou has become the most commonly used stain in cytology partly because it was the first stain in the field.

its application for screening smears (see ***Box 13.6***). Other stains are useful for following up abnormalities. Staining needs to be well controlled and reliable since explaining any observed changes in staining is an important contribution to inform a diagnosis. Most laboratories use automatic staining machines to ensure a constant staining routine and to cope with the large numbers of smears being stained.

Papanicolaou stain

The Papanicolaou stain was devised in the 1940s and has been the main cytological staining method ever since. To a certain extent, the clarity of this method has been a feature in the acceptance of smears as a diagnostic method. The Papanicolaou method provides good nuclear detail and gives a translucent or transparent cytoplasm that is useful if cells overlap. The ability to see through the cytoplasm of a cell means that it is often possible to diagnose an abnormal cell even if there is another cell overlapping it. The cytoplasmic colour gives some indication of the maturation of the cells and of cellular activity.

The dyes involved are:

- **Haematoxylin.** Harris's or Gill's are the commonest formulae. Harris's requires differentiation, which may result in some unevenness if the smear is thick, but many workers think it gives the sharpest detail when well stained. Gill's is less critical in differentiation and is well suited to automatic staining.
- **OG6.** This is orange G in alcohol with tungstophosphoric acid. This stains the superficial cell cytoplasm a distinct but transparent orange colour.
- **EA (eosin azure).** There are several formulations of this stain and they are designated by numbers EA36–EA65. They all contain eosin, light green and tungstophosphoric acid and the slightly different formulae vary in the light green concentration and whether or not they contain Bismarck brown and lithium carbonate. The stains do need careful preparation and the commercially available solutions are often used since they give consistent quality.

Haematoxylin and eosin (H&E)

This is by far the most popular stain for histological preparations and is used by some cytopathologists as it is a familiar technique and matches more easily with histological appearances. Generally the cytoplasmic differentiation is poorer. It is sometimes considered better for specimens prepared by filtering the cells from fluid samples since it is less likely to give a heavy background colour with the filtered material.

Romanowsky methods

These are more commonly used for blood smears and as special stains in histology and can be used as a stain for cytology specimens. Giemsa, Jenner, May–Grünwald and Leishman are all examples and the formulation used is mainly personal preference. The nuclear detail with the Romanowsky dyes is not as good as with a haematoxylin nuclear stain but differentiation, particularly of non-epithelial cells, is good. Romanowsky smears are usually air-dried and fixed in methanol, rather than the usual wet smear fixation in ethanol.

Shorr's stain

This was the preferred stain for hormone assessment of vaginal smears but is rarely used now as it gives poorer nuclear detail and the cytoplasm is not as transparent as with the Papanicolaou stain. It does, however, distinguish well between the different layers of epithelial cells.

Stains for special purposes

There are many other stains that can be used for special purposes. The Perls technique will demonstrate haemosiderin and asbestos bodies; it is used in sputum cytology rather than cervical cytology. The Gram stain, Ziehl–Neelsen and others may be used to detect and identify micro-organisms. Periodic acid–Schiff is used to detect glycogen and fungi.

Liquid-based gynaecological cytology

The problems of efficient cell recovery directly from a spatula or brush have led to the development of a new technique for cervical sampling which, although still employing the scraping or brushing technique to remove cells from the cervix, uses a fluid to recover the cells. Several different commercial techniques of liquid-based cytology (LBC) are available often using a great deal of automation, which gives a more reproducible result; SurePath and ThinPrep systems are the two currently in use in the UK. Cells are collected from the cervix using a brush. The brush is transferred to a vial containing a liquid preservative transport medium. Cells are then dislodged from the brush manually into the transport fluid by simple rinsing into the fluid, and the vial containing the cells is then despatched to the laboratory. An alternative is to break off the head of the sampling brush into a vial of fluid and then despatch the vial and the head of the brush to the laboratory where mechanical vortex mixing can be used to dislodge the cells.

The suspended cells can then be separated from the fluid using centrifugation, with or without the use of density gradients – this is used in the SurePath system. This results in a cleaner preparation, as mucus and blood cells do not end up on the final slide and obscure the epithelial cells. An alternative is the ThinPrep system, which uses gentle suction through a filter to transfer the cells to a slide.

Because the cells have been suspended in a thin fluid they are more effectively separated than in a simple 'scrape and smear' preparation. The preparation disaggregates the cell clumps and this gives better separation between cells on the slide, which is almost a monolayer. Liquid-based cytology also removes most of the unimportant debris, which makes diagnosis easier (see *Fig. 13.2*). These two advantages of liquid-based cytology make the use of automated recognition possible. Although the preparation is generally clearer, liquid-based cytology does still leave some cells clumped to form 'hyperchromatic crowded cell groups' (HCCG). Most of these are likely to be normal or benign rather than

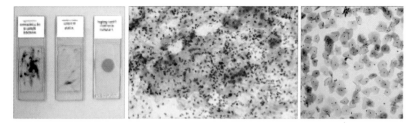

Figure 13.2
The photograph on the left shows three slides prepared for screening. A direct smear that
is too thick is shown on the left, a more evenly and thinly spread direct smear is in the
centre and a liquid-based cytology specimen is on the right. The first photomicrograph is
taken from the over-thick specimen and shows large clumps of overlapping cells and was
considered unsuitable for diagnosis. The photomicrograph on the right is taken from the
LBC slide and shows well-separated cells with little overlapping.

malignant, as malignant cells are less strongly adherent and therefore easier to separate.
However, HCCG cannot be ignored and may hide abnormal cells.

Both systems were investigated by the then National Institute for Clinical Excellence
(NICE) in the UK, which concluded that the benefits of liquid-based cytology were
sufficient for them to be used as replacements for the conventional technique. The
introduction of liquid-based cytology has been accepted and nationwide implementation
was completed in 2008.

Liquid-based cytology is hoped to improve the service by reducing the inadequate
smear rate (a pilot study showed a reduction from 9% to 1–2%) with no loss (and even
possible gains) in detection rates.

Normal gynaecological smear constituents

There are many materials and cells that can be found in normal gynaecological smears. Not
all of them will be present in every normal smear, so the absence of certain constituents
is usually not significant. The presence of some of these constituents is normal only if
they are present in small amounts or small numbers but if they are present in larger
amounts then it may be abnormal even though the constituent is considered normal for
the purposes of this list.

Stratified squamous epithelial cells

These cells arise from the stratified squamous epithelium of the vaginal or ectocervical
epithelium. This is the same type of epithelium that is found in skin. The epithelium is
quite robust and resistant to abrasion (see ***Box 13.7***). The epithelium is multi-layered
with the cells being produced in the basal layer, maturing in the middle layers and being
shed from the superficial layer (see ***Fig. 13.3***). It is this constant renewal from below
that makes the epithelium so resistant. At the base of the epithelium, there is a layer of
dividing cells. Following mitosis, one of the daughter cells will be separated from the
basement membrane and begins to change. The cells become flatter as they mature and
they also move gradually upwards through the epithelium as more cells are produced
below them. The surface cells are quite flattened and will be lost from the surface with
any mild abrasion. It is the upper layers that are sampled in cervical smears and although
terms such as 'basal cells' are sometimes used, they do not correspond with the very base
of the epithelium where the cells are firmly attached and would not be sampled by the
relatively mild abrasion with a spatula.

> **Box 13.7 // Squamo-columnar junction**
>
> The cervix protrudes into the vagina and is lined with columnar epithelium on its outer surface and in the canal leading from the cervix to the uterus (endocervical canal). The junction between the two varies in life. Before puberty (pre-menarche), the squamo-columnar junction is inside the endocervical canal. At puberty, the cervix enlarges and the squamo-columnar junction is now seen in the vaginal region. This results in a gradual, and sometimes patchy, replacement of the columnar epithelium with stratified squamous epithelium. The stratified squamous epithelium is a more resistant epithelium and is better suited to the vaginal environment. This replacement of the columnar epithelium is called **squamous metaplasia**. The appearance of the squamo-columnar junction in the vaginal region has been referred to by various terms including **ectopy, ectopion, erosion** and **cervical discontinuity**. This is not a disease but a normal phenomenon and can give rise to inflammation and bleeding in adult females of all ages.

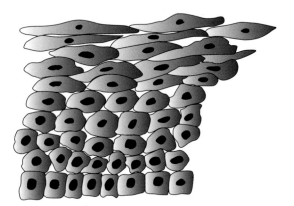

Figure 13.3
Diagram of the layered epithelium of the cervical surface. The cells are produced in the lowest (basal) layer and gradually move upwards

Skin-type epithelium can become keratinized if the upper layers accumulate large amounts of the protein keratin. Although stratified squamous cells in the cervix do have the potential to become keratinized, this does not usually happen in humans, although it does in other animals. The distinctions between cells and layers are not rigid since cells gradually change as they mature and the classes of cell are only convenient labels for identifying the stages in cell maturation. Every superficial cell has passed through all of the stages and the cells do not suddenly alter but change gradually so that some cells are in between categories. Some workers include extra types of cell (e.g. pre-cornified) but these are not included in this list.

Superficial cells. These are large (50–60 μm) cells. They have a very small central nucleus that stains very darkly. This condensed nuclear appearance is called **pyknosis** so the nucleus is called a **pyknotic nucleus**. Superficial cells also have an acidophilic translucent cytoplasm (pink or orange with the Papanicolaou stain), which is often angular or 'squared'. This is the most common type of cell in most smears (see ***Fig. 13.4***).

Intermediate cells. These are smaller (40–50 µm) cells. They have a larger central nucleus with finely granular chromatin. The cytoplasm is pale green with the Papanicolaou stain and the overall cell shape is angular or 'squared'. These are quite a common type of cell in many smears (see *Fig. 13.4*).

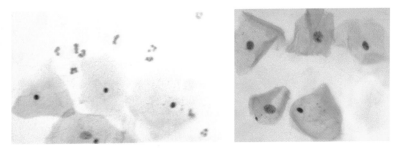

Figure 13.4
The left hand photograph of a cervical smear shows three superficial cells with pink cytoplasm and pyknotic nuclei. At the bottom is an intermediate cell with greenish cytoplasm and a larger nucleus. Also shown are a few neutrophil leukocytes with lobed nuclei. The two cells in the upper left of the right-hand photograph are intermediate cells with green-staining cytoplasm and large round nuclei. The cell at the bottom right shows a similar colour of cytoplasm but the nucleus is pyknotic, whilst the cell at the top right has pink-staining cytoplasm but a more open nucleus. This illustrates that the classification and labels are not absolute and cells go through all stages; the changes in staining and nuclear morphology are not instantaneous and do not necessarily occur together.

Parabasal cells. These are even smaller (20–30 µm) cells. They have a larger central vesicular nucleus with a coarser granular chromatin. The cytoplasm is a denser green with the Papanicolaou stain and the whole cell is rounded or ovoid. This is a less common type of cell in most smears (see *Fig. 13.5*).

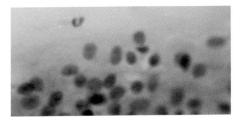

Figure 13.5
Parabasal cells stained with the Papanicolaou stain.

Basal cells. These are similar to the parabasal cells in most respects. They are even smaller (15–20 µm) cells but have a larger nucleus with a coarser granular chromatin, which may be central or slightly eccentric. The cytoplasm is a dense green with the Papanicolaou stain and overall the cell is rounded or ovoid. This is an uncommon type of cell in most smears since it is not normally shed from the surface.

Uterine columnar cells

Endocervical cells. These cells come from the endocervical canal, which is lined with ciliated columnar epithelium. Endocervical cells in smears often vary in how well they are preserved. Cells that are poorly preserved may cause confusion and difficulties in identification, especially for less experienced workers. They are smaller than the squamous cells (15–20 μm). The nucleus of these cells varies quite considerably in size, even in normal cells, so variation in nuclear size on its own should not be taken to indicate malignant change. The nucleus is often eccentric, usually at the base of the cell. The cytoplasm stains a green/grey colour with the Papanicolaou stain and may contain droplets of mucin in the upper part of the cytoplasm.

The top border of the cells may be ciliated. The cilia themselves can be seen in well-preserved cells but more often only the terminal plate below the ciliated surface is seen. These cells may occur in clumps. In side view, the clumps appear as a palisade of cells but in vertical view they appear as a honeycomb pattern. Degenerated cells may appear as little more than bare isolated nuclei.

Endometrial cells. These arise in the endometrial lining and are columnar cells but are usually degenerating since they cannot be sampled directly in a simple cervical scrape and may have been shed several days before and then travelled slowly down to the cervix before being collected and smeared. Endometrial cells are small (10–12 μm). The nucleus is quite large and is usually eccentric with a well-defined nuclear border; often one or two large chromatin clumps are seen. There is very little cytoplasm and this usually stains a green/grey colour with the Papanicolaou stain but may occasionally be pink. The cells may overlap or be bunched together. Endometrial cells are only considered as a normal constituent of cervical scrapes around menstruation; significant numbers at other times in the menstrual cycle should be noted and investigated. Endometrial cells may be more common in women fitted with a coil or other intrauterine device.

Blood cells

Erythrocytes. These are easily recognized, as they are strongly eosinophilic small cells (7–8 μm) without any nucleus. Erythrocytes are common around menstruation but may indicate bleeding if seen at other times or even from the trauma of sampling.

> **Box 13.8 // Significance of blood cells in smears**
>
> The explanations for finding blood cells are complex since they can represent menstruation, which is a normal condition. They can, however, also represent disease states such as inflammation and bleeding. Deciding which it is can be difficult and so smears from around the menstrual period should not be used for diagnosis.

Neutrophils or polymorphs. The full name of these cells is polymorphonuclear neutrophil leukocytes but this is usually abbreviated to neutrophils, polymorphs or even just polys. The name indicates their multiple-lobed nucleus (*poly* = many, *morph* = shape), which may have between two and four lobes. They are very easy to recognize by their small size and strongly basophilic lobed nuclei. Polymorphs are about 9–15 μm in diameter with a pale green/blue cytoplasm. Small numbers are normal but larger numbers are usually associated with inflammation and infection (see *Fig. 13.6*).

Macrophages (histiocytes). Macrophages are motile cells so they can be very variable in shape and this makes them difficult to identify. Macrophages are larger than other leukocytes (16–20 μm) with a large lightly stained eccentric nucleus. The colour of their cytoplasmic staining depends on how active they are and in particular whether they are

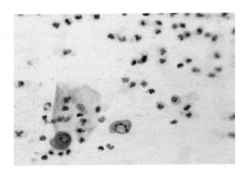

Figure 13.6
Smear showing signs of inflammation. There are large numbers of neutrophil leukocytes
with lobed nuclei and two plasma cells with characteristic clumps of chromatin at the
margin of the nucleus.

actively phagocytosing material and the nature of any phagocytosed material. The most
common cytoplasmic colour is a light green. The cytoplasm is often vacuolated and may
appear foamy if there are large numbers of vacuoles. The vacuoles may contain debris
(bacteria, degenerating cells, etc.) left over from phagocytosed material. Usually only a
few macrophages are present in a normal smear and large numbers are associated with
inflammation.

Lymphocytes. Lymphocytes are smaller cells (8–12 μm) with a large round basophilic
nucleus that occupies most of the cell. They have only a small amount of cytoplasm, which
is basophilic. They are rare in normal smears and if found in large numbers indicate
chronic inflammation. **Plasma cells** are produced from lymphocytes after they have been
stimulated by antigen. Plasma cells have a larger amount of basophilic cytoplasm and
an eccentric nucleus with prominent chromatin granules scattered around the nuclear
border. This may be referred to as a 'clock-face nucleus' (see *Fig. 13.6*).

Other constituents
As well as the cells coming from the epithelia and blood of the woman, there are a number
of other materials and contaminants that can occur.

Mucus. Mucus is secreted to keep the epithelium moist and prevent the cells becoming
desiccated. Mucins appear as translucent strands when stained with the Papanicolaou
technique but may stain more strongly with other stains. Only small amounts are usually
found in smears and because the small deposits and thin films are translucent the presence
of mucin does not interfere with cell recognition. If larger amounts of densely staining
mucus occur, it will stain more darkly and may obscure some cells making diagnosis
difficult. Large amounts of mucus may occur if there is inflammation present.

Fibrin. Fibrin is an insoluble fibrous protein and is produced from the soluble plasma
protein fibrinogen by the coagulation cascade. Fibrin appears as pink strands with
the Papanicolaou stain and is usually associated with bleeding. It is normal around
menstruation; at other times it usually indicates a haemorrhage but it can also indicate an
inflammatory exudate.

Doederlein (Döderlein) bacilli. These are lactobacilli and are a common commensal
organism in the female genital tract. They break down the glycogen that occurs in epithelial

cells and release lactic acid as a product of their metabolism. If bacteria are present in large numbers, they can cause cytolysis of intermediate cells, which therefore appear irregular or 'ragged'; these are often termed 'clue' cells and correspond to bacterial vaginosis. The lactic acid produced by these bacilli lowers the vaginal pH to around pH 4–4.5 and this helps to inhibit the growth of pathogenic organisms (see **Box 13.9**). Lactobacilli appear as small rod-shaped organisms (1–2 μm long), which may form chains. They stain pale blue with the Papanicolaou stain and are Gram-positive. Lactobacilli are common during secretory-phase smears and in pregnancy when glycogen is abundant in the intermediate smears.

Spermatozoa. These are simple to recognize as they have a dark-staining head with a pale cap (acrosome); the tail fibres are not always easy to distinguish. The presence of large numbers may make a smear unsatisfactory for diagnosis.

Box 13.9 // Lactobacilli in the vagina

The term commensal means that the organism is not causing disease. The presence of these bacteria is often beneficial in this case, so it is almost a symbiotic relationship. The loss of this **normal flora** may have serious consequences. If the woman is treated with a broad-spectrum antibiotic, it may kill off all of the lactobacilli and result in the growth of more-pathogenic organisms such as *Candida albicans*. This fungus causes the disease thrush and one treatment for thrush is to use a living culture of lactobacilli to re-establish a healthy and acidic environment. One such culture is live yoghurt, so yoghurt can be used to treat thrush. It must of course be a live yoghurt and not a pasteurized one. Although this 'organic' treatment can help, it is always wise for the patient to consult their doctor and not self-treat in isolation.

Normal changes in gynaecological smears

The appearance of cervical smears alters with time and with the physiological state of the woman. The main controlling feature of the changes is the levels of the hormones oestrogen and progesterone (see **Box 13.10**).

Oestrogen and progesterone levels in the blood vary on two time scales:

- **Short-term changes.** These occur in normal menstruating women with roughly a monthly cyclic change. Although the menstrual cycle is nominally a 28 day cycle, it can vary considerably in some women.

Box 13.10 // Therapeutic hormones

The effects of hormones are not limited to the natural hormones and will be mimicked by artificial hormones such as those in the contraceptive pill and hormone-replacement therapy. The normal menstrual cycle is only nominally 28 days in women and can vary depending on conditions. The use of the contraceptive pill usually results in a regular 28 day cycle because of the exogenous hormones so the cycle becomes constrained. Pheromones are scents that can control aspects of physiology and behaviour in another animal. Women sharing a flat as students often become entrained and will menstruate at the same time and this is believed to be due to pheromone influences. To get into step with each other, the length of the cycles must change.

- **Long-term changes.** Lifetime changes are associated with the onset of puberty (menarche) and the menopause, whilst a pregnancy interrupts the normal menstrual cycle for 9 months.

Cytology of the menstrual cycle

There are four main phases and two intermediate phases that can be identified:

1. Proliferative (oestrogenic) phase.
2. Ovulatory phase.
3. Secretory (progestagenic) phase.
4. Menstrual phase.

Menstruation is classed clinically as the beginning of the menstrual cycle since the onset of bleeding is the easiest point in the cycle to recognize. Menstruation is, however, really the last phase of the cycle when the body is removing the unwanted remains of the previous cycle and preparing the uterus for regrowth of the epithelium. The start of epithelial growth is a more logical start to the menstrual cycle but the simplicity and reliability of the menstrual blood flow as a marker makes timing and counting the days from this occurrence the best method for clinical purposes. Thus, the first signs of menstrual blood flow are counted as the first day.

Proliferative phase

The proliferative phase occurs from the 9th to the 14th day of the menstrual cycle. It is a period of increasing oestrogen levels in the blood and the oestrogen stimulates the epithelium of the cervix and uterus to grow and thicken. The stratified squamous epithelium includes more mature cells. A smear from this phase shows predominantly superficial cells (60% or more in the later stages) and there is a low number of polymorphs and histiocytes. The squamous cells separate well and so appear isolated and they usually lie flat on the slide with very little wrinkling or folding of the cells. This type of smear or the next type (ovulatory) is preferred for diagnosis as they are easier to screen because of the 'clean' smears produced.

Ovulatory phase

The ovulatory phase occurs around days 15 and 16 of the cycle and smears taken at this stage consist almost exclusively of superficial cells. Polymorphs and other background materials are mostly absent giving a very clean smear. This appearance indicates a high level of oestrogen in the blood but, although referred to as ovulatory, it cannot be taken to mean that ovulation has actually occurred.

Secretory phase

Following ovulation, oestrogen concentration in the blood declines and progesterone becomes the dominant influence from day 17 through to about day 24. The effect of this progesterone on the cells in smears is to decrease the number of superficial cells and increase the number of intermediate cells in the sample. The intermediate cells have more glycogen and are more likely to be folded and wrinkled than superficial cells. The epithelial cells often occur in clumps rather than being well separated. Lactobacilli become more noticeable both in numbers and in their cytolytic effect on the cells. Secretory phase smears also have more polymorphs, possibly due to the increased numbers of bacteria. The later stages of the secretory phase, from day 25 until the onset of bleeding on day 28, form one of the minor phases termed the pre-menstrual phase. The progesterone levels are beginning to fall. The appearance of smears is similar to those in the rest of the secretory phase although possibly more pronounced. In the last stages, some superficial cells may begin to reappear if the falling concentration of progesterone is accompanied by an early increase in oestrogen secretion from a developing Graafian follicle.

Menstrual phase

This covers days 0 to 5 of the menstrual cycle and is characterized by bleeding and sloughing of the epithelium of the uterus. During the menstrual phase, the appearance of the smear is dominated by red blood cells and white cells from the menstrual blood flow. Also carried with the blood are endometrial cells and mucus from the sloughed endometrium. Diagnostic examination of these smears is difficult as the cells are clumped, degenerating and obscured by the high numbers of red cells. Generally they are unreliable for diagnosis and a repeat smear should be done.

The 5th to 8th days of the cycle are sometimes distinguished as the post-menstrual phase where the numbers of red cells diminish, although white cells are still present in quite high numbers. Endometrial cells are more common in this late menstrual phase than in most other phases and the predominant squamous epithelial cells are intermediate cells.

Hormonal assessment of smears

Since the types of cell vary with the hormone content, it is possible to assess the levels of circulating hormone from the pattern of cells seen in the smear. The relative amount of hormone present can be assessed by counting the number of each cell type. Although modern biochemical tests can give a more accurate measure of the actual concentration of hormones, counting of cells can still be useful in correlating with hormone levels and may detect cases where, although there is a high concentration of oestrogen, the cells are not responding to the oestrogen. This may occur if the hormone receptors are defective in some way.

Three indices have been used in hormone assessment, as outlined below.

Maturation index

The parabasal, intermediate and superficial cells are counted and reported as a percentage of the total, e.g.:

Parabasal	Intermediate	Superficial	
4	27	69	= oestrogen effect
45	42	13	= low oestrogen

Eosinophilic index

This counts the number of eosinophilic cells, i.e. superficial cells. This measure is only accurate in the absence of inflammation and is very dependent on predictable and reliable staining of cells. In the example above, the eosinophilic index would be 69% and 13%.

Karyopyknotic index

The karyopyknotic index uses nuclear morphology to count superficial cells rather than their staining characteristics. The nuclei counted should be dark, homogeneous and less than 6 μm in diameter. This is also reported as a percentage.

Lifetime alterations

Pre-menarche

The onset of menstruation at puberty is referred to as the **menarche** so **pre-menarche** smears refer to smears taken from sexually immature patients. Children are not randomly screened so pre-menarche smears are only taken as diagnostic smears and only when the clinician believes there is a good reason for the investigation. Girls have very little sex hormone secretion before the menarche, so there is little stimulation of the epithelium and the lack of growth is termed atrophic. The appearance of an **atrophic** smear is one in which parabasal cells are the most common epithelial cell type, although leukocytes and commensal bacteria may also occur in the smears. This pattern persists until the onset of the menstrual cycle when intermediate and superficial cells begin to appear. The full regular cyclic activity may not appear, however, for a year or more following the menarche and the appearances of the smears may temporarily revert to a more atrophic state if a particular cycle is missed.

In the newborn during the first few weeks of life, the appearance of smears may be quite different due to hormones coming from the mother and placenta acting on the foetal epithelium whilst the child is in the uterus. This effect will disappear rapidly as the hormone levels fall in the first couple of weeks of post-natal life. The taking of smears from such young babies is extremely rare.

Pregnancy

The predominant hormone is progesterone so the pattern of cells seen in a smear resembles a late secretory phase smear but these appearances are more prolonged and the pattern becomes more established. The cells in the smear are often clumped and folded, with intermediate cells as the predominant type of cell present. The cells may show a **navicular** form. This refers to elongation of the cells with curling edges giving a boat-shaped cell with an eccentric nucleus.

Some superficial cells may be present, depending on the amount of oestrogen being secreted, but the numbers of superficial cells should be low in a normal pregnancy. If more than a third of the cells that are seen in a smear taken in the first trimester (the first third of the pregnancy) are superficial cells, then this is significant and should be reported to the clinician as it may indicate an early sign of abortion (miscarriage). This is most likely to occur when the production of progesterone switches from the corpus luteum to the placenta around weeks 10–12 of pregnancy. In the second and third trimester, the critical number is only 10–15% superficial cells and if a larger proportion of cells is seen, then it should be reported quickly.

Many of the intermediate cells in pregnancy have a high glycogen content and this encourages the growth of lactobacilli, which are therefore common in such smears. This may lead to large amounts of cellular degradation (cytolysis). Parabasal cells are uncommon in smears taken during pregnancy except in cases where there is inflammation. Trophoblastic cells from the placenta may sometimes occur in smears and are another possible indicator of the risk of a miscarriage.

Post-natal smears

Smears taken from a mother during the first week or so following delivery show the presence of erythrocytes, leukocytes and groups of **decidual** cells shed from the remains of the placenta. Parabasal cells are more common than the intermediate cells characteristic of pregnancy. A couple of weeks after the end of pregnancy, the appearance of the smear reverts to the more usual patterns and, depending on the hormones present, the appearance may vary from a secretory pattern to a late proliferative one.

Post-menopausal smears

The menopause is when the usual monthly menstrual periods stop. During this period, there is a gradual cessation of hormone secretion by the ovary but this decline may take months or years. The pattern of cells in the smear varies from a typical proliferative pattern to an atrophic pattern. Most smears will show a more transitional state with many intermediate cells. If a strong progesterone pattern is seen in a reportedly post-menopausal smear, the possibility of pregnancy should be considered.

The atrophic pattern developing shortly after the menopause is often associated with vaginal discomfort and the patient may need hormonal treatment to produce a thicker epithelium that is more resistant and less easily irritated.

Post-menopausal hormone treatment will of course alter the cellular pattern (see **Box 13.11**) so any hormone treatment needs to be known. Highly oestrogenic smears after the menopause in the absence of hormone treatment should always be investigated further.

Box 13.11 // Therapy can alter the appearances of smears from post-menopausal women

Hormones are also capable of acting on cells elsewhere in the body. The epithelial cells in the mouth and trachea also react in the same way as the vaginal and cervical cells but to a lesser extent. The loss of hormones during the menopause can occasionally cause severe problems with thinning of skin and drying of the mucosa and the person may need to have hormone-replacement therapy. The patient can be returned to the same condition as a pre-menopausal woman although the lack of ova prevents natural pregnancy. Women can become pregnant, however, by *in vitro* fertilization. Women into their sixties have become pregnant in this way. This use of hormones and other treatments can complicate the interpretation of smears from older women.

Inflammatory changes in smears

Inflammatory changes occur in a wide variety of conditions and diseases. The inflammation may vary in intensity from mild to severe and can be either short-lived (acute inflammation) or long-lived (chronic inflammation). Sometimes the cause of the inflammation may be directly visible and identifiable in the smear or there may simply be an unidentifiable inflammation present, in which case it would be reported as non-specific inflammation. The classical changes associated with inflammation in tissues are pain, redness, heat and swelling, but these are not, of course, visible in smears so inflammation is recognized by changes in the cells of the smear.

The changes that occur in smears can be considered as either destructive changes or reactive changes.

Destructive or degenerative changes

These occur when the causative agent of the inflammation induces degradation of the cells or of the tissues.

Cytoplasmic degradation. The destruction can be directly due to the bacteria or fungi causing inflammation, with the cells showing direct cytolytic effects; there may be loss of the cellular border giving a ragged edge to the cell (sometimes described as looking 'moth-eaten'). The cytoplasm may contain vacuoles and if there is perinuclear vacuolation it will give the appearance of a pale halo around the nucleus. In extreme cases, degradation may lead to all of the cytoplasm being lost, leaving bare nuclei free in the smear.

Cytoplasmic staining is often altered, with cells that would normally stain as basophilic now showing eosinophilia (e.g. parabasal cells may be red/brown). This eosinophilia is the most likely finding, although other altered staining can occur.

Phagocytosis may occur with the appearance of the engulfed material inside the cells.

Nuclear degradation. There may be pyknosis or the nucleus may fragment into small dense particles of chromatin (a process called karyorrhexis).

Reactive changes

Reactive changes are due to the inflammatory reaction in the tissue, rather than the direct effect of the bacteria. The most important reactive change is the production of inflammatory exudate. In inflammation, there is an increase in the permeability of blood vessels in the inflamed area. The increase in permeability allows protein-rich fluid to pass out of the blood vessel. This is the cause of the swelling (oedema) that is found in acute inflammation. The constituents of the exudate can vary. The most frequent finding is an excess of leukocytes, particularly neutrophils and macrophages (see *Fig. 13.6*). If there is bleeding as well as exudation, then erythrocytes will be seen even in non-menstrual smears. Fibrin strands may be present either from haemorrhage or from a protein-rich exudate.

Chronic inflammation differs slightly in that it is a much longer lasting change. The fluid exudate is usually much less than in acute inflammation and the cells will be mainly lymphocytes, plasma cells and macrophages. The macrophages may form large multinucleated phagocytic cells called giant cells.

In addition to the exudate, there may be reactive changes in the epithelium resulting in altered cell maturation and desquamation (shedding of cells). The normally non-keratinized epithelium may become hyperkeratotic (increased keratin) in response to continued irritation and this can result in white patches on the surface (leukoplakia). In hyperkeratosis, the cells appear as large flat cells without a nucleus although they may have a clear area in the centre of the cell where the nucleus has lysed. The cells are polygonal in shape and the cytoplasm is acidophilic and stains orange with the Papanicolaou method.

The agent causing the irritating can be physical, chemical or may be one of several infective agents (see below). The infection may be protozoan, fungal, viral or bacterial in origin. Many of these infections are considered to be sexually transmitted diseases, although this is not necessarily the only method of transmission (see *Box 13.12*).

Infective agents

Trichomonas vaginalis ('Trikes' or TV) is a unicellular flagellate that is found in up to 20% of females. In infections, it is associated with a yellow/green discharge that can be frothy and has an unpleasant odour. TV can be transmitted as an STD but infection from inanimate objects such as toilet seats is also possible. It can also occur in males so if an infection is found then both partners need to be treated. There is no longer thought to be a strong correlation between TV infections and a higher risk of cervical malignancy. Early reports of such a link were probably due to confusing factors such as promiscuity (which increases the risk of both TV and cervical malignancy). However, there is an established

> **Box 13.12 // Sexual transmission**
>
> Many of the diseases of the female genital tract can be transmitted by sexual intercourse. Such diseases were called venereal diseases but the term sexually transmitted disease (STD) is now preferred. Although a disease can be transmitted sexually, this may not be the only route of infection. Infection from inanimate objects such as towels, toilet seats, etc. is possible in some cases. Contaminated inanimate objects that can transmit disease are sometimes called **fomites**. Other diseases are almost totally restricted to direct sexual transmission and cannot be caught from inanimate objects, as the organisms are unable to exist away from the human body. An example of this is syphilis, which is almost impossible to catch except by direct, usually sexual, contact. It requires tact to tell someone or their partner that they have a disease that can be sexually transmitted as there can be the hidden suggestion that one or other partner has been involved sexually with a third party. This can lead to emotional problems for the people involved. It is essential that the possibility of non-sexual transmission is explained.

link with ectopic pregnancy and sterility caused by obstructed Fallopian tubes and TV infection.

TV stains rather poorly with the Papanicolaou stain and the flagellae are rarely visible. They appear as pear-shaped faint grey/green structures (8–20 μm) with an eccentric nucleus ('Mongol's eye'). The organisms can be seen more easily in wet preparations using phase-contrast microscopy where they can also be seen to be motile. If necessary the organism can also be cultured to confirm its presence.

The presence of TV can often be inferred even when the organism itself is not seen, since it usually causes a characteristic red staining of the cytoplasm of the epithelial cells. Vacuolation and perinuclear haloes are also common. The inflammatory changes in the smear are often severe. Some cytopathologists have reported the reactive changes as 'appearances suggestive of *Trichomonas* infection'.

The changes seen with TV may mimic some of the signs of early malignant change so the patient should be treated to get rid of the infection and a repeat smear taken so that dyskaryotic changes are not missed.

Candida albicans (monilia, thrush) is an eosinophilic yeast that may appear as long eosinophilic threads ('pseudohyphae'). Squamous cells often appear to cluster around these pseudohyphae in smears. *Candida* is a common finding in smears and may behave as a commensal or a pathogen depending on the conditions. In heavy infections, there should be treatment of both the female and any male partners. It occurs mainly in reproductive life, more frequently in pregnancy and following antibiotic therapy. It does not seem to be associated with the modern low-oestrogen pill formulations but high-progesterone pills may increase the risk of infection.

Herpes simplex virus is a virus that infects humans and has several variants. Herpes simplex virus 1 (HSV-1) is usually associated with oral infections (cold sores or fever blisters), whilst HSV-2 is more common in genital infections and is considered an STD. It causes ulceration of the cervix. If infection occurs during pregnancy, it can be transmitted to the foetus with fatal results (foetal malformation and miscarriage). This usually only applies to women acquiring the infection during pregnancy; once infected, the antibodies in the maternal blood usually protect the foetus. There is a higher incidence of anti-herpes antibodies in women who develop cervical malignancy than there is in the general population. There is no direct evidence that herpes virus is actively involved in malignant change and there is often no sign of herpes DNA or any viral antigen in the malignant cells.

Smears show characteristic changes during the active phase of a herpes infection but simple morphology is not as reliable as cultures. The most characteristic cell (syncitial) seen during an active infection is a large multinucleated (up to 30 nuclei) giant cell (50–200 μm). The chromatin is often most prominent around the margins of the nucleus, which makes the nuclei well defined and results in a 'ballooned' appearance with the empty centre of the nucleus having a ground glass appearance. The virus may cause eosinophilic viral inclusion bodies in the nucleus but this depends on the stage of infection and is not always present. Confusion with malignant change is possible, so accurate recognition of the changes caused by herpes virus is important. Immunological techniques using anti-HSV antibodies to confirm herpes infection may be useful in some cases.

Human papilloma virus (HPV) is the cause of genital warts and is also passed on through sexual intercourse. Having multiple sexual partners increases the risk of infection and thus the risk of developing malignancy. There is a long delay, sometimes in excess of 20 years, between persistent infection and the development of malignancy, so the rise in the incidence of cervical cancer to a peak in 1985–87 has been attributed by some to the more sexually permissive attitudes following the availability of the contraceptive pill. From this peak the rates of cancer have declined and this has been attributed to the introduction of the nationwide cervical screening programme, and its detection and effective treatment of precancerous conditions (see *Box 13.13*). There are many variants of HPV (more than 100 have been recognized), with each type having a characteristic pathological condition.

Box 13.13 // Vaccination controversy

Cervical HPV infection only occurs by infection from an infected sexual partner. Before the introduction of vaccination, there was controversy as to whether it was needed and if it would lead to an increase in promiscuity and earlier sexual activity. Religious groups and some parents argued that promoting faithful monogamous sexual relationships was a more appropriate response to minimize cervical cancer. This echoes the controversy when effective cures became available. Resistance to vaccination has proved to be a minority view, with the uptake of vaccination of girls exceeding 80% in some areas.

The suggestion that vaccination should also be offered to boys was equally controversial. In males, HPV infection was thought to cause only genital warts and not high levels of malignancy. Male vaccination would reduce the rate of infection in women by increasing herd immunity. Provided that female uptake of vaccination was high, the further reduction in rates of infection by male vaccination would be small. Therefore male vaccination was not introduced at the same time as female vaccination.

More recently it has become apparent that the incidence of male throat cancer has more than doubled since the 1990s. The proportion of cancers linked to HPV infection has risen from 35% to over 70% in the same period. HPV infections in the oropharynx are strongly linked to oral sex. The incidence of oropharyngeal cancer is predicted to exceed cervical cancer during the 2020s. This has prompted renewed calls for male vaccination and this would protect both sexes.

Fifteen types have been found in the vaginal environment and some may predispose the person to cervical malignancy. Two forms have been identified with a high risk of cervical cancer (HPV-16 and HPV-18), whilst some others are less strongly linked but still possible factors (e.g. HPV-31 and HPV-33). HPV-16 is more strongly associated with squamous epithelium and HPV-18 with glandular epithelium of the cervix. Both are associated with head and neck malignancies.

HPV infections usually disappear after 6–9 months but 10–20% of infected women remain infected for extended periods. It is the persistent infection that is linked to malignancy. Thus, a single positive test is not significant in itself but is cause for an earlier recall. Persistent HPV infections have been found in >90% of cervical cancers. In the absence of infection, disease rates are very low. Vaccines for the high-risk types of HPV have been developed and approved for use. These vaccines use only the viral coat (L1 protein) without any viral DNA and so are not capable of infecting cells. The UK Joint Committee for Vaccination and Immunisation recommended that HPV vaccination should be available for all girls before they become sexually active and in 1988 vaccination was made available to girls from the age of 12.

HPV testing. The presence of HPV infection can be determined by detection of viral DNA using, for example, hybrid capture techniques. More details of DNA detection can be found in *Chapter 16*. HPV testing for several different HPV strains is now used as a triage step in the diagnosis and management of cervical malignancies. This was adopted following several trials to determine the benefits of HPV testing. Women who show abnormal morphology are tested for HPV. If the HPV test is positive, the patient is referred for immediate colposcopy (see *Box 13.14*). Depending on the colposcopy results, the patient is treated or monitored with a 12-month follow-up recall. If the patients are HPV negative they are returned to routine 3- or 5-year screening.

The full protocol for the HPV triage can be found on the website of the NHS screening service at www.cancerscreening.nhs.uk/cervical/hpv-triage-test-flowchart.pdf

Some studies suggest that molecular HPV testing may be more reliable than morphology alone. The sensitivity of HPV tests may be as high as 95% (current morphology targets are 85% sensitivity) and the negative predictive value is >90%. In a sample of 11,000 women,

---

Box 13.14 // Colposcopy

Following an abnormal smear the patient may be referred for colposcopy. This involves a colposcopist examining the cervix microscopically. The patient lies on her back and the vagina is opened using a speculum. A specialized microcope (the colposcope) is used to examine the surface of the cervix for visible abnormalities. The colposcope has a very long working distance and the examination does not require the colposcope to be inserted into the vagina. Magnifications used are typically ×2–×25, which allows for good visual examination. A dilute solution of acetic acid (aceto-white) is used to help visualize abnormalities; an iodine solution may also be used.

If an abnormality is seen, a small biopsy may be taken using biopsy forceps. Alternatively a LLETZ (large loop excision of the transition zone) may be made, in which a loop of heated wire is used to remove a cone of the cervix. A larger cone biopsy, using a scalpel rather than hot wire, may be done under general anaesthetic.

Cone removal may be curative, with no further treatment needed, or following biopsy more aggressive therapy may be indicated.

HPV testing picked up 97% of anomalies, whilst standard cytology detected only 76%. Combined negative predictive values for morphology and HPV testing approached 100%. Both the standard morphological testing and the HPV screening can be done from the same liquid-based cytology sample, making the process simpler and less stressful for the patients.

The most characteristic feature of HPV infections is the presence of **koilocytes**, which have a distinct perinuclear halo. There is usually some abnormal keratinization associated with the infection. Following a diagnosis of HPV warts, it is probably wise for the patient to have annual cervical smears because of the increased risk of cervical malignancy. Again, like HSV, the use of antibody techniques and in some cases nucleic acid hybridization methods may be useful.

Many other infections may occur from bacteria (*Leptothrix, Gardnerella, Neisseria, Treponema* and *Mycobacteria*) and other infectious agents including actinomycetes and chlamydia.

Malignancy

The diagnosis of malignancy has two related aspects. One is the recognition that the cells have changed from their normal growth pattern and become malignant. This change is often called transformation. The other diagnostic task is to recognize exactly which type of malignancy is present. The cytological recognition of malignant transformation is most obvious in the nuclear abnormalities of the transformed cells. Cytoplasmic changes are frequently present in malignant cells but are less useful in recognizing that malignant change has occurred. The appearance of the cytoplasm is more important in determining which type of malignancy is present in the smear.

The changes that are found in gynaecological smears are the same changes as those described in **Chapter 2**, but they are reiterated and expanded in this chapter.

Nuclear changes (see Fig. 13.7)

Disturbed chromatin distribution. The pattern of dark heterochromatin and pale euchromatin is usually regular and distinctive. Malignant cells usually have a less regular and coarser chromatin pattern in the nucleus. This finding is one of the more significant findings and would be expected in most malignant cells.

Alterations in size. Nuclear enlargement is common but some tumours with small malignant nuclei do occur. Simple alteration in size can be difficult to spot but it is often easier to see the change in the relative proportions of the nucleus and the

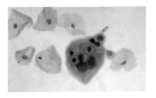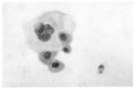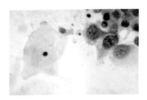

Figure 13.7
Cells from cervical smears showing abnormalities indicative of pre-malignant or malignant change.

cytoplasm, so the altered size is often referred to as an increased nucleocytoplasmic ratio. Humans are much better at recognizing relative changes than identifying absolute size changes.

Increased DNA content in the nucleus. The increased amount of nuclear material (chromatin) may not always be apparent if the nucleus is greatly enlarged. The enlargement of the nucleus means that the chromatin is more spread out giving a large pale nucleus. The increased DNA content comes from two sources. One is that malignant cells are usually actively dividing and, as part of the mitotic activity, the DNA in the cell must be duplicated before the cell divides. Thus, many of the cells will be synthesizing new DNA prior to division and as a consequence they may have twice as much DNA as other cells. The malignant cells may also have extra chromosomes present over and above the normal diploid 46.

Prominent nucleoli. Although abnormal and not always present in malignant cells, they are quite significant. The nucleoli may be enlarged or irregular in shape and there may be more than one present in a single cell. The increased mRNA production of these abnormal nucleoli is indicative of the increased protein production characteristic of cell proliferation.

Abnormal nuclear outline. The nuclear border in malignant cells is often quite irregular and may be less well defined than in normal cells. With a pale nucleus, an irregular and an ill-defined border, the nucleus may appear to merge with the cytoplasm rather than being distinctly separated from it.

Multinucleation. The presence of more than one nucleus in a cell is a significant finding but they must be distinguished from non-malignant cells that can be multinucleate such as giant cells.

Mitosis. Cell division is usually more frequent in malignant tumours than it is in normal tissues and mitosis may occur in superficial cell layers that are normally non-mitotic. This may mean that cells in mitosis may appear in the smear. If an irregular mitosis is seen, e.g. a tripolar mitotic figure, this is even more significant.

Abnormal staining. The malignant cells often have quite odd colours. The cells may be darker than normal (hyperchromatic), paler than normal (hypochromatic) or stain in different colours to normal (polychromatic).

Cytoplasmic changes

Cytoplasmic changes are less important in recognizing malignancy but the sorts of change that occur include variation in the total cell size, variation in shape of the cell, abnormal cytoplasmic vacuolation and abnormal cytoplasmic staining. Malignant cells may also lose the characteristic features that make them identifiable as a particular cell type. This loss of differentiation is an important feature. The less well differentiated the cells are, the more likely it is that the tumour is an aggressive tumour and is becoming invasive and spreading.

Different descriptive terms used to describe the changes

The recognition and reporting of malignant changes is complicated by various factors and varies from centre to centre (see **Box 13.15**). There is no absolute test of the degree of malignancy and the diagnosis is made on the basis of experience.

> **Box 13.15 // Problems with missed diagnoses**
>
> The missing of abnormal cells is a serious problem. The target rate for detecting abnormalities with conventional smears was set at between 85 and 95%. In 1996, there was a case (The Times, 6 Feb 1996) where one laboratory fell below this expected standard, resulting in the rescreening of 70,000 smears at a cost of £140,000. It was expected that 350 previously undiagnosed abnormalities would be detected although none of these was expected to be full malignancy. However, the scale of the problem at that laboratory may have been underestimated. There are reports of five invasive carcinomas, 15 hysterectomies and five deaths in women involved in the screening programme. Many of these women are now involved in litigation against the hospital and the hospital is no longer screening smears.
>
> In 1997 (The Times, 6 Feb 1997), a woman was awarded £60,000 following a missed diagnosis that resulted in her losing a child when she became pregnant.

Malignancy does not happen as a single change but as a gradual development. There is a continuous spectrum of changes from mild changes (which may never progress to a full malignancy) to invasive tumours (which are life-threatening). There are no sharp breaks in this spectrum of changes, so deciding into which category a particular smear should be classed can be difficult.

The reporting phrases also differ even though they refer to the same condition and this may be confusing. The main categories of description are:

- **CIN.** This stands for cervical intraepithelial neoplasia and the degree of severity is indicated by numbers CIN1 to CIN3 (with three being the most severe). The only way to be sure it is intraepithelial is to examine a histological section, so this is a histological description rather than a cytological one.
- **Dysplasia.** This is also a histological description and relates to the changes in cell maturation within the epithelium. Severity is usually indicated by words (mild, moderate and severe).
- **Dyskaryosis** (originally introduced by Papanicolaou). This is a cytological description of the nuclear changes and indicates severity either using the same phrases as dysplasia (mild, moderate and severe) or by relating it to the cytoplasmic appearance of the cells (superficial cell dyskaryosis, intermediate cell dyskaryosis and parabasal cell dyskaryosis). In addition, smears may be classified as negative, borderline, ?glandular or ?invasive. This is the classification favoured by the British Society for Clinical Cytologists (BSCC).

In the USA, the Bethesda system 2001 is used to classify the changes into only two categories: low-grade squamous intraepithelial lesions (LSIL) or high-grade squamous intraepithelial lesions (HSIL). The classification also includes negative and atypical squamous cells (ASC), which may be qualified as 'atypical squamous cells of undetermined significance' (ASC-US) or 'cannot exclude HSIL' (ASC-H).

Within the limits already mentioned, the use of CIN, dysplasia and dyskaryosis correspond reasonably as follows:

CIN1	Mild dysplasia	Mild dyskaryosis
CIN2	Moderate dysplasia	Moderate dyskaryosis
CIN3	Severe dysplasia	Severe dyskaryosis (and carcinoma *in situ*)

The Bethesda 2001 classification is less directly comparable to the system above. The LSIL classification covers HPV cytopathic changes and the CIN1/mild dysplasia category, whilst HSIL covers the CIN2 and CIN3 and carcinoma *in situ* categories.

The cytological changes corresponding to the three categories are outlined below (also see ***Fig. 13.7***).

Mild dyskaryosis. Cells resemble superficial or intermediate cells by having abundant cytoplasm with angular outlines. The nucleus may show slight irregularities but occupies less than half of the cell.

Moderate dyskaryosis. The nucleus is more enlarged and may stain irregularly and the nuclear border may be less distinct and irregular. The enlarged nucleus occupies one-half to two-thirds of the cell.

Severe dyskaryosis. Only a thin rim of cytoplasm is usually present. The cell is largely occupied by an irregular nucleus. Sometimes the cytoplasm is larger but there is extra cytoplasm present. The appearance can resemble the nucleocytoplasmic ratio in mild dyskaryosis but the nucleus in severe dyskaryosis is actually much larger than in mild dyskaryosis.

The severity of the disease is judged purely on morphological grounds. The number of cells is not significant. A single cell is just as significant as hundreds, provided the cytological changes are the same. The number of cells seen to be abnormal is as much an indication of the sampling efficiency as of the severity of the disease.

The term **carcinoma *in situ*** refers to the stage when the cells are showing all of the cytological signs of full malignancy but have not yet begun to invade or metastasize through the underlying basement membrane. This is a histological diagnosis since it is not possible to tell whether invasion has occurred from isolated cells.

The final stage of the diagnostic spectrum is **invasive carcinoma** where the cells have begun to spread beyond the epithelium and into other tissues. This again is a histological diagnosis but a smear will often have cells that indicate that invasion is occurring, so the smear can be reported as 'suspected invasive carcinoma'. The change from the pre-malignant state to the frankly invasive malignancy is usually marked by the appearance in the smear of bizarre cell shapes of which the **tadpole** and **fibre cells** are the most distinctive. Tadpole cells are named from their shape and have a malignant nucleus in the 'head' of the tadpole. Fibre cells are excessively elongated and often have an elongated hyperchromatic nucleus. The cytoplasmic staining is irregular in both cell types.

Other types of malignancy apart from squamous cell carcinoma may also occur (e.g. adenocarcinoma) but their appearance will not be dealt with in this account.

The other report that can be made is for an 'inadequate' or 'unsuitable' smear. An inadequate smear is one that cannot be satisfactorily diagnosed. The inadequacy may be due to one or more of the following:

- Too few cells in the smear to be reliably diagnosed as negative.
- The presence of red cells that are obscuring detail.
- The presence of pus that is obscuring detail.
- Air-drying or inadequate fixation. If the cells have been badly treated during preparation, a reliable diagnosis is not possible.

Assessment of the suitability or otherwise of a smear is subjective and again varies with the laboratory. If the smear is unsatisfactory, then the reason why it is unsatisfactory should be reported back to the requesting doctor and a repeat smear obtained as soon as possible. It is hoped that the introduction of the liquid-based cytology systems will result in a much lower rate of inadequate smears.

Suggested further reading

Bancroft, J.D. and Gamble, M. (eds) (2002) *Theory and Practice of Histological Techniques*, 5th edn. Churchill Livingstone.

Coleman, D.V. and Chapman, P.A. (1989) *Clinical Cytotechnology*. Butterworth.

Hopwood, J. (1995) *Background to Cervical Cytology Reports*. Schering Health Care Ltd.

National Institute for Clinical Excellence (2003) *Guidance on the Use of Liquid-based Cytology for Cervical Screening*. Available at www.nice.org.uk/guidance/TA69

Shambayati, B. (2011) *Cytopathology* (Fundamentals of Biomedical Science Series). Oxford University Press.

Self-assessment questions

1. What are the characteristics needed for a screening method? Outline the properties of exfoliative cytology that make it useful for screening.
2. What staining method is used for gynaecological smears? Why is this method preferred to other staining methods?
3. Which of the following are normal smear constituents, which are abnormal constituents and which are normal but only under specific circumstances? (a) Superficial cells; (b) parabasal cells; (c) *Trichomonas vaginalis*; (d) dyskaryotic cells; (e) koilocytes; (f) erythrocytes; (g) plasma cells.
4. What is meant by a navicular cell and when would you expect to see them in smears?
5. Which stage(s) of the menstrual cycle are best for taking smears and which are not suitable?
6. What effect does acute inflammation have on a vaginal smear? How does this differ from chronic inflammation?
7. Name four common infectious organisms that can cause inflammation.
8. What do the initials CIN stand for?
9. List the characteristics that would make a smear unsuitable for diagnosis.

14 Non-gynaecological cytopathology

Learning objectives

After studying this chapter you should confidently be able to:

- **Describe the methods for preparing fluid samples for cytopathology**
  Cells can be harvested and concentrated by simple centrifugation, the use of a cytocentrifuge or by filtration. The cells can be prepared as a smear sample or processed to give a cell block.

- **Discuss urinary cytopathology and describe the cell types encountered**
  Urine samples can be used to detect bladder and other cancers. Urine has a low cellular content so large amounts need to be used for concentration.

- **Discuss respiratory cytopathology and describe the cell types encountered**
  Samples from the respiratory tract can detect infections, primary malignancies of the respiratory tract and secondary tumours which have metastasized to the respiratory tract. Samples can be obtained as sputum samples which may be used for direct smearing or concentrated following thinning of the mucus. Samples can also be obtained using a bronchoscopy brush and cells prepared by either direct smearing or liquid-based cytology (LBC) techniques.

- **Discuss effusion cytopathology and describe the cell types encountered**
  Effusions arise in the body's serous cavities (peritoneal, pleural and pericardial cavities) and the fluid can be collected and treated as for other fluid samples.

- **Outline the use of fine needle aspirates**
  Fine needle aspirates are used to obtain fluid samples by inserting a fine needle into a fluid-filled cavity and aspirating the fluid.

14.1 Preparation techniques using fluid samples

As well as the direct scraping technique there are a variety of other samples, mainly non-gynaecological, that can be made into cytology smears. The following are a few of the more common ones.

Fluid samples

Many body fluids contain exfoliated cells and can be used to make smears and/or LBC slide preparations. Fluid can be collected by inserting a fine needle into a fluid-filled area (see *Box 14.1*). Fluid samples include pleural and peritoneal aspirates, urine, CSF, amniotic fluid, breast cyst fluids and breast (nipple) discharges.

They vary in volume from about 1 ml to more than 1 litre. They can vary in consistency from watery urine to more viscous discharges and they vary in their cell content from very sparse to heavily cellular. A significant number of epithelial sites can also be accessed and sampled using gentle washing (lavage) with sterile PBS solution. The procedure is controlled endoscopically in order to identify a definite site of abnormality.

> **Box 14.1 // Fine needle aspirates**
>
> Fine needle aspirates are not really 'exfoliative' cytology since they involve introduction of a needle into tissue under local anaesthetic to withdraw cell samples from small nodules that have been detected by palpation or some form of imaging (X-ray, ultrasound or magnetic resonance imaging). The needle can be guided by ultrasound into the lesion. The use of fine needle aspirates is increasing and is a great advance in the management of breast tumours but is also used for thyroid and lymph node lesions. The cells recovered by these techniques can be examined by the same methods that are used for other smears, so the specimens are examined by cytologists along with the more conventional exfoliated cell smears.

Not surprisingly, then, there are several different ways of preparing smears from such varied fluid samples.

Heavily cellular specimens can be directly smeared but most fluids require some method of concentrating the cells. Concentration of fluids can be achieved by centrifugation or filtering.

Centrifugation

Centrifugation may be done in one of two ways:

Simple centrifugation. This involves putting the fluid sample into tubes and then centrifuging them (e.g. 2500 rpm for 5 min). Centrifuge tubes can be quite large so reasonably large quantities can be concentrated in this way. Following centrifugation, the supernatant is carefully decanted off, leaving the deposited cells in the bottom of the tube. The tubes are drained and the cells resuspended in a few drops of fluid. The concentrated suspension can then be smeared using a wire loop or a pipette and the smears are fixed in the usual way. Very watery fluids may not have sufficient protein to allow the cells to adhere and may need the addition of protein such as serum or albumen to aid adhesion of the cells.

Cytocentrifugation. This involves a centrifuge (see ***Fig. 14.1***) which takes a specially designed Cytofunnel that spins small (about 0.5 ml) samples of fluid directly onto the slide (see ***Fig. 14.2***). A larger version using a Megafunnel allows volumes up to 6 ml to

Figure 14.1
A cytocentrifuge is shown on the right with the rotor and fluid chambers on the left.

be processed. This has proved very effective, although early techniques did not always give good cell yields because cells were lost before being spun. The discrete pellet of cells produced by the cytocentrifuge makes screening easy and cell yields are now very good. Again, watery fluids may need additional protein to aid adhesion. With the cytocentrifuge it is critical to have the correct amount of cells in the fluid sample. If there are too many cells in the sample, they will produce several layers of cells instead of a monolayer, whilst if there are too few cells in the small volumes used in the cytocentrifuge, the smear will be very sparse. The samples may need to be diluted or concentrated before use. Concentration would normally be done by simply centrifuging a larger sample, as outlined above. Multi-channel sample holders allow several cell samples to be processed simultaneously, and this produces up to eight discrete areas of cells on a slide. The prepared slide can then be stained as required and as all the samples are on a single slide, the same standardized result for all the samples is achieved.

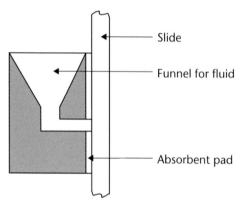

Figure 14.2
Cytocentrifuge chamber. The centrifugal force would be from left to right forcing fluid out of the funnel on to the slide

Filtering

This involves using a specialized plastic membrane filter which can be made translucent if immersed in xylene. The fluid containing the cells is placed in a funnel with a filter and the fluid is gently sucked through the membrane using reduced pressure. Filtration is suitable for concentrating large volumes of fluid containing very few cells because all the cells are trapped on top of the filter.

The cells are usually stained whilst still on the filter, which becomes reasonably transparent in the mounting process. Staining and processing the filters is comparatively simple and the filters are mounted on a slide under a coverslip. The filter does obscure the cells slightly and cells may be distorted by the filter, especially if too much pressure is applied (see *Fig. 14.3*).

Figure 14.3
Filter with a cell trapped on the upper surface

This gives very high recovery rates for cells and can be carried out with fixed cells. This is useful if specimens need to travel some distance to the laboratory because the specimen can be stabilized and rendered less of a health hazard by adding fixative to the fluid sample before despatch. The filters effectively trap epithelial cells but smaller materials such as bacteria and blood cells may pass through the small pores in the filter. The filters are not suitable for Romanowsky staining methods which tend to stain the background. The filters are not suitable if there are many cells because with a high cell content the filters will quickly become clogged. Filtration is an essential part of some liquid-based cytology, which is replacing traditional gynaecological scrapes.

Thick fluid samples

If a fluid sample is very thick, it may be thinned to allow easy separation of cells. Thinning can be done by enzyme digestion, chemical thinning (e.g. dithiothreitol), HCl (8%) or mechanical disruption (e.g. by ultrasonic treatment).

Specimens that are heavily contaminated with blood may be similarly cleaned up using acetic acid to lyse the red blood cells or enzymes such as streptolysin, but separation by careful centrifugation is probably better as it produces fewer problems with staining and later cell identification. The red blood cells are denser than other cells and so get pushed to the bottom of the tube with a thin layer of other cells on top (the 'buffy layer') allowing separation of the red blood cells from the other cells. The separation can be enhanced by **density gradient centrifugation** using a flotation material such as Ficoll or Histopaque. The flotation material is prepared to have a density between that of red blood cells and the epithelial cells. When it is used in a centrifuge, the red cells, being denser than the medium, go to the bottom of the tube but the other cells are less dense and float in the medium instead of being deposited along with the red cells. The cells from the buffy layer can be removed by careful pipetting.

Cell blocks

Liquid samples containing cells can be used to prepare paraffin wax embedded blocks which can be sectioned and used for simple H&E morphology or for immunohistochemistry. After preparing smears using one of the previous techniques, the unused cellular material can be processed into a block. Centrifugation of the fluid will result in the cells forming a pellet at the bottom of the centrifuge tube. The cells can be simply fixed in the tube using a standard histological fixative. The pellet is then gently removed and wrapped in paper and processed in a cassette by routine processing. Rapid processing schedules are particularly appropriate as the pellet is so small. This only works well if there is sufficient protein in the specimen to keep the cells together as a pellet.

The cells can be resuspended in a protein matrix before processing to ensure the cells remain as a single block. The protein is added to the pelleted cells and then solidified before fixing. Commercial kits are available or a plasma fibrinogen solution can be used and then coagulated with thrombin, as in blood clotting. The resulting fibrin clot can then be fixed and processed. Warm gelatine or agar solutions can similarly be used and will set into a firm gel on cooling. A suitable agent is one that provides solid support after fixation and during sectioning but does not interfere with processing or produce significant background staining that would confuse or obscure areas of interest.

In the following sections there will be a brief outline of the application of cytological techniques to different systems. It is not possible to fully consider all the different diseases found in these cases. If more information is needed, specialized cytopathology books and articles should be consulted.

14.2 Urinary cytology

Urine may contain cells that have been spontaneously shed from the urogenital tract. Although cells can potentially come from a variety of sources such as the kidney, ureter, prostate and seminal vesicles, the major cell type and the main malignant tumours recognized in urine samples come from the transitional epithelium of the bladder. Bladder cancer is the fourth commonest cancer in men but only the eleventh commonest in women; men are 2.5 times more likely to develop it than women, with a total of over 8000 new cases each year.

Table 14.1 Cellular and non-cellular constituents of urine samples

Origin	Cell type	Appearance
Transitional epithelium of bladder and urethra	Urothelial cells (superficial and basal)	Single cells (clumps less common) with dense cyanophilic cytoplasm.
		Basal cells are smaller and have a higher nucleo-cytoplasmic ratio.
Stratified squamous epithelium	Squamous epithelial cells (superficial, intermediate types)	As for cervix (more common in female smears).
Male urinogenital glands	Glandular cells	Large dark-staining nuclei but show a variety of forms (pleomorphic). Cytoplasm may contain lipofuscin granules.
Prostate		
Seminal vesicles		
Testis	Spermatozoa	Small homogenous densely stained heads (nucleus). Long flagellae which may be twined together in clumps.
Blood-derived cells		
	Red blood cells	Anucleate biconcave discs
	Neutrophils	Multilobed nucleus
	Macrophages (histiocytes)	Round or indented (bean-shaped) nucleus. Usually single nucleus but occasionally multinucleated.
Non-cellular materials		
Renal tubules and collecting ducts	Casts	Cylindrical structures formed inside the renal tubules and then passed into the urine.
	Acellular	
	Erythrocytic casts	Acellular are normal but others may indicate disease.
	Leukocytic casts	
	Epithelial cell casts	
Urine solutes (crystallization depending on pH, temperature, etc.)	Crystals	Typical crystal angular shape depending on chemical nature.

The bladder is at risk from many carcinogenic chemicals, as the kidney filters the blood and concentrates the urine. Any chemical passing through the kidney will be more concentrated than elsewhere in the body and as urine can be stored in the bladder for several hours, the urothelial cells have a high level of exposure. The biggest single factor in bladder cancer is smoking but industrial workers are exposed to a wide variety of chemicals which may be carcinogenic, e.g. β-naphthylamine in the rubber industry. Lower risk factors include infections such as schistosoma parasites (the cause of bilharzia).

Urinary cytology is used to investigate symptomatic cases but has also been used in industry as a screening test to monitor workers.

Urine collection

The most obvious way to sample is simply to collect voided urine, and the vast majority of samples are collected in this way. For bladder infections (cystitis) a mid-stream urine sample is recommended, but for cytology the first and last samples may be higher in cellular material. Many of the cells in an early morning urine sample will have been lying in the bladder for several hours and may show degradation. The second urine sample of the day may be more representative and give better cellular morphology. Unless the urine is being processed immediately it should be mixed (approximately 50:50) with preservative. Catheters can be inserted into the bladder via the urethra to drain off urine. This is rarely done purely for sampling but many patients are catheterized as part of their treatment, with urine collecting in a plastic bag outside the body. Urine can simply be sampled from the bag but again cell deterioration is likely. Bladder washings, where fresh fluid is injected via a catheter and then aspirated, are possible but are rarely done. Using a brush to collect cells from endoscopy specimens is a specialized technique and is less common.

Preparation of urine samples

Urine is too dilute to smear directly so the sample must be concentrated. Large scale centrifugation of large volumes of urine and then decanting off most of the supernatant gives a more concentrated sample. The sediment can then be prepared directly as a smear but will need an adhesive coated slide or the addition of an adhesive to the sample to prevent cell loss (less watery samples from cervical scrapes, sputum samples, etc. contain proteins which act as adhesives).

The sediment can alternatively be processed by the techniques of liquid-based cytology, as developed for cervical cytology, or by cytocentrifugation to produce a neat pellet on the slide.

Pathology of the urinary tract

Infections

Inflammation can be caused by a number of diseases including bacterial cystitis (e.g. *E. coli*) and urethritis (e.g. chlamydia and gonococcus). Viral infections include herpes, HPV and cytomegalovirus. Infection with the candida fungus occurs more commonly in women by infection from the vagina. Parasitic infections include those from trichomonas, which have increased due to urethral infections, and also from schistosoma. The latter is, however, unlikely in the UK except in immigrants and occasionally travellers from the Middle East and Africa, where the disease bilharzia occurs. The schistosoma parasite is a trematode fluke and infection occurs through skin exposed to contaminated water.

Neoplasms

The major cancer seen in urine cytology smears is the transitional cell carcinoma ('bladder cancer'). Other tumour cells are much less common and include squamous carcinoma, adenocarcinoma, prostatic adenocarcinoma and secondary metastases from primary tumours elsewhere in the body. Occasionally shed renal cell carcinoma cells can be detected.

14.3   Respiratory tract cytology

The respiratory tract is a very vulnerable system as it is constantly exposed to the environment. Breathing continuously pumps the atmosphere in and out over the delicate epithelial linings of the tubes and lungs, so exposure is not simply passive but forced. As a consequence, the lungs are prone to a large number of diseases including malignancies. Surgical biopsies of the lower parts of the tract are fraught with difficulties and hazards. By contrast, cytological sampling can be simple and convenient, as with sputum samples where the patient is only required to cough deeply and then expectorate the resulting sputum into a pot (see *Box 14.2*). Endoscopy samples, although more intrusive, are not technically difficult to obtain nor excessively distressing for the patient and can provide excellent samples of suspicious lesions. Intra-costal percutaneous needle biopsy of the lung, which was previously favoured, is no longer recommended because of the risk of inducing a pneumothorax (collapsed lung).

Table 14.2   Cellular and non-cellular constituents of sputum samples

Origin	Cell type	Appearance
Oral mucosa – stratified squamous epithelium	Squamous epithelial cells (superficial, intermediate types)	As for cervix.
Bronchi – pseudostratified ciliated columnar epithelium	Ciliated cells	Similar to cervical ciliated cells but cilia are often easier to visualize.
Bronchi – pseudostratified ciliated columnar epithelium	Mucus-secreting goblet cells	Similar to cervix.
Blood derived cells		
	Red blood cells	Anucleate biconcave discs.
	Neutrophils	Multilobed nucleus.
	Macrophages (histiocytes)	Round or indented (bean-shaped) nucleus. Usually single nucleus but occasionally multinucleated. May contain ingested carbon particles.
Non cellular materials		
	Mucus	May form Curschmann's spirals.
	Asbestos bodies	Perls positive beaded rod structures.
	Food debris	Fibres and cells of plant or animal origin. Variable appearance.

Specimen collection

There is no current cytological screening of healthy patients in England so all respiratory tract samples will have been taken from symptomatic patients. The samples should

> **Box 14.2 // Cough samples**
>
> An early morning deep cough specimen is easily explained to the patient. However, human nature and the intrusion of etiquette can restrict the actual execution. Patients are often reluctant to produce a deep 'hacking' and productive cough as it is 'impolite'. Spitting viscous, discoloured or smelly sputum is similarly seen as impolite. Yet this is exactly what is needed. A polite little 'Ahem!' cough will produce little more than saliva and will be useless for diagnosis of the lower respiratory tract. Taking samples over several days may improve the samples, as the patient will become less sensitive to the impoliteness.
>
> Almost everyone inhales carbon particles from diesel fumes, fires and cigarette smoke. The carbon particles are ingested by alveolar macrophages (histiocytes). The presence of carbon-laden histiocytes is a strong indication of a deep cough specimen. The absence of these alveolar macrophages indicates that the smear is inadequate, as it has not been sampled representatively.
>
> Using vigorous massage / chest thumping has been suggested as a way of dislodging cells. This seems excessive as cells are shed regularly and such vigorous interventions in a patient already showing respiratory compromise cannot be recommended.

therefore be treated as infected with very noxious organisms, as the cause of the symptoms is still unknown.

Sputum samples

The sample is best collected as an early morning specimen, when the mucus has lain undisturbed in the airways for several hours and will include cells exfoliated over the course of the previous night's sleep. By using an early morning specimen the presence of food particles or toothpaste, which may obscure details and make the sample inadequate, can be avoided. A deep cough specimen may contain cells from very deep regions of the respiratory tract and lungs. The quality of the sample is entirely under the control of the patient and adequacy for diagnosis is unpredictable (see *Box 14.2*). Repeat sampling is fortunately easy and is more likely to produce at least one adequate specimen.

The sputum can be directly sampled simply by pouring it into a flat Petri dish and looking for cellular clumps or blood flecks. Samples can then be taken from these areas and smeared onto a slide, fixed in alcohol and then stained. Sputum is very viscous so if centrifugation or liquid-based cytology is to be used, the mucus must be thinned. This may be done chemically by mucolytic agents or mechanically using homogenizers or ultrasound.

Bronchoscopy samples

Bronchoscopy involves the use of a flexible optical device (using either fibre optics or small video CCD chip) to view the respiratory passages. A small brush can be inserted into a channel in the bronchoscope. The brush can be extruded from the channel if a suspicious area is seen with the bronchoscope; this area can then be gently stroked with the brush and the brush then drawn back into the channel. The brush can be retrieved from the bronchoscope and direct smears can be made by gently rolling it onto a slide. Alternatively, the brush can be rinsed in preservative for preparation by liquid-based cytology. The bronchoscope can also be used to collect washings of the surface tissue. Saline is used to wash the surface and dislodge cells. The fluid is recovered by suction through the endoscope and treated as for other liquid samples.

Pathology of the respiratory tract

Infections

The respiratory tract (including the oral cavity for sputum samples) can have a wide range of commensal and pathogenic bacteria (e.g. TB organism). Most are of little significance in cytology but can be important and need to be investigated microbiologically. Viruses which affect the appearance of cells include the herpes virus. As well as candida, other fungi occur such as aspergillus (causing aspergillosis) and *Pneumocystis carinii* (renamed *P. jirovecii*, but the older name is still common) also occur but only cause disease if the patient is immunocompromised.

Minerals

Inhalation of non-degradable materials can result in pneumoconiosis. Several types of pneumoconiosis occur. Asbestosis is found mainly in individuals who have worked with asbestos (the use of asbestos is now banned but it still remains in the environment). In the lungs asbestos fibres become coated with an iron-rich protein to form asbestos bodies which stain Perls positive.

Neoplasms

The commonest malignant tumours in the respiratory tract are carcinomas with fairly similar incidences for all four types (15–40% of tumours in each case). They are usually categorized and named from the appearance of the cells. Squamous cell carcinoma is the tumour most closely linked to smoking and can be well, moderately or poorly differentiated. Small cell carcinoma ('oat cell carcinoma') often metastasizes early, resulting in secondary tumours which can be much larger than the primary tumour. Large cell tumours are undifferentiated (hence the term large cell undifferentiated tumour (LCUC)). Adenocarcinomas can form from pneumocytes (type 2), mucus-secreting cells or club cells (Clara cells). The lung is also a common site for metastatic spread.

14.4 Aspirated effusion samples

Effusions are the leakage of fluids into a 'potential' cavity. A potential cavity is where there is discontinuity between an organ and the surrounding tissue. Normally the gap is small and contains only small amounts of fluid which act as lubricants. The function of these lubricated potential cavities is to allow the organ to slide easily over the surrounding tissues. The major cavities are:

- The **peritoneum** in the abdomen which holds the alimentary canal and the sliding of which allows the normal movements of the gut.
- The **pleura** which surround the lungs and allow the lungs to slide against the chest wall during breathing.
- The **pericardium** which allows the heart to move as the heart beats.

The cavities are lined with a thin flat epithelium called mesothelium, and this epithelium together with its basement membrane and supporting connective tissues are called the serosa. The mesothelium is a semi-permeable membrane which allows diffusion of small molecules but is impermeable to large molecules such as proteins. The fluid within the cavities is formed by fluid passing out of capillaries and through the serosal membrane into the cavity. The rate of production is dependent on the hydrostatic pressure in the

> **Box 14.3 // Dialysis using the serosa**
>
> Because the serosal membrane is semi-permeable it can be used to treat patients with renal failure as an alternative to a dialysis machine. If the peritoneal cavity is filled with a dialysis medium, the serosa will allow diffusion of small molecules in a very similar way to the glomerulus. Regular and frequent draining of the now used dialysis fluid and replacement with fresh fluid gives an effective dialysis without the patient being restricted to a bed attached to a machine. The patient can walk around and get on with ordinary everyday life, hence the term continuous ambulatory peritoneal dialysis (CAPD) and the procedure is an important alternative during early renal failure.

capillaries and the osmotic pressure difference due to proteins (colloid oncotic pressure). Normally there is a dynamic balance between these factors on different areas of the cavity, resulting in continuous production and reabsorption which gives a slow circulation of the serous fluid.

If there is any disturbance of the dynamic equilibrium then serous fluids may accumulate, causing swelling. Peritoneal effusions are probably the best known and the larger peritoneal effusions may be referred to as ascites and the fluid as ascitic fluid. An archaic term for the accumulation of peritoneal fluid was dropsy. Peritoneal effusions result in a distended abdomen but the fluid pressure has little other effect. In the case of pleural effusions, the fluid presses against the rigid ribcage causing pain and interfering with breathing movements and, in the case of pericardial effusions, the pressure inside the limited pericardial cavity can seriously interfere with the heart contracting and is referred to as a cardiac tamponade. The fluids expressed into the cavity can be low in protein (exudations) or high in protein (transudates).

Sampling

Fluids can be collected from needles inserted into the cavity and this is often controlled by ultrasound guidance. The fluids will drain out gradually or they can be gently aspirated with a syringe. Between 20 and 50 ml can be taken for several techniques. Cells survive well in serous fluids without deteriorating rapidly, so preservatives are not essential if the fluid is processed fairly quickly. If a proteinaceous serous fluid is rich in fibrinogen, it may produce clotting. The clot can be prepared as a histological block but cannot be used for routine techniques. Clotting can be prevented with anticoagulants such as citrate. Serous fluid samples for cytology can be concentrated by the methods outlined earlier for fluids.

Diseases causing serous effusions

Ascites has many causes and not all are malignant. Some of the wide range of causes of ascites are given in *Table 14.3*; the diagnosis for many of these will not be aided by cytology.

Ascites can be caused by organs lying in or adjacent to the serosal cavity, such as the ovary in peritoneal ascites and breast cancer in the pleural cavity. Unlike many other cytological samples, the majority of the tumours are not of the serous cavity itself, with the exception of mesothelioma, which is a true serous tumour of the mesothelium. The majority of malignant cells (see *Table 14.4*) are from other organs and are therefore metastases or secondary tumours. Metastatic cells survive well in ascitic fluid and continue to grow and may produce nodules of cells on the wall of the serosal cavity (see *Box 14.4*).

Table 14.3   Causes of ascites

Exudative ascites	Transudative ascites
Bacterial peritonitis	Liver cirrhosis
Tuberculous peritonitis	Liver portal vein obstruction
Mesothelioma	Right heart failure
Metastatic (secondary) malignancies	Malnutrition (kwashiorkor)
Systemic lupus erythematosus	Meigs' syndrome (benign ovarian tumour plus ascites and pleural effusion)

Table 14.4   Cellular constituents of effusions

Origin	Cell type	Appearance
Serosal membrane	Mesothelial cells	Round, usually central nucleus. Green with Papanicolaou. May show mitosis.
		Groups show a clear space between cells.
Damaged or inflamed serosal membrane	Reactive mesothelial cells	Variety of changes and may be confused with malignant cells.
Blood derived cells		
	Red blood cells	Anucleate biconcave discs.
	Neutrophils	Multilobed nucleus.
	Macrophages (histiocytes)	Round or indented (bean-shaped) nucleus. Usually single nucleus but occasionally multinucleated.

Box 14.4 // Ascitic fluid and cell growth

Ascitic fluid has been used as a supplement in culture media and mouse peritoneal cavities have been used to culture cells otherwise difficult to grow.

The good survival of cells means that unlike many other fluids in cytology, there is good cell preservation in ascitic samples.

Pathology of serous membranes

Infections

Acute inflammation of the serous membranes (peritonitis, pleuritis and pericarditis) with the involvement of infective organisms can be very dangerous and a medical emergency. Peritonitis can be caused by rupture of the alimentary canal (e.g. gastric ulcer, acute appendicitis).

Tumours

The only true tumour of serous membranes is malignant mesothelioma. Malignant mesothelioma cells in cytological preparations look similar to normal reactive mesothelial cells so definitive diagnosis relies on immunohistochemistry. There is no single antigenic marker that is specific but by using a panel of antibodies a more definitive diagnosis is possible.

Although many of the tumour cells are metastases from primary tumours elsewhere, the identification of the original tumour type can be important. The original tumour may not have been detected and the first sign is the serous effusion. The cells will show the usual signs of malignancy but the tissue of origin may be unrecognizable by simple morphology. By using immunocytochemistry it is usually possible to positively identify the source of the primary tumour.

14.5 Fine needle aspirates

Fine needle aspiration (FNA) uses a thin hollow needle to reach deep within the body to organs and lesions that previously were inaccessible except by major surgery. The use of FNA alongside ultrasound, computed tomography and angiography and mammography has made sampling and diagnosis much simpler, more accurate and less traumatic for the patient.

In FNA the sample is a fluid which can be simply drained through the needle by internal pressure or by the gentle use of suction. The technique can be used for cytology (fine needle aspiration cytology, FNAC). The same technique used on a solid lesion will produce a very thin core of tissue akin to a very small biopsy; this is the fine needle aspiration biopsy (FNAB) technique.

Superficial lesions can often be felt by palpation (palpable lesions) so the clinician taking the sample can hold the lesion steady with one hand whilst manipulating the needle into the lesion with the other hand. The needles can be standard clinical syringe needles (e.g. 23–25 gauge needles, external diameter 0.6–0.5 mm, internal diameter 0.34–0.26 mm). Once inserted, the needle can be gently rocked to help dislodge cells whilst the syringe plunger is gently withdrawn to produce a slight suction pressure. It is not intended that large amounts of fluid are withdrawn, just enough to fill the needle bore. When fluid can be seen at the upper end of the needle the plunger is pushed back to the base to stop the suction. The needle is withdrawn from the patient and the cells can be expelled by gentle positive pressure (this involves removing the needle from the syringe, pulling the plunger up slightly to suck in air, reattaching the needle and expelling the contents of the needle). The contents of the needle can be smeared immediately and fixed, or for liquid-based cytology preparations the material is expelled into LBC transport medium with gentle rinsing of the needle.

For deeper (non-palpable) lesions a longer needle and ultrasound guidance are needed. Once the needle and its contents are withdrawn, procedures are the same to expel and treat the sample.

Sites which are regularly sampled by FNA include:

- Breast lumps
- Lymph nodes
- Thyroid gland
- Gastrointestinal tract
- Respiratory tract
- Chorionic villus sampling.

A wide variety of cells can be present in FNA samples depending on the site being investigated. Similarly, the infections and malignancies will vary with the sites being sampled. Details of the cells and pathologies found are beyond the scope of this chapter and the interpretation of FNA smears requires a great deal of experience.

Gastrointestinal tract cytology

In the past cytological examination of the GI tract was a useful method for diagnosis. Gastric lavage was an early technique but with the development of flexible endoscopes, brushing techniques became more usual. Nowadays, endoscopy with biopsy forceps or collecting biopsy specimens has proved more useful. The other development has been the use of FNA guided by endoscopic ultrasound, which means that cytological specimens are more commonly provided by FNA and would be processed as for other FNA samples.

Direct brushing of the buccal cavity (cheek) is relatively simple and can be used for diagnosis but is not used for diagnosis on a regular basis. Buccal smears are used to determine sex chromosome status, as explained in *Chapter 15* (see also *Box 14.5*).

Box 14.5 // Other use of buccal smears

In the scientific literature there are reports of the use of buccal smears in the diagnosis of a large range of diseases from diabetes to leishmania infections. Most of these are not cytological but rather genetic. The ease of sampling the buccal mucosa as a source of the patient's cells is the main reason for using buccal smears.

Buccal swabs are also a preferred method of obtaining cells for forensic DNA profiling ('genetic fingerprinting').

Cerebrospinal fluid

The preparation and analysis of cerebrospinal fluid (CSF) is somewhat of an orphan technique and may be found residing in any one of the haematology/chemistry, microbiology, cytology or histology specialist laboratories – it usually depends on who owns and can operate the cytocentrifuge. CSF analysis is partly observational, with a yellow and more viscous appearance indicating increased protein and a reddish tinge a degree of haemorrhage. Both are indicative of an inflammatory response, usually due to infection, whereas the presence of a significant red blood cell component is more likely to have resulted from trauma or a cerebral vascular accident (CVA). Samples can be very small (sometimes as little as 500 µl from a paediatric case) but are extremely valuable in demonstrating the presence of reactive inflammatory cells, e.g. eosinophils, PMNs or plasma cells. Equally infectious organisms can be demonstrated directly, for example, meningococcal meningitis or toxoplasma.

Bone marrow aspirates (and core)

A bone marrow trephine may be indicated for two main reasons, either abnormalities with haemopoiesis or the presence of malignancy. The malignancy may relate to the white cell component of the blood, i.e. lymphoma/leukaemia, or may be a metastatic deposit within the marrow space. Again, there is overlap with other specialisms; in this case sets of slides are often required for both haematology and histology. Practice does vary according to the expertise and involvement of clinicians but often it is useful to roll a marrow trephine core along the length of a slide. This is sufficient to transfer a representative sample of the cell types present that can then be stained very rapidly (5 min) using a commercial

Romanowsky (Giemsa) staining kit. An early indication of the likely pathology will permit informed clinical decisions to be made before the results from the bone marrow core are available, given the time needed for decalcification and processing.

Suggested further reading

Bibbo, M. and Wilbur, D (eds) (2008) *Comprehensive Cytopathology*, 3rd edn. Saunders.

Cibas, E.S. and Ducatman, B.S. (2009) *Cytology: Diagnostic Principles and Clinical Correlates*, 3rd edn. Saunders.

Gray, W. (2002) *Diagnostic Cytopathology*. Churchill Livingstone.

Orell, S.R., Sterrett, G.F. and Whitaker, D. (2005) *Fine Needle Aspiration Cytology*, 4th edn. Churchill Livingstone.

Shambayati, B. (2011) *Cytopathology* (Fundamentals of Biomedical Science Series). Oxford University Press.

Self-assessment questions

1. Outline how simple centrifugation, the use of a cytocentrifuge and filtration can be used to concentrate fluid samples. Briefly explain when each method would be applicable.
2. List the cell types found in normal urine and outline the origin and significance of transitional cell carcinoma.
3. Give four examples of infective agents that can be found in urine samples.
4. Briefly describe three methods of obtaining cell samples from the respiratory system.
5. List the cells that can be found in respiratory samples.
6. What is the significance of asbestos fibres in sputum samples and how can asbestos fibres be identified?
7. Outline what 'effusions' are and name the places where effusions occur.
8. Briefly describe the use of fine needle aspirates in cytopathology.
9. In which other biomedical specialisms are CSF samples frequently seen?

Learning objectives

After studying this chapter you should confidently be able to:

- **Outline the diagnostic applications of cytogenetics**
  Cytogenetics can be used to diagnose genetic disease caused by alterations in chromosome number and structure. This can be either pre-natal or post-natal diagnosis. It can also be used in some acquired conditions such as leukaemia where there may be a change in the chromosomes.

- **Describe how chromosomes are prepared for examination and describe a normal human chromosome complement**
  Chromosome examination requires cells to be grown in culture to ensure they are in mitosis, which is the only time when chromosomes are visible. Chromosomes need to be spread using hypotonic treatment of cells. Fixed smears can then be stained to show the bands and this allows recognition of altered number or structure. A normal human cell contains 46 chromosomes including 22 paired chromosomes (autosomes) and two sex chromosomes, which differ between the sexes (two X chromosomes for females and one X and one Y chromosome for males).

- **Describe the effects of alterations in chromosome number on spontaneous abortions and live births**
  Spontaneous abortions are more common in chromosomally abnormal pregnancies. Chromosomal abnormalities are the commonest cause of spontaneous abortions. Hydatidiform moles are abnormal euploid conceptuses that implant and mimic normal pregnancies but do not contain a viable foetus.

- **Briefly discuss and name the syndromes that occur in humans from altered chromosome number**
  Chromosomes may be altered either in complete sets of the 23 human chromosomes (euploid variations), giving an embryo or foetus with 23 or 69 chromosomes, or the change may only affect one chromosome (aneuploid variations). All euploid variations are inviable. Most autosomal aneuploid variations are also inviable with only Down's syndrome surviving to birth in significant numbers. Aneuploid variations in the sex chromosomes are less vulnerable and several syndromes occur, notably Klinefelter's, Turner's, supernumerary X and double Y syndromes.

- **Outline the effects of alterations in chromosome structure**
  Chromosome structure changes can result in less-severe forms of genetic disease than loss or gain of complete chromosomes. Deletions act as incomplete monosomy, whilst insertions act as incomplete trisomy. Translocations may not cause problems if there is no loss or gain, only simply rearrangement of genetic material (balanced translocation). However, translocations may cause problems to the offspring of carriers as the meiotic rearrangement of genetic material can result in an unbalanced chromosome structure. Inversions may also cause problems during gamete formation but by a different mechanism.

- **Interpret an abbreviated karyotype**

Genetics is the study of inheritance and encompasses three areas of study, two of which involve aspects of cellular pathology and histology:

- Classical genetics is the study of pedigrees and the pattern of inheritance from one generation to the next. The basis for classical genetics was laid in the nineteenth century by the Austrian monk Gregor Mendel whose name is associated with this type of study as Mendelian inheritance.
- Cytogenetics is the part of genetics involved in the study of genetic changes that occur in cells and can be observed microscopically. These include changes in chromosome number or structure. The chromosomes can be stained so that they show **banding**. The banding is unique to each chromosome and individual chromosomes can be recognized. By sorting the chromosomes into order, a person's **karyotype** can be determined. The karyotype shows whether any chromosomal abnormalities are present. It is even possible to recognize changes within an individual chromosome such as loss of certain bands. The techniques involved include smear preparation and staining and are quite closely allied to histology and cytology.
- Molecular genetics studies the changes in the sequence of DNA bases. These changes can be visualized in sections or smears by using molecular biological techniques and the role of these techniques is rapidly increasing, including their use in histology.

15.1 Cytogenetics and chromosome studies

Chromosomes are the structures that contain the DNA of the cell and normal humans have 46 chromosomes. Abnormalities in the number and structure of these chromosomes are important in many fields of medicine.

Chromosomal abnormalities

Chromosomal abnormalities are a common cause of abortion

Abnormalities of chromosomes are often present from the instant of fertilization and remain until the death of that individual since there are no cures for chromosomal abnormalities. Many chromosomal abnormalities will spontaneously abort and only become apparent as miscarriages in the pregnancy (see *Box 15.1*). Although diagnosis of a miscarried foetus may seem like a low priority in medicine, it can be important to the parents by indicating that the dead foetus would never have been able to lead a full and normal life. Knowing this often helps parents come to terms with their grief.

Diagnosis of the cause of foetal death may also be important to rule out other causes of abortion. Some unfortunate couples suffer repeated abortions and recognition of the cause of the miscarriage is often a necessary first step in helping to overcome the problems.

A somewhat similar argument also applies to children and adults who are born with chromosomal abnormalities. Although at present little can be done to help people with these conditions except for treating individual symptoms or aspects of the syndrome, it can be helpful if the parents and the individual understand the nature of their abnormality. Knowing the cause of a problem can also eliminate the need to search further for a diagnosis.

> **Box 15.1 // Abortion and miscarriage**
>
> The term abortion is often misused by the general population. In medical terms, an abortion is any premature termination of pregnancy before the foetus is viable.
>
> There are several different forms of abortion. **Spontaneous abortions** are where the pregnancy terminates because of physiological reasons without any intervention or traumatic cause. **Induced** or **artificial abortions** are where the pregnancy is terminated medically and a **criminal abortion** is an induced abortion carried out by someone other than trained, licensed medical personnel. **Therapeutic abortions** are induced abortions performed to save the life of the mother.
>
> To many people, the term abortion means an induced abortion and they would use the term **miscarriage** for a spontaneous abortion. The term miscarriage in medicine strictly only applies to spontaneous abortions in the second trimester of pregnancy. Pregnancy is divided into three trimesters, each lasting 3 months, so the second trimester refers to the 4th, 5th and 6th months of a pregnancy.

Pre-natal diagnosis may allow termination of the pregnancy

Diagnosis of chromosomal abnormalities in the foetus during the early part of pregnancy allows more active intervention. This may involve a medical termination or abortion of the pregnancy if the defect is severe and the prospective parents do not wish to have a severely affected child. Pre-natal diagnosis is controversial because of this link with induced abortions.

Diagnosis of the sex of the foetus can also be used as a method of preventing certain diseases, even though the presence of the disease itself cannot be identified directly. Diseases that can be controlled in this way are those carried on the X chromosome. Such X-linked or sex-linked disorders only affect males. Females can carry the recessive aberrant gene but are not affected. Sex-linked abnormalities include Duchenne-type muscular dystrophy, adrenoleukodystrophy and haemophilia. Although these diseases cannot currently be recognized in the foetus, it is possible to prevent the birth of affected children by only allowing female foetuses to go to full term and be born. Females are not affected so this will effectively prevent the disease occurring in these individuals, although they may still be carriers. It does mean, however, that all male foetuses need to be aborted. Half of these male foetuses would be expected to be normal. This again is ethically controversial.

It may be possible to diagnose abnormalities before pregnancy begins

The increasing use of *in vitro* fertilization also allows the early diagnosis of abnormalities. The prospective mother is stimulated with fertility drugs causing her to hyperovulate and produce several viable ova instead of the usual single ovum produced in a normal ovarian cycle. The fertile eggs are collected by a minor surgical procedure using ultrasound to detect and aspirate mature Graafian follicles that contain eggs. The eggs are then mixed with the father's sperm in the laboratory and grown in culture until they form a small ball of cells (called a **morula**). At this stage, every cell is identical and indeed each cell is capable of forming a whole human being and it is separation of the cells at this stage that leads to identical twins. Removal of a single cell from the morula will not affect the foetus or its development adversely. The removed cell can then be used to produce a diagnosis. Normal embryos can be used for implantation and abnormal embryos would be discarded. By diagnosing before conception, the problems of abortion can be avoided. This technique is being used on a trial basis but is not yet routinely available.

Chromosomal abnormalities arising after birth may be a cause of malignant transformation

There is another use for cytogenetic investigation and this is in the study of chromosomal abnormalities that have arisen after birth in one or a few cells. The majority of the body's cells will be quite normal but the altered cells may cause disease. The commonest disease involving altered chromosomes is malignancy. Alteration of the normal chromosome complement is frequent in malignant tumours and some abnormalities are characteristic of certain diseases. In chronic myeloid leukaemia (CML) there is a large abnormal chromosome called the Philadelphia chromosome formed from a translocation between chromosomes 9 and 22. About 90% of CML cases have this abnormal chromosome and only 10% do not. The presence of the chromosome gives a better prognosis (median survival of about 4 years) compared with those without (median survival of about 1 year).

Preparation of chromosomes for karyotyping

Chromosomes are only visible during cell division

In order to recognize chromosomal abnormalities microscopically, the chromosomes must be visible. The chromosomes are not normally visible within the cell since the DNA is, for most of the time, in the form of euchromatin, which is very spread out in the nucleus. The chromosomes are only visible at mitosis and meiosis when the DNA becomes highly compacted by coiling and supercoiling into relatively short rods. In humans, meiotic divisions only occur in the gonads and are not used in normal cytogenetic investigations. Mitosis occurs in a large number of cells and tissues during normal cell division and is much more easily investigated and so forms the main target of cytogenetic investigations.

Chromosomes can be studied during prophase and metaphase

The chromosomes begin to appear during the first stage of mitosis (**prophase**) as long tangled threads and become shorter and less tangled as prophase progresses into **metaphase** where the chromosomes line up along the central plane of the cell and attach to the microtubules of the mitotic spindle. It is at metaphase that the chromosomes are at their shortest and thickest. The spindle fibres then pull the chromosomes apart at **anaphase** and the chromosomes begin to lengthen again. The two new nuclei develop a membrane and the cytoplasm divides during **telophase** and **cytokinesis**. Thus, the only time that chromosomes can be seen is in these stages. To ensure seeing cells in mitosis, it is necessary to grow them in culture and even then they need treating to make the chromosomes separate and allow them to be easily visualized.

The technique of chromosome preparation can be divided into three stages: (i) collection of the cells; (ii) growth of the cells in culture; and (iii) the preparation of stained smears.

Collection of cells

The cells must be capable of mitosis so not all samples are equally useful. The most common cell samples in the adult are blood lymphocytes, which can be stimulated into mitosis, small skin biopsies from which fibroblasts can be grown and bone marrow cells. Naturally exfoliating cells are not suitable as they are usually dead or at least non-mitotic.

In pre-natal testing of the foetus, the samples are usually taken either from amniotic cells floating in the amniotic fluid, which can be sampled by amniocentesis, or from chorionic villus cells (see *Box 15.2*). Foetal blood samples can be used to produce mitotic samples but are less common.

> **Box 15.2 // Pre-natal collection of cells**
>
> Amniocentesis is the collection of amniotic cells from the amniotic sac. It is carried out by inserting a needle through the abdominal wall into the uterus and through the membranes surrounding the foetus and drawing off some fluid and cells. This is a fairly safe procedure and rarely (less than 1%) causes an abortion. Amniocentesis can only be done after about week 16 of pregnancy. The cells grow rather slowly so it may take 2–3 weeks to get a diagnosis. This makes it quite late in pregnancy before the results are known and a decision on termination taken.
>
> Chorionic villus sampling is usually done through the cervix and can be performed as early as 8 weeks. The cells grow more quickly so the results can be determined much earlier. Chorionic villus sampling has a higher risk of inducing an abortion than amniocentesis. If it is performed very early (before 10 weeks) there is also the risk of some foetal deformations being caused.

Several different sites may need to be sampled if the disease is believed to be a **mosaic** where not all of the cells of the body are affected (see section on Mosaicism in this chapter).

Growth of cells in culture

Growing cell and tissue cultures is a difficult and painstaking procedure. Cells in the human body are supplied with fresh nutrients, oxygen, hormones and other signalling molecules by the bloodstream. The same vascular system then removes carbon dioxide and other waste products. The vascular system is dynamic and modifies its pattern of blood flow in response to signals from the cells, e.g. if oxygen levels become low. This ideal situation cannot be reproduced exactly in culture and instead the cells are grown in a static fluid that is only occasionally renewed. Nevertheless, some cells can be cultured successfully, provided they are given certain requirements. The method of culture varies depending on the type of cell sample available and the reason for culturing the cells.

Cell cultures are grown from cell suspensions

Cell cultures start from cells that have been separated into individual cells before being grown. The cells are prepared as a suspension and can be handled easily using pipettes. This type of culture is easy when the original samples are already suspensions of cells, as would be the case with blood samples and samples of amniotic fluid. It is slightly more complex when the cells are firmly attached to each other, as they would be in a skin sample or a chorionic villus biopsy. The cells must first be separated by treatment with chemicals or enzymes. Trypsin, a proteolytic enzyme, and ethylene diamine tetra-acetic acid (EDTA), which is a calcium-chelating agent, are useful in disrupting tissues without killing or lysing cells. Mechanical disruptions with pipettes or dissecting instruments also help. The cells are then seeded into a plastic culture bottle (see *Fig. 15.1*), which has a large flat base on which the cells can settle and grow.

Human cells will not grow easily whilst in suspension and must attach to a surface of some sort before they will settle down and divide. The density of seeding is important. If too many cells are added, the cells will be crowded and not grow. If too few cells are added, then they may find that they cannot adapt to the culture medium and will fail to thrive. It is possible to grow cells in suspension as spheroids by using a viscous growth medium with methyl cellulose added, or to eliminate gravitational effects by constantly rotating the culture vessel. Spheroids are a better model and substitute for *in vivo* experiments.

Figure 15.1
Tissue culture bottle containing red growth medium (includes phenol red as pH indicator).

Tissue culture involves growing cells from a small piece of tissue

Tissue culture involves placing a small intact piece of tissue into a culture flask. The cells can then grow and spread from this explanted piece of tissue to form a layer of cells covering the bottom surface of the flask. Small 'punch' biopsies of skin can be grown in this way and fibroblasts will readily spread from the connective tissue layers of skin. Fibroblasts are one of the easier cell types to grow, so this is a good method for obtaining cells. The advantage of this technique is that there is no need to use enzymes or chemicals that may damage the cells.

A further type of culture, organ culture, is described in ***Box 15.3***.

Primary cultures are taken directly from tissues

Along with the way cells are obtained, there is also a difference between the cultures depending on how long they have been cultured. Cells grown directly from the body are termed **primary cultures** and are the most reliable in genetics since they represent cells that have been in artificial culture for only a few divisions. When cells are grown for long periods, they need to be subcultured periodically to keep them growing.

Subculturing cells helps to keep them growing

Subculturing is needed not only to replace the nutrients by replacing the culture medium but also to separate the cells. When cells divide in culture, they spread across the surface

---

Box 15.3 // Organ culture

Although not relevant to chromosome studies, organ culture is a method of culturing cells in which more than one cell type is cultured in the same flask. Various tricks are used to encourage the cells to interact and produce small clusters of cells arranged in the same way as tissues in the body. This is a research method for investigating how cells interact rather than a method of providing a source of cells for diagnosis. The long-term goal of such research is to be able to grow whole organs from a small sample of cells. The growth of small amounts of simple tissue such as skin and bladder has been achieved and used clinically, but the growth of complex organs such as kidney is still not possible. Organ culture would be a valuable way of producing new organs to replace organs that are failing in a patient. Cells taken from a patient in renal failure could be used to produce a new kidney, which would then be transplanted back into the patient. The use of an organ grown from the patient's own cells would avoid the problem of organ rejection. Although currently in the realms of science fiction for complex organs, such a technique may one day be possible.

and will eventually form a complete layer of cells. The cells then stop growing. This stoppage is called **contact inhibition** and can be overcome by gently stripping the cells off the plastic using trypsin or chelating agents and then dividing them into several new flasks. Subculturing ensures that the cells have more room to grow and keeps the cells dividing rapidly. Once cells have been subcultured in this way they are known as **primary cell lines**. Primary cell lines are still useful as the cells should still be identical to the original cells. Normally, primary cell lines will stop growing after a few subcultures as they become senescent and die. Human cells can divide between 10 and 50 times before this senescence occurs. The limited number of divisions is called the Hayflick limit and is believed to be a fundamental property of the cells in the body. In particular, it is related to the length of the terminal telomeres of individual chromosomes and is not an artefact of the culturing technique.

Established cell lines

When cells are grown through several subcultures, sometimes a few of the cells may change their characteristics. This is called transformation (see **Box 15.4**). The transformed cells may have gained quite different growth characteristics and their genetic make-up alters so they are not suitable for genetic analysis. These transformed cells can divide without limitation and are called **established cell lines**. Established cell lines can be important in biological and medical research but not for cytogenetic diagnosis of the original donor.

Box 15.4 // Transformation

Transformation to form established cell lines is believed to be due to mutation within the cell, possibly caused by viruses, radiation or chemicals. Transformation is very similar to the alterations seen in malignant change in the body and primary cultures from malignant tumours often grow immediately in the same way as established cell lines. This has made the process of transformation an important tool in investigating carcinogenesis. The most famous of the tumour-derived established cell lines is the HeLa cell line (the name HeLa comes from the name of the patient with the original tumour) derived from a cervical cancer in an American woman in the 1950s. This cell line is still rapidly dividing today and is one of the easiest cell types to culture. The cells grow so readily they have been able to spread to other cultures and many cell lines have become infected by this aggressively growing cell.

Requirements for cell growth

Cells require several critical factors for their active growth. These factors are now largely understood and we know the specific growth requirements for many cells, including those used in chromosome studies. The growth media are usually made from many individual components and can be quite complex.

Nutrients

This is the most obvious requirement since nutrients provide the energy needed for metabolism and the building materials needed to synthesize new proteins and other cell constituents. Unlike bacteria and plants, most human cells require a complex mixture of nutrients and cannot synthesize many of the basic building blocks of proteins. Cells require a variety of nutrients to be supplied including (see **Box 15.5** for a formulation of Medium 199):

Box 15.5 // Constituents of culture medium

The complexity of a cell growth medium can be seen from the following formula for 1 litre of Medium 199.

Amino acids:
L-Alanine 25 mg
L-Arginine HCl 70 mg
L-Aspartic acid 30 mg
L-Cysteine HCl 0.1 mg
L-Cystine 20 mg
L-Glutamic acid 75 mg
L-Glutamine 100 mg
Glutathione 0.05 mg
Glycine 50 mg
L-Histidine HCl 20 mg
L-Hydroxyproline 10 mg
L-Isoleucine 20 mg
L-Leucine 60 mg
L-Lysine HCl 70 mg
L-Methionine 15 mg
L-Phenylalanine 25 mg
L-Proline 40 mg
L-Serine 25 mg
L-Threonine 30 mg
L-Tryptophan 10 mg
L-Tyrosine 40 mg
L-Valine 25 mg

Vitamins:
Ascorbic acid 0.05 mg
Biotin 0.01 mg
Calciferol 0.1 mg
D-Calcium pantothenate 0.01 mg
Choline chloride 0.5 mg
Folic acid 0.01 mg
Inositol 0.05 mg
Menadione 0.01 mg
Niacin 0.025 mg
Nicotinamide 0.025 mg

p-Aminobenzoic acid 0.05 mg
Pyroxidol HCl 0.025 mg
Pyridoxine HCl 0.025 mg
Riboflavin 0.01 mg
DL-α-Tocopherol phosphate 0.01 mg
Thiamine HCl 0.01 mg
Vitamin A 0.1 mg

Other components:
Sodium acetate 50 mg
D-Ribose 0.05 mg
2-Deoxy-D-ribose 0.5 mg
Adenylic acid 0.2 mg
Adenine 10 mg
Guanine HCl 0.3 mg
Hypoxanthine 0.3 mg
Thymine 0.3 mg
Uracil 0.3 mg
Xanthine 0.3 mg
ATP 10 mg
Cholesterol 0.2 mg
Tween 80 5 mg
$Fe(NO_3)_3.9H_2O$ 0.72 mg

Hanks' salts:
$CaCl_2.2H_2O$ 185 mg
Glucose 1000 mg
KCl 400 mg
KH_2PO_4 60 mg
$MgSO_4.7H_2O$ 200 mg
NaCl 8000 mg
$NaHCO_3$ 350 mg
Na_2HPO_4 (anhyd.) 47.5 mg
Phenol red 17 mg

- **Carbohydrates.** All cells need a supply of energy and carbohydrate is the most readily used energy source. Human cells do not usually need any specific sugars and most media simply use glucose as the energy source.
- **Amino acids.** These are the basic building blocks of protein and a wide variety of amino acids needs to be supplied.
- **Vitamins.** Again, human cells cannot synthesize many vitamins that are often essential enzyme cofactors. Individual cells do not necessarily require the full range of vitamins that a complete human being would, e.g. vitamin K is not essential in culturing cells as it is needed for the production of blood clotting factors rather than cellular materials.

- **Minerals, trace elements and anions.** These are sometimes needed both as nutrients to be included in structural or enzymic proteins and as part of the osmotic and buffering system of the culture medium.

Gases

It is usually appreciated that cells need oxygen since this is the driving force of energy production in the cell but it is less obvious that carbon dioxide is also needed. Human cells produce carbon dioxide as part of their metabolism and it is thought of as a waste product, but of course in the immediate environment of the cells, the concentration of carbon dioxide is much higher than in the atmosphere as a result of the build-up from metabolism. The cells are adapted to this level of carbon dioxide and use it as part of their buffering system. Carbon dioxide levels therefore need to be maintained in the medium. Sometimes the medium is formulated to include carbon dioxide as bicarbonate ions and the medium can then be used in ordinary air, but some media require the atmosphere in contact with the medium to contain 5% carbon dioxide and cells grown in such media require a special atmosphere of gases.

pH

Cells are very sensitive to pH. They must be kept very close to neutrality and will only grow at about pH 6.8–7.4. The pH in the medium gradually changes as nutrients are used and waste products such as lactic acid accumulate. The growth medium is always buffered to minimize these changes using phosphate and bicarbonate buffers, but the pH will still fall slowly. In order to show the changes in pH, the growth medium usually includes phenol red as a pH indicator. Phenol red changes from red (at pH 7.4) to orange and then yellow. Thus, tissue culture media are red at the beginning of the culture and then gradually change colour. Once the orange/yellow colour develops, the medium is exhausted and must be replaced (see *Fig. 15.2*).

Figure 15.2
Two tissue culture bottles. The bottle on the right is still at a reasonable pH, but the more orange medium on the left shows a lowered pH. The colour change is due to the phenol red indicator incorporated in most media. NB phenol red can have oestrogen-like properties and this may affect cell growth characteristics.

Osmotic pressure

Osmotic pressure (sometimes called tonicity) again alters during the culturing of cells mainly due to the gradual loss of glucose and other nutrients. The original osmotic pressure is calculated to include these materials so as they are broken down or incorporated into proteins the osmotic pressure gradually falls. Changes in osmotic pressure of the medium are rarely a problem in practice since the medium needs to be changed more quickly for pH changes than osmotic pressure changes. Thus, provided the pH is acceptable, the tonicity will also be acceptable. Osmotic pressure is usually kept in the range of 280–320 mOsm.

Complex molecules

In addition to the direct nutrients needed for growth, human cells also need hormones, growth factors and other complex molecules to grow properly. These play no structural role in growth but act as stimulants for the cells. Without such stimulation, most cells are not mitotic and will not divide or grow. These complex molecules can be added in pure form but it is more common to add a natural source of these as serum (e.g. foetal calf serum) or embryo extracts (made from fertile hen's eggs), which are rich sources of these growth-stimulating substances.

Even then, some cells will not grow adequately and will need further stimulation. Adult human lymphocytes will not enter mitosis unless they are first stimulated by one or more special mitogens. The mitogens used include the lectins phytohaemagglutinin and pokeweed mitogen, which can be added to the culture medium and will stimulate cell division.

Temperature

Cells are very sensitive to slight rises in temperature and are easily killed at even a few degrees above normal body temperature. Lower temperatures than normal are better tolerated by cells but will greatly slow growth. It is quite usual to handle cells at room temperature for replacing medium, subculturing, etc., but it is important to grow cells in a well-controlled incubator. Although cells will tolerate temperatures down to about 10°C reasonably well, they are killed at very low temperatures, e.g. freezing is lethal. It is possible to freeze cells and keep them alive but this requires the use of an antifreeze, e.g. dimethyl sulphoxide (DMSO), to protect them.

Sterility

The usual cell culture media are not only ideal for growing human cells but are also ideal for growing bacteria, fungi and other micro-organisms. Bacteria and fungi will usually grow more rapidly and aggressively than human cells. To prevent micro-organisms adversely affecting the growing cells, it is important to keep the culture medium free from contaminants. Simple sterile procedures are the best defence but are usually backed up by including penicillin and streptomycin in the medium to inhibit bacterial growth. Mycoplasma pose a serious challenge as they are often present in sub-lethal numbers and survive on or within cells without killing them. Mycoplasmal infections will compromise experimental results (e.g. growth characteristics), making it essential to use detection kits at set intervals to ensure that infection is not an issue.

Because of the complexity of the formula and the difficulties of maintaining sterility during preparation, most laboratories buy their medium ready made up and sterilized, rather than preparing it themselves. This is the easiest and most reliable way to get well-balanced and reliable growth media.

Preparation of stained smears

Cells can be stopped in metaphase to obtain more mitotic cells in the smear

The purpose of growing cells in culture is to obtain cells in mitosis; peripheral blood monocytes are often used because of their ease of sampling and manipulation. Most cells, even in culture, will not be in mitosis itself but in interphase when the chromosomes are not visible. Mitosis normally only lasts between 1 and 2 h, and even in culture many cells only divide once every 24–48 h, so as few as one in a hundred cells may be in mitosis at any one time. Some laboratories like to enhance the number of cells in division by adding a mitotic spindle inhibitor to the medium some hours before the smears are prepared. This stops the cells in metaphase, when the chromosomes are aligned along the cell equator. Cultures will have initially been given a growth stimulus by the addition of the natural plant lectin phytohaemagglutinin (PHA). This is followed by the addition of colchicine for the final 24 h of culture, ensuring that all the cells that have entered mitosis will be held in metaphase. This increases the number of mitoses seen but the chromosomes are at their shortest and thickest in metaphase, which limits the resolution of banding patterns. Thus, other workers prefer to leave the cells alone so that all stages of mitosis will be present. Metaphase is the best stage to prepare chromosome spreads for simple counting, but prophase, when the chromosomes are longer, allows better banding resolution.

Cells are made to swell to separate the chromosomes

Just before smears are made, the cells are stripped from the bottom of the flask and treated with a hypotonic solution. Cells in hypotonic solution will absorb water and swell. This causes the chromosomes to separate and makes counting and identifying the chromosomes easier. The hypotonic treatment needs careful timing and control to prevent the cells from bursting since once the cell membrane bursts individual chromosomes may be lost, making accurate counting impossible. The ideal situation is maximum swelling without rupturing of the cells.

Chromosome spreads are fixed in methanol/acetic acid

The swollen cells are fixed in suspension using methanol/acetic acid. This fixative is a good nuclear fixative but is a poorer cytoplasmic fixative so the chromosomes are well preserved but some of the cytoplasmic protein is lost. This loss of protein is an advantage for this application since it makes the final preparation less obscured by protein. The spread preparations can then be made by placing a single drop of the suspension of lightly fixed cells on to a slide and air-drying. Drying is accelerated if the slide is warmed on a hotplate before dropping on the suspension. The dropping distance is crucial: it should be just sufficient to rupture the cytoplasmic and nuclear membranes. Increased height above this distance will spread the chromosomes too widely on the slide, making imaging difficult.

Banding patterns are stained using Giemsa

Staining to produce the bands can be done by a variety of techniques but the commonest is using Giemsa staining, often with extra treatments such as enzyme digestion or salt treatment to enhance the banding pattern. The banding pattern produced with Giemsa is referred to as G-banding. If the proteolytic enzyme trypsin is used before Giemsa staining, the banding pattern can be referred to as GTG (G-banding after Trypsin–Giemsa). The substituted nucleotide 5-bromodeoxyuridine (BRDU) can be used during the cell culture and this slightly alters the pattern and is referred to as GBG banding (G-banding after BRDU–Giemsa). Fluorescent banding techniques, e.g. using quinacrine dyes, produce a slightly different pattern of banding and are referred to as Q-banding. These fluorescent probes attach to the adenine/thymine-rich regions of DNA (see **Chapter 9**). Specific techniques are also available to stain the centromere.

> **Box 15.6 // Brief history of chromosome counting**
>
> Chromosomes were recognized in the 19th century but the correct number of human chromosomes was not determined until 1956 (by Tjio and Levan). Until the discovery of banding in the late 1960s, chromosomes could only be grouped into seven groups and could not be recognized individually.
>
> Q-banding was the first to be introduced in 1968. Giemsa banding was first used in 1971 with warm saline treatment to enhance the bands. Trypsinization was also introduced in the same year and is now the most popular method.

Chromosomes are counted and arranged in order

After staining, the slides are observed under very high magnification and the chromosomes are counted and measured. By taking photographs, it is possible to cut out each chromosome from the photograph and pair it with its partner chromosome. The pairs can then be lined up in order (see *Fig. 15.3*). This is known as the **karyotype**. The chromosomes are numbered to identify them. The larger chromosomes are numbered first so chromosome number 1 is the largest human chromosome. Cutting out the photographs is slow and tedious so computer-aided techniques are preferred. These not only make it easier to assemble the chromosomes but also help in identifying the chromosomes by using image analysis techniques (see *Chapter 18*). The appearances are vague and need an experienced eye to ensure that all abnormalities are detected. This is especially true of structural abnormalities.

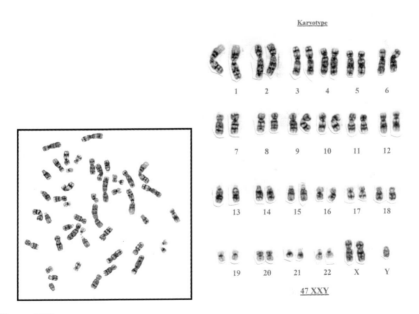

Figure 15.3
Karyotype of a patient with Klinefelter's syndrome. The chromosomes are cut out of the photograph on the left and arranged in order to give the karyotype

15.2   The effects of abnormal chromosome number

The normal human chromosome number is 46, which consists of 23 pairs of chromosomes. One set of chromosomes comes from each parent and this double set is referred to as the **diploid** number. Abnormalities of chromosomes occur quite frequently in human reproduction with about 7.5% of conceptions being chromosomally abnormal. This high level of abnormality falls to about 0.6% of pregnancies at birth since many of the abnormal embryos cannot survive and spontaneously abort. Chromosomal abnormalities are the commonest cause of spontaneous abortions, with about 60% of spontaneous abortions showing abnormal numbers of chromosomes. Abnormalities can occur in a number of ways, each of which has its characteristic set of syndromes. These changes vary in their severity; the most severe is dealt with first here.

Euploid variation

Euploid variation involves a change in a complete set of chromosomes with the total number of chromosomes following the pattern: total number of chromosomes = $n \times 23$. Thus:

- $n = 1$ is **haploid** and the total number is 23;
- $n = 2$ is **diploid** (normal) and the total number is 46;
- $n = 3$ is **triploid** and the total number is 69;
- $n = 4$ is **tetraploid** and the total number is 92.

Haploid and polyploid foetuses do not survive

All euploid abnormalities are inviable in human beings although they occur frequently in plants. Although humans as a whole cannot survive abnormal numbers of chromosomes, some human cells are often polyploid (e.g. liver). More than 99% of triploid conceptions abort during early pregnancy and the few live births that have occurred did not survive long after delivery. No tetraploid conceptions have survived to birth.

Euploid variations may be due to abnormal divisions or abnormal fertilization

There are probably several different causes of euploid variation. Haploid states could occur due to the development of an unfertilized egg but seem totally inviable even as embryos and do not occur even in aborted embryos. Polyploid states may be due to non-separation of the chromosomes in gamete formation resulting in diploid gametes. Normally the meiotic division reduces the number of chromosomes in the gametes to a haploid state. About 10% of polyploid conceptions occur due to the extra set of chromosomes coming from diploid ova and about 24% of the extra chromosomes come from diploid sperm.

An alternative way of getting polyploid conceptions is by double fertilization. There is normally a mechanism in the ovum to prevent more than one sperm head penetrating into the ovum but if this fails then several different sperm may each supply a complete set of chromosomes resulting in polyploidy. Double fertilization seems to occur in about two-thirds of all triploid conceptions. Triploid abortions may be recurrent in some women and seem to run in some families, suggesting a genetic susceptibility to this type of abnormality. Euploid variation is easily diagnosed from chromosome spreads since it is obvious that there are large numbers of extra chromosomes. Euploid variations will normally only be seen in aborted foetuses.

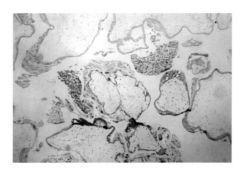

Figure 15.4
Section through a hydatidiform mole

Euploid abnormalities may produce a molar pregnancy or hydatidiform mole

Although not producing viable foetuses, euploid variations may result in implantation in the uterus and the production of placental structures. The abnormal placenta does not contain a foetus and the mass of placental tissue is called a hydatidiform mole (see *Fig. 15.4*). The patient appears to be pregnant, with enlarged ovary, amenorrhoea and a positive pregnancy test (raised HCG), so it is referred to as a molar pregnancy. Molar pregnancies are more common (up to 1 in 150) in the Far East than in Western countries (1 in 1000). Two types of mole are distinguished and differ in their chromosomal nature and the fertilization abnormality.

Partial molar pregnancies. These are triploid, with the major abnormality in fertilization being a double fertilization by two spermatozoa. The chromosome complement of 69 chromosomes (see *Fig. 15.5*) includes 1 maternal (23X) and two paternal (23X or 23Y) with 3 sex chromosomes and all possible variations are known to occur (i.e. 69XXX, 69XXY and 69XYY). Initially a foetus may be present but does not survive. The placental structure may show some normal placental villi, some abnormal oedematous (hydropic) villi and there may be some foetal remnants. Not all triploid conceptions result in a molar pregnancy and many simply never successfully implant.

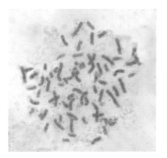 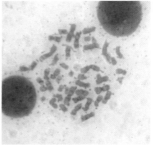

Figure 15.5
A chromosome spread from a triploid hydatid mole with 69 chromosomes is shown on the left, with a spread showing the normal 46 chromosomes on the right for comparison. It is easy to spot the difference in number without counting them

Complete molar fertilizations. These are diploid but have only paternal chromosomes and are typically 46XX. This suggests that the abnormality occurs when an anucleate or 'empty' ovum is fertilized by a single spermatozoa (23X) which then reduplicates the chromosomes to produce 46XX (a mitotic division of the nucleus occurs without any associated cytoplasmic division). Fertilization of an 'empty' ovum by a 23Y spermatozoa with reduplication to 46YY would not produce a viable cell; many important genes are carried on the X chromosome so the absence of an X chromosome would be lethal. In a minority of cases the constitution of the mole is 46XY, suggesting a double fertilization of an 'empty' ovum. A complete molar pregnancy is much larger – all the placental villi are abnormal, and there is trophoblastic proliferation. There is no evidence of any foetal remnants, which suggests very early embryonic death. Why apparently normal diploid karyotypes (46XX) produce an inviable foetus and a hydatidiform mole placenta is not clear, but one possibility is that as all the chromosomes have come from a paternal source they may have 'male imprinting'. Imprinting refers to changes which are not mutations but where genes are selectively switched off by methylation. This may mean that certain key functions for early embryonic development may be missing in male imprinted chromosomes but would be active with female imprinting.

In both types of molar pregnancy the male placental component is invasive; this is more the case in the complete molar version. If the abnormal placental material is not completely removed, a possible complication is choriocarcinoma – about 2–3% of complete molar pregnancies are complicated by this invasive, HCG-secreting malignancy. About half of all choriocarcinomas occur after a complete molar pregnancy, the partial ones having a much lower risk.

Autosomal aneuploid variation

Variation of less than a complete set of chromosomes is termed **aneuploid variation** or **aneuploidy**. Autosomes are those chromosomes that are identical in both sexes and are numbered from 1 to 22. Aneuploidy is usually a variation of only one or two chromosomes at most and multiple chromosome variations are uncommon. The change of only one chromosome results in the chromosome number being 46 ± 1. Thus, 46 – 1 is termed **monosomy** (e.g. 45, –7) and 46 + 1 is called **trisomy** (e.g. 47, +21), with the missing or additional chromosome indicated after the chromosome number.

The presence of an extra set of genes seems to alter the cell's metabolic balance and control. An aneuploid foetus is affected by multiple abnormalities due to the imbalance, affecting a large number of genes on each chromosome. The larger the chromosome, the greater the burden of defects and each chromosome has a characteristic set of metabolic abnormalities related to the genes contained on that chromosome (i.e. a characteristic syndrome; see *Box 15.7*).

Autosomal monosomy is inviable and most affected embryos abort spontaneously, often very early in pregnancy, and are rarely seen even in abortions. This lack of the abnormality ever being seen suggests that the embryo is unable to implant or even that

---

Box 15.7 // Syndromes

The occurrence of a set of symptoms or conditions together is called a syndrome. A syndrome may not always be a single disease but the symptoms commonly occur together. Syndromes are often recognized before the underlying cause is determined that would allow them to be described as a disease.

the zygote is totally inviable. Occasionally, foetuses that are 45, –21 have survived through to birth but they did not survive long after being born.

Autosomal trisomy is the commonest abnormality of chromosome number. Most of the affected foetuses do still abort and probably over 95% do not reach the end of pregnancy and instead miscarry. Trisomy is also the commonest form of abnormality in spontaneous abortions with nearly one-third of all abortions being trisomic.

The commonest form of trisomy at birth is **Down's syndrome** where there is an extra chromosome 21 (see *Box 15.8*). Down's syndrome occurs with a frequency of about 1 in 700 live births. Although it is the commonest abnormality at birth, over 60% of Down's syndrome foetuses still spontaneously abort during pregnancy.

Box 15.8 // Dr Down and his syndrome

Dr John Langdon Haydon Down was the superintendent at the Earlswood Asylum for Idiots in Surrey in the 1860s. He published a short paper on the 'Ethnic Classification of Idiots' in 1866. He held a view that would nowadays be held to be 'politically incorrect'. Down thought that male Europeans were the highest level of evolution and he looked at his 'idiots' (a term that was then a technical term rather than the more general meaning it has assumed today) as lower forms of humans. Down classified them into different, and in his view lesser, racial types. He identified Ethiopian, Malay and other idiocies. One type, however, has proved to outlast his simplistic and unfounded philosophy of the hierarchy of races and that is his classification of 'Mongolian idiocy'. This real syndrome is still often referred to as 'Mongolism' although it has no real link to oriental races except perhaps the epicanthal fold over the eyes. Ironically, his name has been used to try to replace this 'non-PC' name, yet he was the originator of the ill-thought epithet.

Down's syndrome has many genetic abnormalities

Chromosome 21 is a very small chromosome so it carries fewer genes than most chromosomes. The smaller number of genes is probably the reason that so many Down's syndrome children survive. Yet even with this tiny chromosome the number of defects is large. Down's syndrome patients have characteristic facial features including a generally flattened facial profile, a small low-bridged nose and an extra fold of skin over the eye (epicanthal fold). They also have an enlarged protruding tongue and their mouths are not held closed. They have redundant skin folds around the neck and misshapen, low-set ears.

They have a shorter than normal stature and their limb bones are short. The middle phalanx of their little finger is abnormal making it incurving. The palm of their hand shows only a single crease instead of the more usual double crease. They also have an extra-wide gap between the first and second toes ('sandal gap'). Their joints are more flexible than normal. The tone of their muscle fibres is less than in normal muscles and they have pelvic abnormalities.

Mental retardation is probably the single most important feature of the syndrome and Down's syndrome patients are one of the largest groups of mentally retarded people in institutional care. Their intelligence may in some cases be within the normal range but most have an IQ of less than 70.

Female Down's syndrome patients have reduced fertility but any children are usually normal and not affected by Down's syndrome. Males are usually infertile.

They suffer from an increased incidence of certain diseases: endocardial defects 40%; cataracts 2%; epilepsy 10%; hypothyroidism 3%; and leukaemia 1%. Alzheimer's disease is common in Down's syndrome patients who survive beyond 40 years old.

This is a large list of symptoms so the syndrome is quite complex and involves the imbalance of many genes.

Trisomic foetuses other than Down's syndrome rarely survive

Other trisomic foetuses that sometimes survive to birth are Patau syndrome, which is a trisomy of chromosome 13 and occurs with a frequency of around 1 in 5000 births, and Edwards' syndrome, a trisomy of chromosome 18 with a frequency of 1 in 3000 live births. Most other trisomics abort and do not reach birth. The incidence of most trisomies shows a link with maternal age (see *Box 15.9*). Other factors predisposing to the incidence of trisomy include exposure to radiation, hypothyroidism, viral infections and familial tendency.

Box 15.9 // Effect of maternal age

A number of chromosomal abnormalities including Down's syndrome have a strong link with maternal age. The incidence rises with the age at conception. Mothers over the age of 40 have a 20- to 40-fold increased risk of having a Down's-affected child compared with a 20-year-old mother. This indicates a non-disjunction in ovum formation possibly due to the long meiotic prophase.

Girls are born with all of the ova in their ovary already partially in the first meiotic prophase. The meiotic division only goes to completion during ovulation. This means that some eggs may remain in the middle of cell division for 40 or more years. It is hardly surprising that occasionally this results in chromosomes sticking together and not separating. Because of the non-disjunction, one of the daughter nuclei is effectively trisomic and the other monosomic.

In Britain, the number of women delaying the birth of children until later in life is increasing. This results in an increase in chromosomally abnormal conceptions. However, the availability of pre-natal diagnosis and elective abortion of affected embryos means that the number of affected live births is falling. This means that published figures for the frequency of chromosomal abnormalities often differ depending on whether they consider only live births and the years that are included in the survey.

Sex chromosome aneuploidy

The sex chromosomes X and Y have unique features that make them more likely to produce viable aneuploidy syndromes than most of the autosomes.

The Y chromosome is a very small and insignificant chromosome that has no function except for determining sex. The Y chromosome converts the embryonic gonad into a testis and does very little else. The testis then produces testosterone that controls the main body structure and produces either a male or a female type of body. The Y chromosome is, however, dominant and if present will turn the foetus into a 'male' through the switching on of androgen (testosterone) production at puberty. This occurs regardless of how many X chromosomes are present. With its regular shortening in length it is entirely possible that the Y chromosome will eventually become redundant. This is an event that has happened in other animal species, resulting in another chromosome taking over its primary function.

The X chromosome is large and important and contains many genes totally unrelated to gender. However, in adults only one X chromosome is ever active in any cell. Any extra X chromosomes above the single one that is needed for normal cell activity are 'switched off' during early embryonic life between day 12 and day 16. This inactivation of any supernumerary X chromosomes is called Lyonization. This results in a visible heterochromatic (dark-staining) structure close to the nuclear border in epithelial cells (see *Fig. 15.6*). This structure is called the **Barr body** and provides a convenient way of determining the number of X chromosomes within a cell (see *Box 15.10*).

Figure 15.6
Two cells showing Barr bodies. In the cell on the left, the nucleus shows a Barr body at the bottom right (at about '5 o'clock'), whilst the one on the right is visible at the top of the nucleus (about '12 o'clock')

If only one X chromosome is present (e.g. normal male), then no Barr bodies are seen. If two X chromosomes are present (normal female), then one Barr body is seen as the unwanted X is switched off. The number of X chromosomes is thus the number of Barr bodies plus one. Abnormal numbers of Barr bodies are found when the phenotypic (body) sex is different from the chromosomal sex, e.g. Klinefelter's, Turner's and 47XXX syndromes.

Barr bodies can be visualized by taking a buccal (cheek) smear and staining with a basic dye such as cresyl fast violet. The Barr bodies are then seen as small dark-staining structures inside the nucleus close to the nuclear border. Not all cells allow the Barr body to be identified but about one-third of cells should show distinct Barr bodies, so examination of about 30–40 cells should give a good indication of their presence.

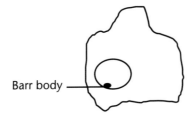

Barr body

Klinefelter's syndrome

This syndrome is associated with an extra X chromosome giving a complement of 47XXY or less commonly 48XXXY. It can also occur as a mosaic in which only some of the cells of the body have an abnormal chromosome number. Klinefelter's syndrome has a frequency of around 1 in 1000 live births so it is slightly less frequent than Down's syndrome.

Klinefelter's syndrome patients appear to be male, since they have the dominant Y chromosome, but they are usually infertile. Investigation of infertility may be the first time they are diagnosed and about 10% of all infertile males turn out to have Klinefelter's syndrome. Klinefelter's syndrome patients usually have a lower than normal testosterone level and have small testes and poorly developed secondary sexual characteristics. They tend to have long limbs and may develop slight breast enlargement, which is called gynaecomastia. They are more likely to develop diabetes mellitus and as group they have

> **Box 15.10 // Sex determination in sport**
>
> The use of Barr bodies to determine chromosomal sex was introduced into athletics to prevent cheating by men masquerading as women. It was first used in Mexico in 1968. This is at first glance a good test for sex since buccal smears are a better and less intrusive test than physical examination of the external genitals of female athletes, which was the previous method for determining the sex of athletes.
>
> What has happened in practice, however, is that the test has not detected any cheating where men (XY) have tried to compete as women (XX), but it has, perhaps unfairly, disqualified athletes who have genuine medical conditions such as XX/XY mosaics and androgen insensitivity. Androgen insensitivity is a condition in which the receptors for the male hormone are abnormal resulting in the inability of cells to respond to testosterone. Although they are Barr body-negative (they have only one X chromosome), they are, if anything, at a disadvantage since they have no response to testosterone (an anabolic steroid) whereas 'normal' women have a low level of testosterone, which would enhance athletic performance.

a lowered average IQ although individuals may be more intelligent than average. There is one (or more) Barr body present, despite the person being apparently male.

There is no major link between the incidence of Klinefelter's syndrome and maternal age; about 60% of the extra chromosomes are of paternal origin and 40% of maternal origin.

Turner's syndrome

In this syndrome there is only one sex chromosome. Thus, they are usually designated as X0 (X nought). Turner's syndrome often occurs as an incomplete form of the disease with about 57% of affected individuals being 45X0, 16% being mosaic with only some cells affected, 10% having deletion of part of the X chromosome and 17% having their X chromosomes fused to form an isochromosome X. Cases of Turner's syndrome are less common than Klinefelter's syndrome with around 1 in 5000 live births being affected.

Turner's syndrome patients appear female but have primary amenorrhoea and never menstruate. They have redundant skin around the neck with a short stature and this gives them a very stocky appearance. Their intelligence is normal and they have a normal life span. Congenital heart disease is more common in Turner's syndrome patients, with up to 20% of patients being affected, and they also suffer more commonly from hypertension (27%). Over 75% have only the maternal X and 25% only the paternal X, suggesting that a defect in sperm formation is more common in this syndrome than ovarian defects. The effects on the patient vary in severity. Mosaic and deletional variants of the syndrome often show relatively mild effects. There is no Barr body present in the buccal smears, despite the person being apparently female, although Barr bodies may be present in some cells of mosaics since the defect only affects some cells in these individuals.

47XXX, superfemale or supernumerary X syndrome

In these patients, there are more than two X chromosomes present and although 47XXX is the most common form, it is also possible to get 48XXXX and 49XXXXX. They occur in about 1 in 1000 births. This syndrome often goes totally unrecorded since the affected female is apparently quite normal but there may be some slight reduction in IQ, especially with XXXX and XXXXX variants, although this would usually be within the normal range for an individual. Affected females are usually fertile and, despite being chromosomally abnormal themselves, their children are quite normal. There does seem to be a maternal age effect. Buccal smears show multiple Barr bodies present depending on the number of X chromosomes present.

47XYY double Y syndrome
The majority of these patients are asymptomatic and are never diagnosed. XYY occurs with a frequency of around 1 in 1000. Although there are no distinctive signs, their IQ may be reduced and they may be more aggressive and have behavioural problems. This is borne out by the reported finding of XYY at a rate of 20 per 1000 of the prison population compared with 1 in 1000 in the general population. Thus, XYY patients were 20 times more likely to end up in prison than a normal male in this study. This link with behaviour is controversial. Patients with XYY are usually fertile and have normal offspring.

Box 15.11 // Syndrome symptoms may be corrected by therapy

Early diagnosis of affected children is becoming more common and treatment can be given to correct abnormalities. Thus, growth hormone treatment of a Turner's syndrome child would mean that the characteristic short stature would be prevented. Similarly, plastic surgery can correct some abnormalities such as webbing of the neck. Thus, the symptoms described are often missing and the descriptions may apply only to untreated patients.

Structural abnormalities of chromosomes
Structural abnormalities involve changes of less than one complete chromosome (see *Fig. 15.7*). Different types of abnormality occur but the overall incidence is about 1 in 200 live births. The number of different possible changes means that a shorthand description is often used (see *Box 15.12*).

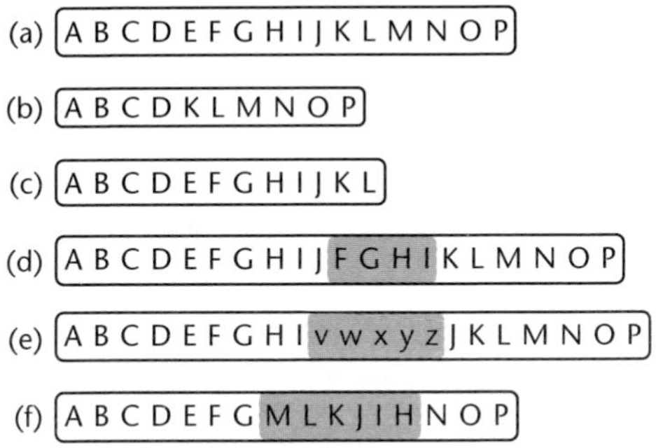

(a) A B C D E F G H I J K L M N O P

(b) A B C D K L M N O P

(c) A B C D E F G H I J K L

(d) A B C D E F G H I J F G H I K L M N O P

(e) A B C D E F G H I v w x y z J K L M N O P

(f) A B C D E F G M L K J I H N O P

Figure 15.7
Structural abnormalities of chromosomes: (a) normal chromosome; (b) internal deletion; (c) terminal deletion; (d) duplication; (e) translocation; (f) inversion

---

Deletions
Deletions involve the loss of part of the chromosome and act like incomplete monosomy. The symptoms are usually similar to the full monosomic syndrome but with a reduced range and intensity of effect. The severity is dependent on the extent of the deletion, e.g.

> **Box 15.12 // Nomenclature of chromosomal abnormalities**
>
> The position and form of chromosome abnormalities are described in a shorthand form such as 46, XX, t(9:22)(q34;q11), which shows the Philadelphia chromosome in a woman. The first figure is the total chromosome count (in this case 46, normal diploid number). The next gives the sex chromosomes (in this case XX, female). The letter 't' indicates a translocation (del = deletion, dup = duplication, inv = inversion, fra = fragile site). The next pair of numbers in brackets indicates the chromosomes involved and the last set indicates where on the chromosomes the changes occur. In this case, chromosomes 9 and 22 are involved with the short arms (q = short arm, p = long arm) being affected at positions 34 and 11 in the banding patterns.
>
> The complications of nomenclature have increased since the technology has improved the resolution of chromosome bands. Whole books on nothing but nomenclature have been written. Details can be found in *ISCN 2005. An International System for Human Cytogenetics Nomenclature.*

deletion of part of chromosome 5 allows survival to birth as Cri du Chat syndrome, whilst monosomy of chromosome 5 is lethal. Smaller **microdeletions** show less widespread defects and are linked to single diseases (e.g. malignancy) rather than a syndrome of symptoms.

Duplications

Duplications are small increases in the amount of chromosomal material and act as incomplete trisomy. Again the range of symptoms and severity vary with the size of the duplication.

Translocations

Translocations involve transfer of genetic material from one chromosome to another. If the translocation goes in both directions, it is described as a **reciprocal translocation**. If there is no loss or duplication of material, then the effects are minimal as they constitute a **balanced translocation**. Although balanced reciprocal translocations are not deleterious, they can result in problems during meiosis when the balance can be upset and the offspring may be affected by **unbalanced translocations**.

Robertsonian translocations involve the fusion of the long arms of two chromosomes, often with the total loss of the short arms. Translocation Down's syndrome is typical of such translocations. **Isochromosomes** are formed by the fusion of the long arms of both isologous chromosomes.

Translocation Down's syndrome. This involves the translocation of part of chromosome 21 on to another chromosome. The commonest form (about 4% of Down's cases) involves a Robertsonian transfer of the long arm of chromosome 21 on to the long arm of chromosome 14 giving a large abnormal translocation chromosome.

The loss of the short arms does not seem to cause any major problems and people with the chromosome can appear quite normal, i.e. they have a carrier status. The carriers are, however, easily detected cytogenetically as they have only 45 chromosomes and the abnormal chromosome is much larger than normal. Carriers can transmit the condition to their children depending on how the chromosomes transfer into the gametes. The chromosomes pair, with chromosome 14 dominating the pairing. There are four gametes that are likely to be produced and these are shown in *Fig. 15.8.*

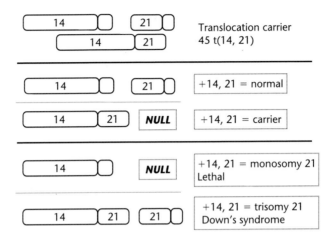

Figure 15.8
Inheritance of translocation Down's syndrome. This shows the chromosomes of a carrier at the top with a translocation 14, 21 chromosome and a normal 14 and a normal 21 chromosome. The two lower sections show the gametes produced depending on the pairing of chromosome 21. If the normal chromosome 21 ends up in the same nucleus as the normal chromosome 14, then after the addition of a normal set of chromosomes from the other gamete, the embryo will be diploid. The other nucleus formed from this pairing will contain only the translocation chromosome and after fertilization with a normal gamete will produce a translocation carrier. If the normal chromosome 21 ends up in the same nucleus as the translocation chromosome, then one nucleus would produce a Down's syndrome-affected child and the other would produce a monosomy of chromosome 21 that is inviable

The Down's affected patients with translocation have 46 chromosomes (unlike the 47 of trisomic Down's syndrome) and have the large abnormal chromosome. Those with the translocation Down's syndrome show a familial pattern of inheritance and it is not connected to maternal age.

The risk of a carrier having an affected child is not a simple 1:4 Mendelian ratio as the Down's syndrome foetus is less viable than a normal foetus and abnormal gametes are likely to be less fertile (especially abnormal sperm). If the male is carrying the 14:21 translocation chromosome, then the risk of having an affected child is about 1%, whilst if the female is the carrier the risk rises to 15%.

Inversions

Inversions occur when a chromosomal segment becomes reversed. Provided there is no loss or gain of chromosomal material in the inversion, then the person will be normal. Problems only arise during gamete formation and only then if a crossover occurs at meiosis.

If a crossover occurs within the inverted segment, then abnormal gametes will be formed. The severity depends on whether the inversion includes the centromere (called a pericentric inversion) or is entirely within one arm of the chromosome and does not include the centromere (paracentric inversion). Paracentric inversions will give recombinant chromosomes, one of which is acentric (no centromere), whilst the other is dicentric (two centromeres) and both nuclei would produce non-viable cells. Crossovers affecting pericentric inversions produce abnormal offspring due to the chromosomes having both a deletion and a duplication.

In female carriers of an inversion, about 8% of the offspring will be affected and in male carriers 4% of the children will be abnormal.

Chromosome 9 is unusual and shows a relatively high incidence (about 1% of the population) of a small pericentric inversion but no abnormal offspring have been reported as being associated with this inversion.

Fragile sites on chromosomes

These are sites on chromosomes that are particularly vulnerable to damage. These fragile sites occur in some inherited diseases.

The commonest disease associated with a fragile site is **Fragile X syndrome** (see *Box 15.13*). This occurs in about 1 in 2000 males and 1 in 4000 females. The cause is an abnormality of the distal end of the X chromosome where there is a multiple repeat of the CGG sequence of bases. In a normal X chromosome, there are less than 50 such repeats but in affected individuals there are more than 200 such sequences. These sequences render the chromosome more fragile than normal and can cause loss of part of the chromosome. The affected individuals show mental and physical handicaps. The abnormality can be detected cytogenetically or by Southern blotting using the restriction enzymes *Eco*RI and *Eag*I.

Other fragile sites also occur in Huntington's chorea and myotonic dystrophy. All show the phenomenon of **anticipation** where the disease gets worse from generation to generation, presumably as the number of trinucleotide repeats increases making the site more fragile.

Box 15.13 // Fragile X Syndrome

Fragile X syndrome is the commonest form of inherited mental retardation with an estimated incidence of 1 in 1000 males. It is sex-linked with the incidence figures being lower for females. The true incidence is not known as there have not been any large-scale population studies. It is possible that around 60–90 % of people with Fragile X syndrome are not yet correctly diagnosed.

Fragile X syndrome patients may show some physical features associated with the disease including a long narrow face and prominent ears, jaws and forehead but these are not seen in all patients. Up to 80% of boys who inherit Fragile X syndrome have some mental impairment. The majority are only mildly to moderately retarded. Girls are less affected with only about 30% having some degree of mental retardation; this is because the faulty gene is supplemented by the other X chromosome.

Up to 20% of those with Fragile X syndrome may show some autistic behaviour, e.g. poor eye contact, hand-flapping, hand-biting and poor personal skills. Behaviour problems and speech or language difficulties are also frequently associated with Fragile X syndrome.

Mosaicism

Most chromosomal abnormalities in foetuses occur due to defective meiosis or fertilization and result in all cells of the embryo being affected. However, it is possible that, in the early embryo, mitosis may be defective resulting in only a proportion of cells being chromosomally abnormal. Such individuals are termed **mosaic**. The effects of mosaicism are similar to the full chromosomal abnormality but in a milder form. The severity of the symptoms is related to the proportion of cells affected.

The extent to which cells are affected depends on how early in embryogenesis the abnormal mitosis occurs. If it is an early division, then many organs can be affected, whilst a later abnormal division may affect only a few tissues. If the abnormal mitosis was the very first division of the fertilized ovum then about half of the cells in the final foetus would be affected. This would give a more severe syndrome than if the abnormality occurred in the second division when only one quarter of cells would be affected.

About 1% of Down's syndrome cases, 16% of Turner's syndrome and 5% of Patau syndrome cases are mosaic rather than the full syndrome. In mosaicisms, several tissues may need to be karyotyped to detect the abnormality since only a few tissues may have developed entirely from the defective cell in the embryo.

Suggested further reading

Gersen, S. and Keagle, M.B. (eds) (2012) *The Principles of Clinical Cytogenetics*, 3rd edn. Springer.

Heim, S. and Mitelman, F. (eds) (2009) *Cancer Cytogenetics*, 3rd edn. Wiley-Blackwell.

Kingston, H.M. (2002) *ABC of Clinical Genetics*, 3rd edn. Wiley-Blackwell.

Pfeifer, J.D. (2006) *Molecular Genetic Testing in Surgical Pathology*. Lippincott Williams and Wilkins.

Strachan, T. and Read, A. (2010) *Human Molecular Genetics*, 4th edn. Garland Science.

Self-assessment questions

1. Explain what the term cytogenetics means.
2. How can cytogenetics be useful in clinical practice?
3. Outline the preparation of chromosome spreads from human lymphocytes. How are the chromosomes made to separate?
4. How can the banding patterns on human chromosomes be demonstrated?
5. Give the karyotype of a normal human female. What would a karyotype of 47 XY, +21 mean?
6. Approximately how common are chromosome abnormalities at conception and how common are they at birth?
7. Why do alterations in the sex chromosome numbers occur more frequently than abnormalities in the number of most other chromosomes?
8. What are the differences between the usual Down's syndrome and the translocation form of Down's syndrome?
9. What is a pericentric inversion and how will it affect the individual?
10. What is a hydatidiform mole and what abnormalities of chromosomal number occur?

Molecular biology and autoradiography

Learning objectives

After studying this chapter you should confidently be able to:

- **Briefly describe the role of *in situ* hybridization (ISH)**
  In situ hybridization allows changes in the DNA sequence to be identified and localized. This can be used to identify mutations in individual cells as well as whole-body changes.

- **Explain the mechanism of DNA hybridization**
  DNA is dissociated (melted) by heating. Labelled probe DNA can then combine with the dissociated DNA and anneal with it when the temperature is lowered.

- **Briefly outline the principle of the PCR technique**
  PCR involves repeating the sequence of melting and replicating DNA. At each round of the sequence the restriction endonucleases will copy the dissociated DNA, forming a new strand. At each round the quantity of target DNA will double. The increase in the target DNA allows easier detection of very small amounts of DNA.

- **List and compare the different methods of labelling DNA probes**
  DNA can be labelled with radioisotopes, fluorescent dyes, haptens or colloidal gold. Radioisotope labelling requires autoradiography and is therefore slow but is very sensitive. Fluorescent dyes and haptens are more convenient for routine use and are quicker. Colloidal gold is useful for electron microscopy.

- **Describe how tissue should be prepared for *in situ* hybridization**
  Routine formalin-fixed wax sections are suitable for ISH. Cryostat sections can be used but need formalin fixation after cutting, to retain DNA. Tissue nucleases need to be inactivated. DNA may need unmasking with a protease.

- **Outline the applications of ISH**
  Applications include research, identification of viruses and bacteria, identification of tumours and indications of the best therapeutic approach, detection of abnormal genes and karyotyping.

- **Outline the principle of autoradiography**
  Autoradiography uses radioisotopes to follow the fate of molecules in the body. The labelled compounds are used in cell cultures or experimental animals and metabolized by the cells. Sections of the tissue can be prepared and the isotopes are located using photographic emulsion. Ionizing radiation released by the isotopes affects the photographic emulsion, which shows blackened areas following development.

- **Name some radioisotopes commonly used in autoradiography and the nature of their radiations**
  These are ^3H (tritium), ^{14}C, ^{32}P and ^{35}S. These are all β-emitters with energies varying from 19 to 1710 keV.

16.1 Basics of molecular biology

The sequence of bases in DNA specifies all the proteins that can be manufactured by cells and is the fundamental mechanism used in the basis of the transfer of genetic information from one generation to the next. The role of the double helix and the specific pairings of bases (adenine pairing with thymine and guanine pairing with cytosine) is well known and the details can be found in any modern biochemistry or genetics textbook. The manipulation and control of this base-pairing mechanism is the basis of most molecular biology techniques. The accuracy of the pairing allows specific sequences of bases in DNA to be identified and localized. This can be used to identify the position of mutations and therefore genetically caused diseases, including hereditary errors of metabolism and acquired mutation-related diseases such as malignancy in cells.

The decision to take a genomic, DNA probe approach for analysis is not solely for the confirmation of abnormal gene expression. It may be that a key cellular product, e.g. a peptide hormone, is rapidly metabolized or transported away from the cell of origin, making the proteomic (antibody) approach either insensitive or inaccurate. In this case the ability to demonstrate either the relevant mRNA or number of gene copies in malignant cells provides a valid alternative mode of analysis. The pairing of a labelled probe with its complementary sequence in RNA is perfectly feasible and allows the localization and characterization of specific mRNA base sequences. The detection of these characteristic mRNA sequences enables the molecular biologist to identify either which cells are using a particular gene or any changes in the activity of particular genes with time.

The combined use of a chromosome label (centromere specific) together with a DNA *in situ* probe has become the gold standard in revealing the nature of some key malignancies, e.g. breast cancer.

Even a brief explanation of the methodological approach requires a reasonable knowledge of DNA behaviour and the enzymes involved, so first we will briefly cover the basic biology relevant to the use of molecular probes. Much more detail can be found in the many molecular biology texts that are available, or from probe suppliers' websites.

Many enzymes are used in molecular biology for the manipulation and identification/characterization of DNA sequences. Key enzymes used in molecular biology include:

- **Nucleases** (deoxyribonucleases) are enzymes which split DNA.
- **Exonucleases** remove bases from the ends of a DNA chain.
- **Endonucleases** split DNA in the middle of the chain.
- **Restriction endonucleases** (see *Box 16.1*) are particularly useful in molecular biology because they cleave DNA at specific sites; they will only bind to a specific sequence of nucleotides. For example, the restriction endonuclease *Hind*lll attaches to the sequence ---AAGCTT--- and splits it between the two adenosine nucleotides in both DNA strands. This cleaves the DNA into two fragments each with a protruding single chain 'sticky' end which can re-anneal as:

  ---A **AGCT**T---
  ---T**TCGA** A---

- **Ligases** are enzymes which 'heal' or 'restitch' breaks in a strand by relinking the nucleotides at either side of the break. In the case above the ligase would relink the two adenosines after annealing of the sticky overhangs.

  Point of ligase relinking
  ↓
  ---A **AGCT**T--- ---A**AGCT**T---
  ---T**TCGA** A--- ---T**TCGA**A---

Box 16.1 // Restriction enzymes

Restriction enzymes are named from their discovery in strains of bacteria that were more resistant to bacteriophage viruses than other strains of the same bacteria. The resistant bacteria were said to be 'restricting' the bacteriophage by cleaving the viral DNA.

Restriction endonucleases are also the basis of 'genetic fingerprinting'. The very precise mechanism for cleaving the DNA has made it possible to compare the fragmentation pattern of a DNA sample collected during a crime investigation with the fragmentation pattern of a suspect. The chances of two humans having identical DNA fragmentation patterns are very small (unless they are identical monozygotic twins) so a positive match is a strong form of forensic evidence of guilt.

- **DNA polymerases** are important in DNA replication and repair. The polymerase will form a new DNA strand onto a single stranded DNA by adding nucleotides (A, T, G or C) one at a time, matching each nucleotide to those on the existing strand to form double-stranded DNA. The copying can only go in one direction along the strand (5′ to 3′ direction). The polymerase can also only add nucleotides to an existing chain and cannot start a new strand. To start the polymerase activity there must be at least a short section of double helix DNA to provide a free 3′ end to which new nucleotides can be added.
- **Reverse transcriptases** are enzymes which use RNA as a template to produce a DNA strand. These were originally discovered in RNA viruses (retroviruses) such as HIV and they allow a DNA copy of the viral genome to be formed and then inserted into the host cell DNA. Reverse transcriptases are used in molecular biology to produce DNA, which is referred to as cDNA (copy DNA).

Polymerase chain reaction (PCR)

Nucleic acids in sections can be extracted and identified using gel-based blotting techniques (e.g. Southern blotting) or the very powerful **polymerase chain reaction (PCR)**. Together these are two of the main techniques used in molecular biology (see *Box 16.2*).

PCR is a method of cloning DNA very rapidly. It uses a short primer that is known to bind to the DNA to be cloned. The sample is incubated in a solution containing the primer, DNA polymerase enzyme (called Taq polymerase) and the four deoxyribonucleotide triphosphates needed to synthesize DNA (see *Fig. 16.1*). The sample DNA is heated to separate the two strands of the DNA helix and then the mixture of single-stranded DNA, enzymes, oligonucleotide primers and ribonucleotides is cooled to allow hybridization with the primer oligonucleotides. The DNA polymerase then forms a new strand on each of the old strands, resulting in two DNA helices. By repeating the heating and cooling sequence, each of the two will gradually separate to form four new DNA helices. Every

---

Box 16.2 // PCR and Southern blotting

These techniques are not usually done in histology laboratories but in more specialized molecular biology units.

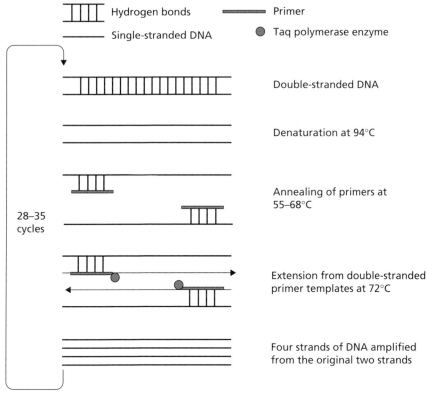

Figure 16.1
Diagram of PCR mechanism.

time the cycle (heating to separate the strands, cooling to allow hybridization and finally growing a new strand) is repeated, there will be a doubling of the amount of DNA. Approximately thirty cycles generates a huge number of copies of the original DNA (about 10^5–10^6 copies).

In histology the PCR cycle will give copies of DNA in the cells but the new DNA molecules will be in solution rather than localized in the tissues. The reaction can be useful to detect very small amounts of DNA that might be missed when using *in situ* hybridization. This might occur in viral infections where perhaps only one or two cells in the tissue might have the viral genome. In this case a positive PCR result will allow the virus to be detected, although it will not indicate which cells are infected.

These techniques use the tissue section simply as a source of DNA and extract it from its true position. This ignores the ability of histological sections to identify exactly where specific sequences are located. The PCR is also so sensitive that extremely high levels of care and control are required to ensure there is no contamination. The benefit of a histological approach is that it will not only detect if a sequence is present but will also locate which cells have the sequence. This is achieved by a range of methods that leave the nucleic acid in position (*in situ*) and are referred to as *in situ* hybridization methods (ISH). The following sequence indicates the main steps of a standard approach for slide preparations.

- Attach sections or cells to coated slides.
- Dewax histological FFPE sections.
- Digest with buffered proteinase at 37°C (helps to free DNA from histones and binding proteins).
- Separate DNA strands (correct melt temperature and time are crucial).
- Incubate with labelled probe (regulate probe binding with salt concentration = stringency conditions).
- Anneal the separated DNA strands (correct temperature and time are crucial).
- Visualize the label attached to the probe.

N.B. this sequence also has buffer washes between each of the main steps.
More specific detail of the methodology is given in ***Section 16.2***.

16.2 DNA hybridization

In order to detect the hybridized DNA, the probe DNA needs to be tagged or labelled to allow easy visualization of the hybrid DNA.

The production of labelled probes requires two components: firstly a sample of DNA (probe DNA) which will hybridize with the target nucleic acid, and secondly a label, either attached to or incorporated in the probe DNA which is detectable by a reliable visualization method. In this account, DNA techniques will be used as an example, but similar modified methods are applicable to detecting RNA sequences.

DNA hybridization involves 'melting' and 'annealing' of two DNA strands

DNA consists of two complementary strands that can unite to form a double helix and this is the usual form of DNA in human cells and in fixed tissue samples. The helix is held together by hydrogen bonds that are sufficiently weak to be broken by raising the temperature ('melting' or denaturation of DNA) or by using mild chemical treatment. Different DNA samples will melt at different temperatures. The more G–C base pairs there are in the DNA, the higher the melting temperature will be, since G–C pairing is stronger as it involves 3 hydrogen bonds. A–T pairs, on the other hand, only have 2 hydrogen bonds. A temperature of 90–95°C will usually melt mammalian DNA. If the temperature is then reduced below 70°C, the melting can be reversed and complementary sequences can realign and reform the double helix ('anneal') using their specific hydrogen bonding (see ***Fig. 16.2***). Sequences that are not complementary will not anneal in this way since the specific hydrogen bonding will not be correctly aligned. Annealing will only occur

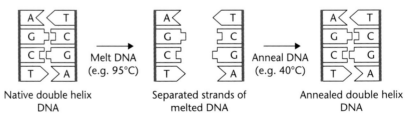

Native double helix
DNA

Separated strands of
melted DNA

Annealed double helix
DNA

Figure 16.2
Principle of DNA hybridization.

when the two sequences are identical or at least very similar. This very specific bonding is made use of in hybridization techniques to identify specific DNA base sequences.

Probe DNA sources

The DNA for probes needs to have an identical sequence of bases to the target DNA. In most cases the sequence will be found in a natural source. To detect viral DNA in a cell, the DNA can be extracted from a cultured virus and then split using restriction enzymes to make probes of an appropriate size; it is then separated by electrophoresis into individual probes. Bacterial DNA can be similarly extracted and fractionated. The probe DNA then needs to be verified to ensure it will hybridize with the target DNA and will not cross-react with other DNA which might be in the specimen. For example, viruses which insert themselves into human DNA can pick up fragments of human DNA, so some of the potential probe DNA may be human in origin; this probe DNA will hybridize with normal human DNA as well as infecting viral DNA.

When the target DNA is a human gene, the problem becomes more difficult. Extracting all of the DNA present in the human nucleus, purifying it and then fractionating and determining the DNA sequence is an immense undertaking. Fortunately this has already been done by the **Human Genome Project** and DNA libraries of all the human gene fragments are available. Probe DNA can be obtained from these libraries and then cloned to provide sufficient DNA probe for the testing.

If the target sequence is known then it is also possible to construct the probe directly from the sequence information *in vitro* using molecular biology techniques. DNA up to several thousand base pairs can be synthesized commercially. Synthesis is only economical for short lengths of DNA but this can be ideal for many probes.

Labelling of probes

The labelling problems and solutions are in many ways similar to the labelling techniques used with proteins, as outlined in *Chapter 12*. There are two different approaches to labelling:

1. Chemical treatment of the probe DNA to attach the label to the probe DNA.
2. Incorporation of labelled nucleotides into the probe DNA during enzymatic DNA synthesis.

Chemical treatment

Chemical methods add the label randomly onto the DNA. This can give a higher degree of labelling and therefore a stronger signal. The disadvantage is that the chemical treatment alters the nucleotide base structure and this can interfere with hybridization. Direct chemical treatment is less popular for ISH than incorporation of labelled nucleotides.

Labelled nucleotide incorporation

Nucleotides are commercially available already tagged with an appropriate label. Different techniques are used to incorporate the label and three are explained below.

Nick translation labelling. Double stranded DNA is treated with a low concentration of deoxyribonuclease 1, which hydrolyses the bonds between known bases leaving a 'nick' in one of the DNA strands and the reaction is stopped. Nucleotides including some labelled nucleotides and DNA polymerase1 are added to the nicked DNA. The polymerase uses the nick to begin synthesizing a new strand in the 5'–3' direction, using the other strand as a template. The polymerase has an exonuclease activity which removes nucleotides; the polymerase then incorporates labelled nucleotides into the new DNA. This results in several labelling sites in each probe, which will give a stronger signal.

PCR labelling. The PCR method outlined above can be used to incorporate tagged nucleotides into the DNA and will increase the number of probe DNA at the same time. The PCR method does of course require an oligonucleotide primer.

End labelling. As well as labelling in the middle of the probe it is also possible to label just the 3′ end using a terminal transferase. This results in a weaker signal as there is less label, because it is only incorporated at the free end of the probe. However, this can be more than one nucleotide long.

Labelling of the reporter nucleic acid
There are a variety of different labelling methods including fluorescent, hapten, metal and radiolabelling. As well as the generic methods which will be outlined below, there are a number of commercial variants which are copyrighted. All of the labelling techniques are still used but some are more popular than others. Some are more suited to particular areas of investigation such as research, routine diagnostic and forensic.

Radiolabelling. This was the original labelling technique in which DNA bases have one of their atoms replaced with a radioactive isotope, such as the replacement of hydrogen with tritium, or carbon with C14. Since the atoms are chemically identical there are no problems in their compatibility with enzymes and hybridization. The radiolabelled molecules are treated the same as the normal materials. The disadvantages are the dangers and problems of radioactive compounds, the slowness of autoradiography for obtaining the results and the cost. It is, however, the most precise localization method and is still used for some research applications. Additionally it can be used in conjunction with standard staining methods to visualize normal microanatomy, e.g. H&E or Giemsa.

Fluorescent labelling. Also known as FISH (fluorescent *in situ* hybridization), this has become very popular. The fluorescently labelled nucleotides are widely available with either fluorescein or proprietary commercial dyes such as AlexaFluor. The results are rapidly available following hybridization. Fluorescence does require a special microscope and staining of normal tissue structures is limited. Fluorescence fits in well with confocal microscopy, allowing precise localization in all three dimensions. The fluorescent signal is sometimes weak, making visualization difficult and the fluorescence does fade, which makes the reviewing of older preparations impossible.

Hapten labelling. A hapten is a small molecule attached to the probe which, though not immediately visible, can be detected using the chromogenic techniques of immunocytochemistry. The chromogenic techniques can amplify the result, can produce coloured precipitates visible with ordinary light microscopy and can be combined with standard staining techniques to show tissue structure. The results are more stable than fluorescence and can even be adapted for electron microscopy. This has made chromogenic *in situ* hybridization (CISH) increasingly popular.

Hapten-labelled nucleotides are widely available using a variety of haptens. The limitations on the hapten are that it must not seriously interfere either with the enzymes used in incorporating the labelled nucleotides into DNA or with hybridization between the probe and the target DNA. These limitations restrict the size of the hapten. Common haptens include biotin, which is detected with the same systems as biotinylated antibodies (e.g. avidin / streptavidin HRP) and digoxygenin (uses antidigoxygenin antibodies as part of an IHC detection). Fluorescent dyes such as fluorescein can also be considered as haptens and can be detected with antibodies (anti-fluorescein antibodies) and one of the standard IHC methods. This will amplify the signal from a weakly fluorescent specimen and also raises the possibility of reviewing old FISH slides which have faded with age, although how useful this might be is open to question.

Silver-enhanced ISH uses the enhancement capabilities of silver techniques combined with the CISH probes to amplify the signal and produce a stable silver deposit, which gives a dense black and allows the use of two probes with very different appearances on the same specimen. The silver grains produced can also be used in immuno-EM.

Colloidal gold spheres. Gold spheres for electron microscopy can be used with DNA probes in a similar fashion to the use of immunogold techniques.

The choice of the probe nucleic acid and size can be crucial. Small probes will react with more than one site in the genome (see ***Box 16.3***).

The choice of method is determined by the speed and sensitivity required, and whether EM detection is needed. Radiolabelling is rarely used in diagnostic ISH because it has several drawbacks (it is slow, expensive and radioactivity is a problem for handling and disposal).

Commercial probes are available for many important genes, viruses and other diagnostic applications and these are often available with a range of different labels. In research, ready prepared probes for a particular purpose may not be available, so custom-made probes may be synthesized and labelled by the laboratory or commissioned from commercial sources.

Box 16.3 // Probe size

There are four nucleotides, so as a rough calculation the frequency of a particular sequence occurring is related to 4^X, where X is the number of bases in the probe. One nucleotide will occur every 4 (4^1) bases. Three nucleotides (i.e. codon equivalent to 1 amino acid) will occur every 1 in 64 bases (4^3). Sometimes quite a small oligonucleotide probe is appropriate (20–30 nucleotides, $4^{25} = 1.125 \times 10^{15}$). The human genome has around 3.2 $\times 10^9$ bases so, assuming random base frequency, only the target complementary site would be expected. Base sequences are certainly not random. Many repeated sequences occur so it may be necessary to check if the probe sequence is repeated in the genome; this can now be done on the internet. As the size of the probe increases, the chances of a false match occurring diminish. A typical probe may have around 250 bases ($4^{250} = 3 \times 10^{150}$) so 'accidental' matches are vanishingly small. Probe sizes may be even larger (up to 1000 bases) but as the probe size increases, DNA penetration into the tissues, and hence annealing, becomes more of a factor.

Cell and tissue preparation

Fixation and processing are often crucially important steps in the success or otherwise of a technique, and whilst the same is true of ISH it is not really a problem as the standard histological approach of formalin fixation and paraffin wax embedding remains the recommended best practice. The formalin fixation preserves the DNA by fixing the attached proteins. The proteins also help to protect the DNA from formaldehyde forming addition compounds with the reactive groups of DNA, whilst the essential hybridization groups of the nucleotide bases are protected from reacting as they are already hybridized and bound into the double helix. Rapid fixation is required, because if tissue begins to deteriorate by autolysis, nucleases are released and will fragment the nucleic acids. For critical applications, perfusion fixation and the use of buffered paraformaldehyde is preferred. The precipitated histones and nuclear proteins serve to act as a barrier to the probe DNA, so tissues normally require protease digestion (e.g. with proteinase K) to unmask the DNA prior to hybridization. The longer fixation is allowed to proceed, the greater the masking effect and this requires more careful unmasking with proteases.

Frozen sections are suitable but need formaldehyde fixation to prevent nuclease activity. The recommended fixative for chromosomal spreads is alcohol; acetic acid with methanol is preferred to ethanol. Other fixatives can be used but are generally inferior.

Paraffin wax processing does not seem to harm the DNA or its ability to hybridize with probes. Normal paraffin wax schedules, including rapid microwave processing, seem appropriate and no special treatment is required.

Since formaldehyde fixation and paraffin wax embedding have been the standard technique for over half a century, it is possible to use old paraffin blocks for ISH. However, even in wax blocks there may be some slight alterations and the standard ISH protocol may need adapting, for example in the choice of enzyme and timings for protease digestion.

Following processing, the main precautions that are needed are good cleanliness and the wearing of gloves to prevent contamination with foreign DNA. Purified reagents, including distilled water, are essential to exclude nuclease activity which could otherwise damage the sample. Sections need to be firmly fixed to the slides as the treatments (e.g. protease digestion) will tend to dislodge the section. The use of charged or coated slides is always advisable. Wax sections need to be dewaxed and returned to water before moving on to the next step.

Pre-hybridization and unmasking of DNA

The major step in pretreatment is the protease digestion to unmask the DNA and facilitate the entry of the labelled probe. Digestion is usually with either proteinase-K or pepsin. Following proteinase treatment, the tissue is often briefly refixed in formaldehyde to minimize losses. In a few specific cases unmasking may require extra treatments, such as proteoglycan removal or removal of lipids.

Blocking of non-specific binding of the probe can be applied here (e.g. acetylation). The tissue is then equilibrated with the hybridization buffer but without the probe DNA.

Separation of the two DNA strands from the DNA helix (denaturing) can be done at this stage before adding the probe DNA (separate or indirect denaturing), with the denaturing of the probe DNA being done separately. Alternatively the probe DNA and target DNA can be denatured at the same time, as part of the hybridization step (simultaneous denaturing). Heating the DNA will cause denaturing ('melting') and can be assisted by adding formamide to the hybridization buffer. The addition of formamide lowers the temperature needed for denaturing. Depending on the presence and concentration of formamide, the melting temperature used is between 65°C and 95°C.

Probe hybridization

In simultaneous denaturation the hybridization probe containing the probe is applied to the section. The total volume added is very small (<50 µl) and evaporation can be a problem, so a coverslip is placed over the section and the edge sealed (e.g. with rubber cement). The section is then placed in a humid environment. The temperature is raised (65–95°C) on a heater block to denature the DNA. Once the DNA of both target and probe have been denatured the temperature is reduced and the tissue held at a lower temperature (25–45°C) to allow annealing. The exact time and temperatures depend upon the probe (particularly probe size) and the constitution of the hybridization buffer. During hybridization the probe DNA and the target DNA will anneal together (see *Fig. 16.3*).

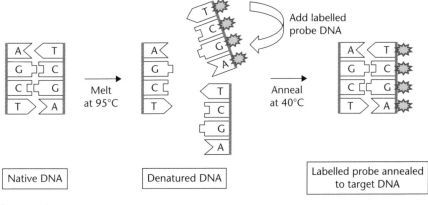

Figure 16.3
Hybridization of probe DNA to target DNA.

Double labelling

It is possible to use more than one DNA probe on a single slide, provided the visualization can produce different colours. With FISH there are several fluorescent dye labels with different colours, so there can be several probes which can all be recognizably different. CISH is slightly less versatile and is often restricted by the labelling to two different probes.

Box 16.4 // Standardization

The methodology of ISH is quite variable, as the conditions needed for good results need to be optimized for particular probes, tissue treatments, and exact requirements. There is no single protocol that will work optimally in all situations. Even within a single laboratory there are likely to be variations. The degree of protein digestion and also the stringency set by salt concentration in buffers both affect melting and annealing temperatures.

A histologist can go into almost any histology laboratory and be able to stain an acceptable H&E, van Gieson or PAS without needing to check for the local protocol, but the same degree of transferability would not apply to ISH.

Nuclease enzymes need to be inactivated

The amounts of DNA being used are very small and can be destroyed by even slight nuclease enzyme activity. The reagents and the sections must be kept free of all nucleases. Ribonuclease is resistant to heat so it needs inactivating chemically. This can be done with the inhibitor diethylpyrocarbonate (DEPC). It is safest to add DEPC to all solutions, including distilled water, and to treat all glassware before use. Nucleases also need calcium ions for their activation so the addition of EDTA to solutions also helps to prevent nuclease activity. Gloves should be worn at all times, as nucleases are present on the skin and even in the water used to float sections before they are transferred to glass slides.

The stringency of hybridization can be controlled using additives

Formamide is a reagent that destabilizes the hybridization complex. Formamide can be used to supplement the effects of heat. The higher the concentration of formamide, the lower the annealing temperature that is needed (about 0.6°C for every 1% formamide). The balance between temperature and formamide concentration is often the critical factor in controlling the formation of precise hybridization complexes.

Sodium chloride can also alter the stringency of hybridization. High salt concentrations will mask the electrostatic repulsion between probe and tissue DNA, allowing less stringent annealing, whilst low salt levels will favour very stringent matching. (This masking of repulsion is reminiscent of the salting-on seen with dyes.)

Fragmented non-specific DNA, usually salmon sperm DNA, can be applied to stop non-specific binding of DNA to tissues in the same way that normal serum is used to block non-specific protein binding in immunotechniques. Dextran sulphate can also be used to block non-specific binding.

Controls

The probes should always be validated for specificity and adequate labelling. These data will normally be available for commercial probes.

Controls are always an important part of any technique and should include as a minimum both positive and negative controls. Positive controls would be a sample known to contain the target nucleic acid that is as close as possible to the test tissue for preparation (fixation, processing, etc.), but this is not always easy to obtain. The simplest negative control for CISH is to omit the labelled probe. Any positive result indicates non-specific binding in the ICC steps. Negative controls can be samples which do not contain the target nucleic acid. For an mRNA target this can be achieved by treating a test section with RNase to destroy the RNA so that any remaining binding of the labelled probe is non-specific background.

An alternative strategy is to use a labelled nonsense probe which has a similar size and GC/AT content as the probe. This will detect non-specific binding by a mechanism other than hybridization. A further way is to use a large excess of unlabelled probe in the prehybridization step as this can bind to the target and block the labelled probe from hybridizing, so any labelling is not by hybridization.

The final steps are much the same as for any other technique, namely to counterstain if required and mount. For CISH and FISH a haematoxylin and eosin counterstain allows normal morphology as well as the ISH, and can be mounted in most routine xylene-based permanent mounting media. FISH requires a fluorescent nuclear counterstain in a contrasting colour, for example with propidium iodide (orange/red) or DAPI (blue), and are usually mounted in buffered glycerol.

Hybridizing for RNA

There are some advantages in identifying RNA. First the target RNA is not a double helix so there is no need to denature before staining. There are often many copies of mRNA so there will be a stronger response than with DNA, where each gene will typically have only two copies. The presence of mRNA, and especially multiple copies of mRNA, indicates gene activity, whilst DNA only shows the gene is present. Similarly if a protein product has a very short half-life because it is rapidly metabolized or exported after manufacture, it may not be detectable by IHC.

Some viruses have RNA as their genetic material, although the RNA may be present in DNA as reverse transcriptase and will produce cDNA from the viral RNA.

RNA is slightly more likely to be lost from the tissue, as the binding to protein is less secure and RNases may degrade it. Fixation in formalin is still the recommended technique to retain the maximum amount of RNA.

Detecting RNA by ISH is very similar to the methods used for DNA and will not be further explained.

16.3 Applications of ISH

There are many individual applications; only a few can be mentioned. The main applications can be considered under the following headings:

- Research
- Clinical – diagnostic
  - virus identification
  - tumour identification and phenotyping
  - indication of therapy strategies
  - detection of abnormal genes
  - karyotyping

An individual probe may actually be relevant in all categories.

Research applications

A lot of the research in molecular biology is done using methods, such as Southern blotting, which are outside the techniques of histology and cytology. One technique using ISH is, however, very useful. This is the use of composite blocks or tissue arrays (TMAs) containing small pieces of different tissues arranged in a grid or array (see *Fig. 16.4*). An array can be made as a 10 × 10 array which would contain 100 individual samples of tissue in a single section. Note that each original sample block is represented by triplicate cores to counteract potential losses caused by variations in core length / orientation. Larger arrays are possible, e.g. 25 × 25 (625 core) but as the number increases, the diameters of each core become smaller. In a single ISH preparation this will therefore identify which of those samples has the DNA or RNA that the probe is targeting. The samples could be

Figure 16.4
Tissue array section showing multiple cores from different tissues used in evaluating techniques

of 100 different malignant tumours and the probe could be of a possible cancer-causing gene such as *p53* or *BRCA1*. The pattern on the array would indicate which tumour types had an association with the gene. Comparisons are therefore quicker and cheaper (only a single ISH slide for 100 samples) and the results easier to grasp. The patterns can be analysed using computerized recognition systems, thus reducing the need for individual human assessment. Arrays for various applications are commercially available, or custom-made arrays can be prepared in house. Direct human confirmation need only be applied to those samples identified as significant by the automated recognition.

Diagnostic applications

Virus identification

The identification of viruses in cells is important in several areas. The role of HPV in cervical screening has been introduced in **Chapter 13**, with conventional screening and HPV status being carried out using the same liquid-based cytology specimen. The detection of HPV is often done using PCR and hybridization rather than ISH, but ISH can be used in tissue samples such as biopsies. The most important HPV strains are HPV-16 and HPV-18. Other viruses associated with malignancy can also be detected by hybridization but these are not used routinely. They are, however, important in research and can be requested in some difficult cases; they include viruses such as Epstein–Barr (Burkitt's lymphoma and nasopharyngeal cancer), herpes virus-8 (Kaposi's sarcoma) and hepatitis B and C viruses (hepatocellular carcinoma).

Tumour identification

Many lymphoma and haematological malignancies can be difficult to diagnose on purely morphological grounds, so methods utilizing immunocytochemistry and molecular biology techniques can be useful adjuncts to conventional diagnosis. There are currently over 30 commercially available probes for a range of haematological malignancies, often those involving chromosome translocations. Detection by these probes can involve PCR or ISH on blood smears. Diagnosis of solid tumours may also be assisted by hybridization techniques; these again are not routine but rather available by request. Tumour types which can be assisted by such techniques include bladder cancer, sarcomas, neuroblastomas and breast cancer.

Rather than being detectable by a single probe, these may require a panel of different probes to identify their tumour phenotype. This can be achieved by preparing labelled probes from the tumour's genes and then using these in conjunction with an array of DNA types fixed on a solid substrate. The tumour DNA can then be seen to hybridize with a subset of the DNA arrays and so the genes active in the tumour cells can be identified and the tumour phenotype determined.

Apoptosis is an important process in a number of pathological processes (see **Chapter 2**) and the large number of 3′-hydroxyl terminal ends produced by the fragmentization can be identified by ISH. The technique uses terminal deoxynucleotidetransferase (TdT) to enzymatically incorporate labelled dUTP at each nicked end, and is referred to as the **terminal deoxynucleotide dUTP nick end labelling** (**TUNEL**) assay. The multiple labelling makes apoptotic cells stand out from normal cells (see **Fig. 16.5**). The TUNEL assay will also detect other cells which have undergone severe DNA damage.

Indication of therapy strategies

The most widely used assessment of therapy is for the *HER2/neu* gene (also known as *c-erbB-2*), which is important in the assessment of breast cancer. This gene is amplified in

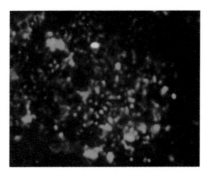

Figure 16.5
TUNEL technique for recognition of DNA fragmentation in apoptosis

approximately 20% of invasive breast tumours. The human oestrogen receptor is a strong indicator of prognosis and can be detected by immunocytochemistry and overexpression of the gene by ISH. The monoclonal antibody Herceptin, used as a treatment for breast cancer, is only effective for this subtype and has toxic effects on cardiac tissue, so the use of Herceptin should be limited to those patients with overexpression of *HER2*. The decision on suitability for treatment depends on the ratio of gene copies (ISH) to the number of marked gene-bearing chromosomes (centromere-specific ISH).

Similar applications are likely to increase as more targeted therapies become more prevalent. These therapies can be very effective but also expensive and possibly with serious side-effects so limiting the treatment to those patients who will benefit is advisable. The use of targeted therapy in which ISH can play a role is already increasing, with antibodies targeting the epidermal growth factor receptor in colorectal cancer and more recently lung cancer. The prognosis for using the antibodies cetuximab or panitumumab can be assessed by ISH for *KRAS* gene status in colorectal cancer, and this will help optimize treatment.

Detection of abnormal genes
Detection of mutated genes is limited, as many mutations are changes in only a single base in the gene (e.g. sickle cell haemoglobin). Such small changes are difficult to detect by hybridization, as a probe will often be able to at least partially hybridize with both the normal and abnormal gene. Larger mutations such as microdeletions or microduplications are more easily detected and several probes are commercially available for known abnormalities.

Karyotyping
The use of FISH, often with two or more different coloured fluorophores is a useful adjunct to G banding for identifying chromosomal abnormalities. This multiple tagging of chromosomes is sometimes referred to as 'chromosome painting' and together with centromeric tagging makes the identification of individual chromosomes with their corresponding deletions, duplications and translocations a much easier process. It is even possible to use chromosome painting to identify some abnormalities in interphase nuclei, not only in mitotic spreads.

Karyotyping can be useful in prenatal diagnosis and ISH can give results in 2–3 days using amniotic cells or chorionic villi samples even without culturing (see *Fig. 16.6*).

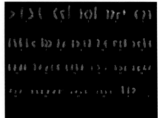

Figure 16.6
Detection of chromosomes using two centromeric probes CEP3 and CEP4 (Chromosome Enumeration Probe) which bind close to the centromeres of chromosomes 3 and 4 respectively.The bright spots indicate where the probes have bound. In the left hand photograph it is possible to quickly identify eight chromosomes with bright spots (four red, four green) indicating four copies of chromosomes 3 and 4, confirming a tetraploid state. This is confirmed on the photograph on the right and confirms a triploid state.

Automation

The complexity of the procedures and the need for precision at each step make ISH an ideal situation for automation. Machines are flexible enough to run several different protocols simultaneously. This not only makes the technique more reliable but in most cases will be more economical with reagents and will complete the full technique in a shorter time. For these reasons most laboratories performing ISH are automated.

16.4   Autoradiography (ARG)

As the name implies, ARG utilizes radioactive compounds that have been incorporated into living systems, making them primary emitters of radiation. This radiation can then be detected and the radio-label is then acting in exactly the same way as any other labelling system that has been considered so far. ARG is a useful method for tagging key structures or molecules in biological systems. Originally devised as a sensitive assay for the tracking of labelled metabolites in whole body sections of animals, it was subsequently realized that the principle could be applied to histological or cytological preparations.

Autoradiography is a useful method for accurately following the fate of chemicals in cells and organs using radioisotopes as labels. It is widely used to trace the routes of metabolism of drugs within animals and so this will be used as an example to show how an autoradiographic investigation is planned and carried out. The applications

---

Box 16.5 // Discovery of autoradiography

The first autoradiograph was done before there was any knowledge of radioactivity and before radioactivity was even suspected. Autoradiography was the first method of identifying radioactivity. Uranium salts were thought to be fluorescent. Antoine Henri Becquerel was testing some salts and left them on top of a paper-covered film in a drawer ready to irradiate them with light the following day. They were left unused for a few days, so rather than irradiating them he decided to just develop the film anyway. He was surprised to see blackening since fluorescence should not have occurred in the absence of light. This was the first indication of material producing unknown rays in the absence of heating or absorption of light.

of autoradiography are very widespread and it is one of the most useful techniques in research.

The fate of a drug in the body involves two phenomena:

- how and where it is distributed within the body; and
- how it is metabolized and excreted.

Ideally we would like to be able to follow both of these processes. The actual concentrations of most drugs within cells and tissues are generally quite low, so detecting the drug using conventional chemical tests would be difficult or impossible. To overcome this problem, radioactive materials are used, which can allow the detection of just a few molecules of drug.

Steps involved in autoradiography

The first step is to prepare some of the drug but replacing one of its normally stable structural atomic isotopes with a radioactive isotope. The modified drug is then administered to an animal or put into a tissue culture medium. The drug is absorbed, distributed in the tissues and may be metabolized. Sections of the animal tissues can then be made. The radioactive isotope will decay and emit radiation, which can be detected using photographic film that reacts to the radiation as it would to light. The exposed film is developed in the same way as for photography and areas that have been exposed to radioactivity will show up as black areas (see *Fig. 16.7*). Only a few atoms decaying radioactively are needed to detect their position in the tissue, so it is extremely sensitive. It is also necessary to be able to locate the film over the original section accurately so that the position of the isotope can be determined. It has to be recognized that some excretory organs, e.g. bladder and liver, will accumulate high levels of activity very rapidly as the drug is metabolized and eliminated. These levels therefore fall off quickly. *Fig. 16.8* shows an example of microscopic autoradiography.

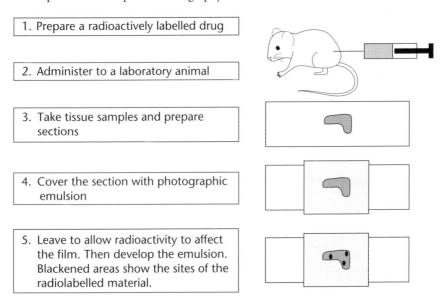

1. Prepare a radioactively labelled drug

2. Administer to a laboratory animal

3. Take tissue samples and prepare sections

4. Cover the section with photographic emulsion

5. Leave to allow radioactivity to affect the film. Then develop the emulsion. Blackened areas show the sites of the radiolabelled material.

Figure 16.7
Outline of the steps in autoradiography

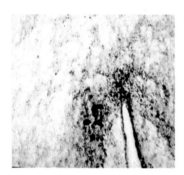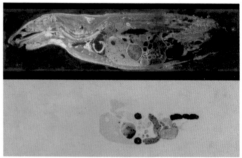

Figure 16.8
Left: Microscopic autoradiograph with the black granules showing the migration of labelled compound from the blood vessel. Right: Macroscopic autoradiograph. A longitudinal section of a whole rat used for testing a radioactive drug. Below is the film used to detect the position of the drug. The drug can be seen to be mainly restricted to the lumen of the gut but with some being absorbed into the liver.

Choice of label and site of labelling

Labelling involves making a sample of the drug with one or more radioactive atoms. This is done by synthesizing the drug using a radioactive precursor. Many standard synthetic reagents are available in radioactive form from commercial sources, so it is possible to make almost any compound. However, it is not simple or cheap and should be done by an expert chemist as radioactivity is dangerous and the preparations are usually done on a very small scale since the radioactive starting compounds are so expensive.

It is important to think carefully about the choice of label before getting the radiolabelled drug synthesized.

Choice of radioisotope

It is not possible to replace every individual element since not all atoms have suitable radioisotopes (see **Box 16.6**). Radioactive decay can produce three forms of radiation, as described in **Table 16.1**.

Table 16.1   Properties of radioactive emissions

Type of radiation	Nature	Charge	Applications
α	Helium nucleus	+	Good for localization but few α-emitters are interesting biologically
β	Electron	−	Localization poorer than α-particles but many of the atoms found in biological specimens have β-emitting isotopes
γ	Electromagnetic radiation	0	Useless for autoradiography as it can travel a long way from the labelled site. This can be a nuisance since γ-rays are often emitted along with β-particles

---

Box 16.6 // Not all important atoms can be radioactively labelled

Some elements have no isotopes that can be used for autoradiography. There are, for example, no useful isotopes of oxygen, nitrogen or chlorine. Thus, labelling cannot be done for every atom in the molecule and the fate of some atoms must be guessed. In some cases, they can be determined by using non-radioactive isotopes and mass spectrometry to determine their fate. This can be done with oxygen, which has a heavy but non-radioactive isotope.

Box 16.7 // Microradiography

Autoradiography is sometimes confused with microradiography. Microradiography involves placing a small piece of tissue or a tissue section on to a piece of film and then irradiating it with X-rays or γ-rays. This is similar to ordinary radiographic methods used in medicine to visualize bones and teeth and will detect radio-opaque materials in the section. Autoradiography is different in using radioactivity within the tissue and using mainly poorly penetrating β-particles to detect radiolabelled materials.

β-particles. β-Particles are usually used in autoradiography. They are relatively light and have the same negative charge as the electrons in the outer shells of atoms. β-Particles are therefore often deflected by the electron shells of atoms. As a result of these deflections, the electrons do not travel through the tissues or the autoradiographic film in straight lines but tend to wander around. Depending on their energy, they may cause blackening some distance away from the isotope's actual position.

If there are many radioactive disintegrations at the same point, then there will be a lot of grains forming a 'cloud' around the site of the radioactivity. The location of the isotope is somewhere within this cloud of dots but its exact position can only be an estimate (see *Fig. 16.9*).

 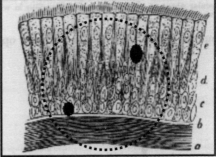

Figure 16.9
The track of β-particles in an autoradiographic emulsion. The left-hand photograph shows the path of the particle, whilst the right-hand one shows the sites of silver grains. Only one or two points on the β-particle's track leave a silver grain. With β-particles, the accuracy of localization is less precise and all that can be assumed is that the radioactivity is somewhere within the circle

β-Emitting isotopes include useful elements such as carbon, hydrogen and sulphur

There are a lot of very useful β-emitters (see *Table 16.2*). Suitable isotopes include hydrogen (tritium), carbon, sulphur and phosphorus. With β-rays, the localization is affected by the energy of the radiation, which is usually measured in electron volts (eV). Tritium (^{3}H) is one of the best isotopes for localization as it has a low-energy emission. Tritium can even be used in the electron microscope to localize to the subcellular level; for example, it can determine whether the drug is in the nucleus or the cytoplasm. With very careful technique, it is even possible to estimate whether the isotope is in a mitochondrion or in the cytosol or in the case of the nucleus, the exact chromosome to which it is bound.

The effective resolution (*Table 16.2*) indicates how accurately the position of the isotope can be determined. Carbon and sulphur are quite good for light microscopy magnifications and can certainly locate the isotope to individual cells. Phosphorus is less accurate for microscopy but is still useful for macroscopic and electrophoretic uses and can be relied on to locate the isotope to small groups of cells.

Table 16.2 Radioisotopes and their properties

	Energy (keV)	Half-life (t½)	Effective resolution (μm)
^{3}H (tritium)	19	12.3 years	0.5
^{14}C	156	5730 years	2–5
^{32}P	1710	14.3 days	5–10
^{35}S	167	87.4 days	2–5

Safety is important as isotopes may concentrate in the body

Most labelled compounds are a health hazard as they can accumulate within the body. Tritium is often rapidly diluted in the body because of the large amount of hydrogen present, particularly in water, but phosphorus concentrates in bone. This location does, however, depend on the form of the isotope. If it is incorporated into a molecule, it is not the nature of the isotope but the nature of the molecule that is important. Weaker radiations (e.g. tritium) do not penetrate far and are easily blocked, even by air. Unless the isotopes actually get into the body, they will not cause any damage since the dead cells on the skin will prevent the radiation effects reaching living cells. If they are ingested, the isotopes will come into direct contact with cells, causing significant breaks in DNA that are not repaired; their cumulative effect is dangerous.

Box 16.8 // Autoradiography is not a diagnostic technique

Autoradiography can never be a diagnostic tool since it involves administering radioactivity to the subject and usually involves killing the subject to recover the tissues. The results of autoradiography can sometimes be confirmed in humans by using positron-emitting isotopes. The positrons can be detected using detectors outside the body and the fate of compounds can be followed. The positrons are detected using tomography equipment that locates the emission in three dimensions. These investigations are called PET scans (positron emission tomography).

16.5   Histological preparation

At the end of the experiment, the fresh tissue obtained from the animal must be sectioned and the processing of the tissue needs to be considered carefully. The main requirement is to ensure that the substance being detected is not lost or redistributed by the processing. Questions need to be asked at every step in the process and careful choices of fixative and processing method (e.g. wax or frozen) are essential. Ideally these questions are all answered long before the experiment begins so that there is no delay in knowing how to treat the tissue once the animal is killed.

It should be noted that autoradiography can also be of value when examining cytological preparations. This is particularly the case at the level of cellular organelles, e.g. tracing the post-translational processing of peptides through the Golgi apparatus, or in the evaluation of cellular activity through the incorporation of a tritiated base into DNA. The general principle for assessing where label has localized is to make a slide preparation of the labelled cells followed by the dipped emulsion technique (see below). This approach is also valid for the demonstration of a radiolabelled *in situ* probe that has been used to evaluate gene expression; however, methods that use radioactive compounds are less favoured in most laboratories because of their associated extra hazards and safety considerations.

Exposure to emulsion

Once the sections are obtained they need to be placed in contact with the photographic emulsion to detect the radioactivity (see ***Box 16.9***). The type of emulsion or film used depends on the accuracy of location required.

With large sections (e.g. whole body of a rat) requiring only general localization to an organ, such as in the blood–brain barrier example above, this can be done using X-ray film. X-ray film is preferred to simple photographic film as it has a more regular and finer grain structure.

Box 16.9 // Autoradiographic emulsions

Photographic emulsions consist of silver halide crystals in a gelatine support. The size of the individual halide grains determines the accuracy of localization. Autoradiographic emulsions have very small grain sizes (fine-grain emulsions) so they can give accurate localization.

When a light ray or β-particle strikes a silver halide grain, it can knock an electron out of the crystal. This results in a silver ion being converted to a silver atom. This one atom then acts as a catalytic site for reduction by photographic developers such as quinol ('hydroquinone'). The whole silver grain turns black giving a visible speck in the autoradiograph. Without that catalytic silver atom, the developers do not cause reduction of the silver grains so they remain as white silver halide crystals.

The unused silver halide crystals must be dissolved out of the autoradiograph before it is examined in the light, otherwise it would slowly blacken in the light. This is done using sodium thiosulphate ('hypo') and is referred to as photographic fixing. This is photographic fixation to remove excess light-sensitive silver salts, not histological fixation in, for example, formaldehyde.

Microscopic sections can be prepared by dipping in photographic emulsion

If the tissue is going to be examined microscopically, a more accurate localization is needed. The emulsion needs to be placed in very close contact with the specimen. The dipping technique uses a low melting point photographic emulsion that can be melted at about 40°C and then held as a liquid in a waterbath. Sections of tissue are brought to water, dipped into the liquid emulsion and then drained and cooled to allow the emulsion to harden. They are then stored to allow exposure. All of these preparations must be done in a darkroom lit only with a safelight and the final preparations must be kept dark. The slides are put into a Coplin jar or staining dish with some silica gel to keep them dry. The whole jar is then wrapped in aluminium foil to exclude light and put into a lightproof box to ensure no light affects them. Finally, they are put into a refrigerator until they have been exposed for a sufficient length of time. This dipping technique is quick, economical and simple to do. However, it has the disadvantages that the emulsion thickness is not uniform and the thickness can vary from one sample to the next making quantitation difficult.

Exposure is difficult to calculate

Exposure of the film is a somewhat hit and miss affair (see **Box 16.10**) in most cases, with several specimens being prepared and each one developed after different exposure times to obtain the best results. Once the first one is done, the results give a reasonable idea of how long the total exposure needs to be to get good results. With a high level of radioactivity present, it may only take a few days to get sufficient exposure, but if the levels of radioactivity are low, then exposures can stretch into weeks or months.

Box 16.10 // The problem of estimating the exposure time

Knowing how much radioactivity is present within the section is not enough to calculate exposure. The radioactivity might all be concentrated in one small area, so the local concentration might be high but the overall concentration in the section could be low. The best way is trial and error. In some cases, it may need more than one length of exposure. The hot spots of radioactivity will be best seen in short exposures, whilst areas with less radioactivity might need longer to produce sufficient silver grains.

Development is carried out in a high-contrast developer such as an X-ray developer and the emulsion is then fixed.

The section can be stained either before or after application of the emulsion. Staining after autoradiography can lead to the gelatine in the emulsion picking up the stain and becoming coloured. Staining the section before applying the emulsion risks the chemicals in the stains chemically affecting the emulsion and giving a false result. After staining, the stained section and overlying emulsion are examined microscopically. Areas of radioactivity show minute black silver grains surrounding the site of the radioactive decay. The small silver grains are more easily seen using a darkground microscope rather than a brightfield microscope (see **Chapter 17**).

Suggested further reading

Alberts, B. *et al.* (2008) *Molecular Biology of the Cell*, 5th edn. Garland Science.

Crocker, J. and Murray, P.G. (eds) (2003) *Molecular Biology in Cellular Pathology*. Wiley.

Hacker, G.W. and Gu, J. (eds) (2002) *Gold and Silver Staining: Techniques in Molecular Morphology*. CRC Press.

Salisbury, J. (1997) *Molecular Pathology*. Taylor & Francis.

Sterchi, D.L. and Astbury, C. (2012) Chapter 21: Molecular pathology. In: *Bancroft's Theory and Practice of Histological Techniques*, 7th edn. (eds K.S. Suvarna, C. Layton and J.D. Bancroft). Churchill Livingstone.

Self-assessment questions

1. Outline the principles of the *in situ* hybridization method.
2. Outline the steps needed to hybridize a probe DNA with target DNA.
3. Give two examples of different labels that can be used for DNA probes and compare their advantages.
4. Briefly explain the principle of the PCR technique.
5. Give an outline of the tissue preparative techniques for ISH.
6. Indicate the applications of ISH.
7. How does autoradiography trace chemicals in the body?
8. A new synthetic prostaglandin has been synthesized. Outline how you would plan an autoradiographic experiment. (The drug is hydrophobic with a low solubility in water and is expected to bind to receptors in blood vessel walls.)
9. Name the radioisotope you could use to label the following compounds
   (a) C_2H_5OH;
   (b) $HSCH_2CH(NH_2)CO_2H$;
   (c) H_2O_2 and
   (d) H_3PO_4. Indicate which atoms would be suitable for stable labelling.

17 Light microscopy

Learning objectives

After studying this chapter you should confidently be able to:

- **Describe the principles of simple and compound microscopes**
  Simple microscopes consist of a single lens, which bends the light from a specimen and thus produces a magnified image. A compound microscope uses two or more lenses held a calibrated distance apart. This combination produces a greater magnification than is possible with a single lens, as the first lens (the objective lens) produces an image (virtual image) inside the tube, which is then further magnified by the eyepiece lens.

- **Describe the importance of numerical aperture and correction of aberrations**
  Numerical aperture (NA) is a measure of the light-gathering power of the lens. The greater the NA, the greater the resolution of the lens. The NA also affects the depth of field of the lens. The NA thus indicates the usefulness of a lens for a particular application.

- **Outline the principles of darkground and polarizing microscopy**
  The darkground microscope uses oblique illumination. This oblique light does not directly enter the objective, resulting in a dark background. Very small objects that diffract light into the objective show brightly against the dark background. Polarizing microscopes use two polarizing filters. One filter is below the specimen and the other is between the specimen and the eye. The two filters are rotated so that no light passes through and only objects capable of rotating the polarized light appear bright in a dark field. This allows the eye to detect rotation of the plane of polarized light, which is not detected by the unaided human eye.

- **Outline the principle of the fluorescence microscope**
  The fluorescence microscope uses two filters. A filter below the specimen absorbs all of the long-wavelength light so the specimen is illuminated only with short wavelengths. The second filter blocks this short-wavelength light leaving a dark background. Only specimens that fluoresce (i.e. shift the wavelength of the light to a longer wavelength) are visible.

- **Outline the role of digitization in microscopy**
  Digitization is used for photographic recording where the speed and low cost of digital technology are useful. Digitization also allows the direct importation of images into computer programs for measurement and analysis.

- **Outline the principle of the confocal microscope**
  The confocal microscope is a form of fluorescence microscope which uses a very small aperture (pinhole) at the back focal plane of the objective to restrict flare from out-of-focus light. This produces a very thin depth of field but requires the specimen to be scanned with illuminating light, rather than the whole of the specimen being illuminated. The image is viewed by reconstruction using digital technology. Different methods of scanning are possible but the use of a laser and deflecting mirrors is the main one.

> **Box 17.1 // A brief history of microscopes**
>
> The term lens comes from the Latin for lentil (*lens culinaris*). The lenses that were first used looked like lentils in shape and the name has stuck. Glass lenses were first used sometime before 1300 but were used mainly as a method for aiding failing vision rather than extending human vision to see minute structures. The idea of a compound microscope was first proposed around 1600 and compound microscopes were used by Robert Hooke (first person to use the word 'cell' in microscopy) in the mid-1600s.
>
> Early lenses were badly ground so for a long time the use of single lenses was often better as they produced fewer aberrations. Antoni van Leeuwenhoek used single lenses with magnifications greater than 200×. The lenses had such short focal lengths that they were used virtually touching the eye and almost touching the object on the opposite side. Lenses were used mainly at lower powers for examining insects and were known as 'flea-glasses'.
>
> In the 19th century, development continued with better correction of lens aberrations and better resolution. Thus, by the beginning of the 20th century routine light microscopes were almost as good as those used today (see *Box 17.4*).

Most human cells are so small that even a very large cell, e.g. 100 μm in diameter, is only just about visible as a tiny speck to the unaided human eye and no internal detail can be seen. Smaller cells, such as erythrocytes or bacteria, are far too small to be seen at all. This means that to observe cells in histology and cytology we must use a microscope to make the image larger (see *Box 17.1*).

17.1 How a microscope works

The size that an image appears to the human eye is controlled by the angle the light makes as it enters the eye. Objects that produce a large angle appear bigger, so we can make an object appear larger by altering the angle of light entering the eye. This happens when we get closer to an object and its apparent size increases. Glass, water and transparent plastics can bend light passing through them (see *Fig. 17.1*) by **refraction** and this bending can be used to magnify the image. A piece of glass with one or two convex curved surfaces

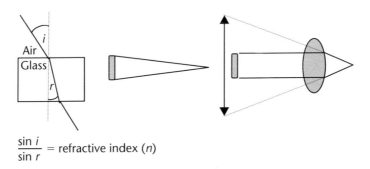

$$\frac{\sin i}{\sin r} = \text{refractive index } (n)$$

Figure 17.1
On the left, the light ray bends on entering and leaving the glass block. This bending is refraction. In the centre, the object is seen as a small structure with only a small angle subtended at the eye, whilst on the right, the lens bends the light and makes it appear larger

can refract the light to a single focal point and this acts as a magnifying lens or **simple microscope**.

The nature of light

Although in most optical diagrams light is shown drawn as a straight line this is just a convenient way to represent it. Light is an electromagnetic wave and as it travels through any medium it vibrates at right angles to the direction in which it is travelling. Light has properties associated with this wave nature that are crucial in the theory of optics including microscopy. The important properties for understanding resolution are the wavelength, frequency and amplitude. In *Fig. 17.2*, the amplitude and wavelength are shown.

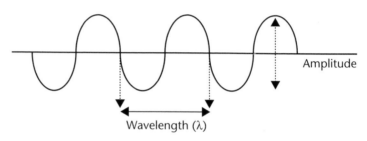

Figure 17.2
The wavelength (λ) and amplitude of a light wave

The amplitude is detected by the human eye as the intensity or brightness of the light, whilst the wavelength is detected as the colour of the light. The frequency of the light is the number of vibrations of the beam in 1 s. Since the speed or velocity of light in a vacuum is constant, the frequency and the wavelength of light are linked:

$$\text{Velocity } (V) = \text{frequency } (f) \times \text{wavelength } (\lambda)$$

In the case of light in a vacuum, the velocity is approximately 3×10^8 ms^{-1}. Although electromagnetic waves can have any wavelength, only certain ones are visible to the human eye. These wavelengths are the visible spectrum of colours and are shown in *Fig. 17.3*.

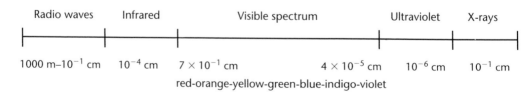

Figure 17.3
The electromagnetic spectrum. The longer wavelengths are on the left and shorter on the right with the visible spectrum in the middle. The diagram is not drawn to scale and the visible spectrum in reality occupies a much smaller part than shown

The wavelength of light varies depending on the medium

When light enters a medium that is denser than a vacuum, it is slowed down but the frequency remains the same. Likewise, light travels more slowly in glass than it does in air. The frequency of vibration remains the same, so in glass the wavelength must become shorter because of the relationship between velocity, wavelength and frequency. For this reason, physical scientists prefer to use the frequency of light rather than the wavelength since frequency is constant, regardless of whether the light is travelling in air, glass, water or any other medium. Most microscopists prefer to use the wavelength of light since the length of the wave relates to characteristics such as resolution, which are also measured as lengths.

The slowing of light causes the light to bend as it passes from one medium to another depending on their densities. The amount of bending is measured by the **refractive index** (RI). Air has an RI only a little higher than 1, water is 1.33 and glass is about 1.5.

Microscope lenses

Simple microscopes are limited to low magnifications

Simple single-lens microscopes are very useful and are widely used for low magnifications such as reading lenses and watchmakers' eye loupes. Single lenses suffer from the problem that they need to have a highly curved surface (which makes the lens very small) in order to get high magnifications and they need to be used close to the specimen and close to the eye. It is inconvenient and impractical to use a simple lens of more than about 20× magnification.

Compound microscopes use more than one lens to increase magnification

To get larger magnifications than is convenient with a single lens, it is possible to magnify the image produced by one lens by using a second lens to produce a higher magnification. This type of system is called a compound microscope. The lens that does the initial magnification and which is close to the object being viewed is the **objective lens**, whilst the lens at the other end through which the eye looks is the **eyepiece**. The lenses are held in a tube that keeps out stray light and holds the two lenses the correct distance apart for optimal performance and allows us to see the image clearly. This is the basis of all compound microscopes.

The overall magnification of a compound microscope is determined by multiplying the magnification of the objective and eyepiece lenses; for example, a ×10 objective used with a ×10 eyepiece gives an overall magnification of ×100.

Other features are usually added to make life simpler when using the microscope or to improve the stability and brightness of the image (see *Fig. 17.4*):

- A **mechanical stage** that holds the specimen and allows it to be moved slowly and smoothly across the field of view.
- A rigid stand **to eliminate vibrations.**
- A **lighting system** to illuminate the specimen and a condenser **to focus the light on the specimen.**
- A **focusing system** to allow the specimen to be focused smoothly and easily.
- Adaptations to allow binocular viewing, or **tilted (inclined) eyepieces**.
- A **turret** to hold more than one **objective**. By rotating the turret, the overall magnification of the microscope can be easily changed.

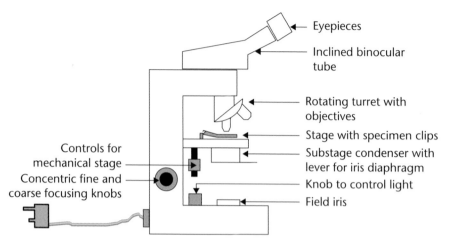

Figure 17.4
Parts of a microscope

There is no reason why more lenses could not be added to magnify the image further to give any magnification required (1,000,000× would be possible in theory). However, there is no point in magnifying much beyond 2000× as there are other limits on the microscope image that make very high magnifications useless. In particular, the **resolution** of the lens limits the usefulness of the image. Resolution is the ability to see two structures that are very close together as separate structures (see *Fig. 17.5*). The resolving power is measured as length (e.g. 1 μm), with smaller distances representing better resolution since objects that are very close together can then be resolved as two separate structures. Thus, a resolution of 0.2 μm is better than a resolution of 1 μm.

Original objects   Objects at limit of resolution   Unresolved image

Figure 17.5
The original objects on the left are quite separate. In the middle, they can just about be resolved as two overlapping images and on the right they are unresolved and appear as a single image

The numerical aperture is a measure of the resolving power of a lens

The resolution of a microscope lens is measured by the **numerical aperture** (NA) of the lens. This measures the light-gathering power of the lens (see *Box 17.2*). The greater the angle of the light, the better the resolution. For microscopists, the NA is more important than the magnification and is always engraved on the barrel of the lens.

In mathematical terms:

$$NA = n \times \sin \alpha$$

where n is the refractive index (RI) of the medium below the lens (air is 1, water is 1.33 and microscope oil is 1.4) and α is the half-angle of the cone of light that can just enter

> **Box 17.2 // Increasing the angle of light captured by a lens**
>
> The resolution relates to the aperture as the amount of light, or more particularly the angle of light entering the lens. The greater the angle of light collected, the better the resolution. This is usually more obvious when dealing with telescopes. Most people readily accept that to see the detail of stars more clearly you need a bigger telescope. A telescope with a 10 cm diameter lens will see more detail than one with a 5 cm lens. The same applies to microscopes but instead of making the lenses larger, the microscope manufacturers usually move the lens closer. This increases the angle of light captured from the specimen without making the lenses too large and cumbersome. This reduces the distance between the lens and the specimen making manipulation difficult. Therefore, specialized microscopes are made with larger 'working distances' for purposes such as fine surgery where microscopes are needed that allow fine manipulation.

the lens. The sine of an angle is always less than 1. Therefore, since the RI is 1 for an ordinary dry lens (one that has air between the objective and the specimen), it follows that the NA of a dry lens is always less than 1. In practice, it is usually less than 0.9. For an oil-immersion lens where the RI is higher (RI of oil is 1.4–1.5), the NA is always less than 1.4, usually 1.25–1.3 for a high-power lens.

The resolution of a lens is given by the mathematical relationship:

$$\text{Resolution} = (0.61 \times \text{wavelength})/\text{NA}.$$

With a NA of around 1.2–1.3, this means that the resolution of an oil-immersion lens is about half the wavelength of light ($0.61/1.2 \approx 0.5$).

The light microscope cannot resolve objects closer than 200 nm

Thus, the ultimate resolution of a light microscope is limited by the nature of light and is roughly half the wavelength of the light used. The shortest wavelength that the human eye can see is about 400 nm (0.4 µm) so the best resolution that a light microscope can achieve is about 200 nm (0.2 µm). No matter how good the microscope, this limit is imposed by nature and cannot be overcome when using visible light. Better resolution can only be achieved by using different types of microscope that use shorter-wavelength radiations such as the electron microscope.

Optimum resolution with 1000× NA

The best resolution for the human eye is achieved when the total magnification of the microscope is about 1000× the NA of the objective. If a much higher magnification is used, the image gets larger but no more detail is seen by the human eye. This phenomenon is termed **empty magnification**. Instead of improving the image, any extra magnification makes the quality gets worse (rather like the way that getting too close to a television or cinema screen makes the image larger but less clear, although the physical reasons for the degradation are slightly different).

Immersion oil improves resolution by increasing NA

Oil is used in microscopy to improve the resolution by acting as a waveguide for light. Without oil, light travelling from the specimen goes through extra refractions and some of the light is lost. Immersion oil has a similar RI to glass and so minimizes the refractions (see *Fig. 17.6*). This allows a greater angle of light to be collected and so

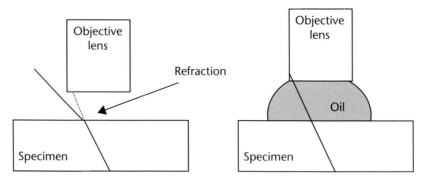

Figure 17.6
On the left in the absence of oil, the refraction as the light leaves the specimen causes it to bend and not enter the objective. On the right, the oil prevents the refraction and the light enters the lens

increases the NA. Oil is only needed between the front of the lens and the top of the specimen and should be cleaned off after use to prevent it hardening. Oil should only be used with lenses designed to use oil (it is always marked on the lens, although it may be in a foreign language, e.g. *Oel* if the microscope is German). Using oil with a dry lens results in degradation of the image without an increase in resolution, as does *not* using oil with an oil-immersion lens (see *Fig. 17.7*).

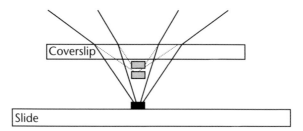

Figure 17.7
Effect of omitting oil with an oil-immersion lens. The lenses are designed to focus light that has not been refracted. If oil is omitted, then refraction occurs and the lens focuses the light as if it was not diffracted. This results in extra images at different focal planes. This smears the image and gives a blurred overall effect. With a dry lens, the opposite would be true as the lens is designed to correct for refraction and again extra images are produced that blur the specimen

Lens aberrations

The objective lenses used in microscopes do suffer from optical problems and these are called aberrations. Aberrations are distortions of the image caused by refraction in the objective lenses. There are seven types of aberration that can occur and that need to be considered when the microscope manufacturer designs the lenses in a microscope.

Chromatic aberration. When light passes through glass at an angle, the different wavelengths of light are not all bent (refracted) to the same extent. This can be seen using a glass prism where a spectrum of light emerges from the prism when a beam of white light is shone on to it (see *Fig. 17.8*). This property of producing a spectrum is

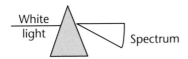

(a) With a prism white light is split to give a spectrum

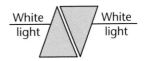

(b) With two prisms there is no spectrum and the light remains a beam of white light

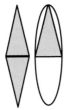

(c) The cross-section of a lens resembles two curved prisms

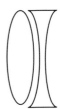

(d) A concave and convex lens doublet can reduce chromatic aberration

Figure 17.8
Chromatic aberration. White light can be split into a spectrum by a prism or by a lens, which, in cross-section, resembles two prisms on top of each other. Two inverted prisms prevent dispersion and a lens doublet can have the same effect

referred to as **dispersion**. Although dispersion and refraction are related phenomena, they are not identical so the simple RI is not a measure of dispersion. In a lens, dispersion results in different colours of light being brought to different foci. Thus, if the red light is focused there are green and blue images that are out of focus. This aberration is usually corrected in good-quality microscopes. The correction is achieved by using two lenses of different materials that differ in their ability to disperse the light into a spectrum. Two levels of correction can be achieved called **achromatic** and **apochromatic**. Apochromatic correction is a better but more expensive option.

Lateral chromatic aberration. Here the effect is that, although all the different colours of light are brought into focus at the same place, the blue image is a different magnification. This results in colour fringes around objects. This is difficult to correct in the objective lens and is usually corrected in good-quality microscopes by using **compensating eyepieces**. The compensating eyepieces have a lateral chromatic aberration built in as part of their design. This aberration is exactly opposite to the aberration in the objective so the two aberrations cancel out. Compensating eyepieces are matched exactly to the objectives and should not be swapped from one system to another or they will introduce chromatic aberrations.

Spherical aberration. This is a consequence of the lenses having a curved (spherical) surface. This aberration is always fully corrected in any decent microscope by the incorporation of multiple lens elements into a single lens. The aim is to bend light incrementally at the surface of each element, to avoid the adverse effects of thick, highly curved single lenses.

Distortion. Distortion results from lenses having slightly different magnifications at the centre and at the edge of the lens. The effect can cause barrel or pincushion shapes from a square object (see **Fig. 17.9**). Usually there is some residual barrel or pincushion distortion in high-power lenses but it is minimized by the manufacturer and is not usually a problem, especially since biological specimens are never exact square shapes so the distortion is not noticed.

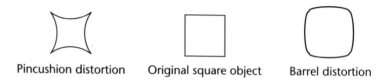

Pincushion distortion   Original square object   Barrel distortion

Figure 17.9
The effect of distortion on a square object

Astigmatism. This is where objects at different orientations are at different focal planes so that radial objects will be in a different focal plane to tangential objects. Astigmatism is always fully corrected in a good microscope.

Coma. This produces a 'tail' coming away from bright objects. Coma is often only minimized in most laboratory microscopes rather than being fully corrected. Coma only affects the periphery of the field of view and is not observable in the centre of a lens.

Curvature of field. Here the edges of the field of view are out of focus when the centre is in focus and vice versa. Curvature of field is usually fully corrected in lenses that are intended for photography but otherwise it is only minimized.

Although it is possible to fully correct each aberration, it results in a more complex lens. Each aberration requires the lens designers to have at least one lens face for each correction. Thus, fully corrected lenses have many optical elements all fitted into one barrel. The more lenses, the higher the cost, so full correction of every aberration is expensive and most microscopes are not fully corrected (but see **Box 17.3**).

Box 17.3 // Better lenses are needed for photography than for visual microscopy

For high-quality photomicroscopy, the lenses need to be of the best quality and as fully corrected as possible but for simple viewing by the human eye this is less essential. This is partly because the human eye and brain are good at ignoring slight defects in the image. This is useful since the lens in the human eye suffers from the same aberrations. The brain, for some reason, does not ignore things in the same way when they are present in a photograph. This phenomenon results in photographs that show people with telegraph poles apparently growing out of their head. They looked all right in real life but look dreadful in a photograph. Thus, photographic corrections need to be of a higher quality than for simple visual microscopy.

> **Box 17.4 // Modern lens improvements**
>
> The best available resolution of the light microscope has not greatly altered since the last century and has possibly even diminished, since electron microscopy can now resolve so well that it is unnecessary to push the light microscope to the very limits of resolution. Instead, lens designers have concentrated on improving other characteristics of the light microscope. The use of special glass and modern plastics has improved the clarity and brightness of the image. There has also been an increase in the field of view seen down the microscope. Older lenses suffered greatly from aberrations at the edge of the field of view so lens manufacturers only made use of the central part of the image. Now better corrections allow more of the image to be acceptable.
>
> One advance has been in the **blooming** or coating of optical components, which reduces flare. Flare is light that passes through the lens but plays no part in the image. The light is scattered over the whole image making it appear as a '**soft focus**'. The same effect is seen with a dirty lens or even a dirty window or windscreen in strong sunlight where there is a lot of light coming in through the glass but it interferes with the clarity as it spreads the light. The coating of lenses means that flare can be reduced to very low levels. The result is a brighter, clearer image but coating of lenses does not increase the resolution.

The distance light travels between the lenses is the optical length

Lenses are usually made to fit a standard type of microscope with a standard optical tube length. The standard is now often 160 mm although other lengths have been used in the recent past. The optical path length is usually inscribed on the barrel of an objective and altering the tube length of a conventional microscope will increase the magnification.

More recently, most high-quality microscopes have been made with optics that have no fixed tube length. Instead, the light travels along the microscope tube as a parallel beam. The tube length can therefore be extended without altering the magnification. Such optics are called **infinity optics**.

The optical path length is often different to the physical tube length since prisms are often used in modern microscopes to make inclined eyepieces and binocular heads. These fold the light path, making the distance the light travels longer than the simple physical distance between the objective and the eyepiece.

Most microscopes allow all of the objective lenses fitted on the turret to be **parfocal**. Changing one lens for another by rotating the turret will leave the lenses approximately in focus and only fine adjustments need to be made. This is very convenient and is the best reason for not mixing lenses from different microscopes. Mixing lenses may mean that **correction collars** are needed to match the lenses and retain the parfocal ability.

Illumination

The illumination system of a microscope is part of the optical system and needs to be correctly set up to allow the best resolution. Below the stage is a condenser that concentrates and focuses the light on the specimen. Condensers, like objectives, have their light-concentrating power rated in NAs. The condenser NA should be matched to the lens NA to allow maximum resolution of the image. The condenser is usually fitted with an iris diaphragm to prevent unwanted light entering the lens. This substage iris should be adjusted to the point where it gives maximum illumination but closing the iris any further would result in a reduction of the light intensity. This is the **threshold of darkening** and gives optimum resolution. Although altering the iris diaphragm can be used to change the brightness of the image, this is not a good idea since it also reduces the resolution by reducing the NA of the condenser. Most microscopes have a knob to control the light source intensity and this is the recommended method to make the image comfortably bright for your eyes. Closing the iris can occasionally be useful with low-contrast objects where the increase in contrast outweighs the loss of resolution. Standard microscopes will have an additional flip-top condenser lens for use at mid-range and high magnifications. This serves to increase the amount of light reaching the objectives.

Most modern laboratory microscopes use a variation of Köhler illumination. The exact method of adjusting it varies with the make and model, but learning how to set up the microscope properly is an essential skill needed to getting the best from the instrument. New generation microscopes favour the use of LED light sources which are bright, provide a truer 'white' light source and operate efficiently at low voltages.

17.2 Special forms of microscopy

The human eye can only detect two changes in light. One is a change in intensity (how bright something appears) and the other is a change in wavelength (the colour that something appears). There are other aspects of light that can be seen by some animals but not humans; however, these can be made visible to the human eye by manipulating the conditions. The microscope with its defined illumination can be used in this way so that, as well as being an instrument to magnify, the microscope can show things that are normally not seen by the human eye. These include very slight differences in RI, phase changes in the light waves and polarization of light. These specialized forms of light microscopy make it possible to see living cells without the need to fix or stain them.

In normal microscopy, the image is seen against a bright background and is referred to as **brightfield microscopy**. In brightfield microscopy, only objects that absorb light can be seen. Usually specimens are coloured (absorb some wavelengths of light) and so can be seen by the change of colour.

Many of the special forms of microscopy involve a dark background with the objects showing up brightly against the darkness. This contrast makes best use of the human eye's sensitivity. It is easy for the human eye to pick out very small amounts of light against a dark background, whilst the same intensity would be invisible if the background was light. The human eye is extremely adaptable and adjusts to the current illumination. Thus, the human eye can detect stars against a dark background, yet the same stars are invisible if the sky is bright. The stars are still there, it is just that the difference in intensity is so slight compared with the overall intensity that the human eye cannot detect the difference; in technical terms the signal-to-noise ratio is too small.

Darkground microscopy

This uses oblique illumination. The light, instead of coming from directly below the specimen and then entering the front lens of the objective, is arranged to come from an angle such that the light will not enter the objective (see *Fig. 17.10*). The oblique angle is achieved by using an annulus in the substage condenser or a special darkground illuminator, both of which produce a hollow cone of light.

Since no direct illumination enters the objective the background remains dark (see *Fig. 17.11*). Only if something in the specimen bends the light will any light enter the objective and be seen. For darkground microscopy diffraction gives the greatest effect, so diffracting objects appear bright against the dark background. This allows the eye to see

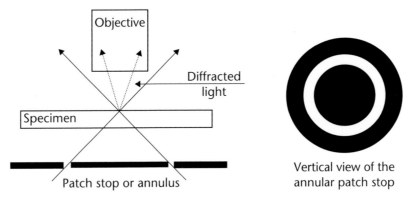

Figure 17.10
Darkground illumination. Only a small circle of light can pass through the clear area of the patch stop and this light is angled to ensure that undeviated light will not enter the objective. If diffraction occurs, then the diffracted light will enter the objective forming a bright image on a dark background

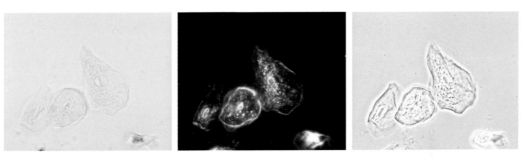

Figure 17.11
This shows three forms of microscopy applied to unstained cheek cells. On the left is routine brightfield microscopy, the middle image is dark ground microscopy and the right hand image is phase contrast microscopy.

them clearly and enables the microscope to produce an image of objects that are smaller than the usual resolution limit (see *Box 17.5* for an alternative illumination technique). The same effect as darkground microscopy can be seen with a bright beam of light such as the projector in a cinema when very small specks of dust can be seen dancing in the projector beam but in even illumination the specks of dust are invisible.

Polarizing microscopy

Polarization is a description applied to the wave motion of light. In polarized light, the vibrations (wave pattern) occur in only one plane (see *Fig. 17.12*). This is unusual. In light produced by most light sources there is a mixture of light with many different directions of vibration. Such unpolarized light can be polarized by passing it through a filter that selectively absorbs light vibrating in every direction but one plane. Such a filter is termed a polarizing filter. Longer wavelengths of radiation (microwaves and infrared) can be filtered using parallel fine wires that are aligned accurately. For visible light, it is usual to use either crystals of calcite (Nicol prisms), which were the original method of polarizing

> **Box 17.5 // Rheinberg illumination**
>
> Rheinberg discs are an alternative to simple darkground illumination. Instead of the annulus being a combination of totally opaque and totally transparent, the two parts are made of different-coloured filters. The annulus can be red, giving oblique red light, and the rest green. Diffracted light will be red and the background green. This gives a 'prettier' image than simple darkground illumination and is popular with illustrators and natural history filmmakers. The colours of the discs can be any pair of contrasting colours so the image can be quite varied. It does not give any extra information about the specimen compared with plain darkground microscopy – it just looks nicer. There are some even more complex variants with different parts of the filter in different colours.

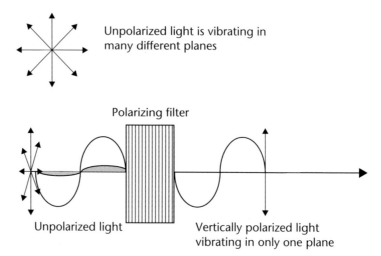

Figure 17.12
Polarization of light. Unpolarized light has its waveform vibrating in all planes. The top view shows a few of the vibrations seen looking along the direction of the light ray. The light would be coming straight out of the page and the wave vibrations would be at right angles to the line of travel. When unpolarized light is passed through a polarizing filter, all the vibrations except one are removed leaving, in this case, vertically polarized light

light, or more commonly nowadays polarizing filters that have precisely aligned crystals (see ***Box 17.6***).

Human eyes cannot see the polarization of light so objects that alter the polarization without altering the colour or intensity of the light appear transparent and lack contrast. The polarization changes can, however, be converted into intensity or colour changes to which the human eye is sensitive by using two polarizing filters. In the microscope, one filter is placed between the light source and the specimen (polarizer) and one between the specimen and the eye (see ***Fig. 17.13***). Thus, the object is illuminated by polarized light. If the analyser is then rotated to a direction at 90° to the polarizer (crossed polars), it will absorb the already polarized light giving a dark background.

Anything rotating the plane of the polarized light will appear bright and coloured against the dark background. This type of microscopy is excellent for crystals (see ***Fig. 17.14***) and regularly spaced materials such as the myofibrils in muscle fibres. Materials

Box 17.6 // Polarizing filters

Polarizing sunglasses use the same principle as polarizing microscopy. The polarizing lens absorbs all the light except that which is vibrating in one particular plane. This absorption of a large fraction of the light has a darkening effect that is useful for sunglasses and has the advantage of not altering the colours. It is essentially a **neutral density filter**.

It also alters the way we see reflected light. Light reflected from a surface is polarized by being reflected. Thus, polarizing sunglasses can apparently remove reflections such as those from the surface of water and allow the viewer to see into the water. Polarizing filters are used in photography for reducing reflections and enhancing colours by reducing the reflections.

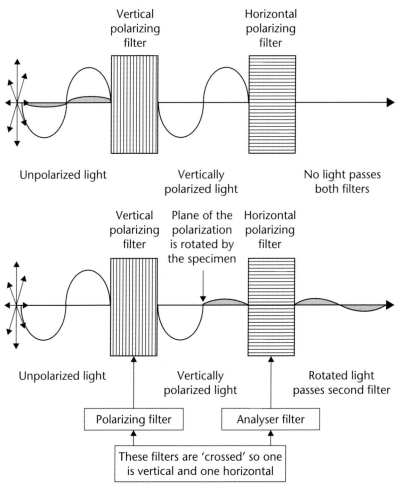

Figure 17.13
Arrangement of filters for polarization

Figure 17.14
Vitamin C crystals using polarized microscopy

that cannot rotate the light are said to be **isotropic** (e.g. muscle I bands = isotropic bands) and those that can rotate the plane of polarization are **anisotropic** (e.g. muscle A bands = anisotropic bands). They are also referred to as birefringent or sometimes called doubly refractile since they have two different RIs depending on the plane of polarization. These terms are not absolutely synonymous to physical scientists but tend to be used interchangeably in biological microscopy.

Fluorescence microscopy

Fluorescence occurs when a material absorbs light and then quickly re-emits it. If some of the energy of the light is lost then the emitted light differs in energy from the absorbed light. The wavelength of light is related to its energy. Lower energy light has a longer wavelength than higher energy light. Thus, when some of the light energy is lost, the emitted light has a longer wavelength than the absorbed light. This change from short-wavelength excitation light to longer-wavelength fluorescent light is called the Stokes shift.

By using two filters it is possible to observe only the longer-wavelength fluorescent light against a dark background (see *Fig. 17.15*). The first filter, the excitation filter, is placed between the light source and the specimen. This allows only short-wavelength light to be transmitted. This short-wavelength light will excite the emission of the fluorescence. The second filter, the barrier filter, is placed between the specimen and the eye (or camera). This absorbs all of the short-wavelength light and transmits longer wavelengths. Thus, it transmits the emitted fluorescence.

It is worth noting that the position of the filters is the same in both polarizing microscopy and fluorescent microscopy. One filter is between the light source and the specimen to control the illuminating light and the other between the specimen and the eye to control the transmitted light.

The filters are complementary and in the absence of a specimen they should stop *all* of the light. Only if fluorescence occurs will anything be seen with the microscope. This technique is widely used in immunofluorescence (see *Chapter 12*), for the detection of neurotransmitters (see *Chapter 10*) and when using ion-selective dyes (see *Chapter 10*).

Filters can be manufactured in two ways:

1.   **Absorption filters.** These are simply dyes in a suitable matrix (gelatine or glass). They are cheaper but give broad transmission peaks since the dyes have limited absorbing ability.

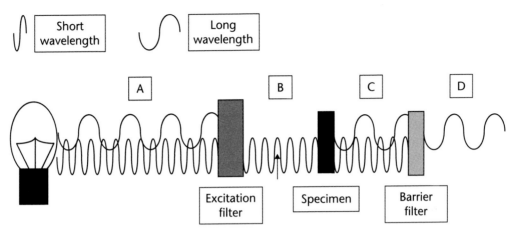

Figure 17.15
Arrangement of filters for fluorescent microscopy. The light from the lamp at A is a mixture of long and short wavelengths. After passing through the excitation filter, there are only short wavelengths of light (B). The fluorescent specimen converts some of the short wavelength light to longer wavelength fluorescent light (C). The barrier filter removes the short-wavelength light leaving only fluorescent light (D)

2. **Interference filters.** These produce their colour by the thickness of a special coating. The thickness and material used in the coating control the wavelengths of light that are transmitted and reflected. This type of colour is seen when there is a thin layer of oil on the surface of water. Neither the oil nor the water need be coloured, yet when the angle of light is right the surface of the oil film appears in rainbow colours.

Interference filters are more expensive than absorption filters but can have very sharp cut-off points in the light that they transmit. Absorption filters are suitable for many simple applications but interference filters are needed for precision fluorescence. Interference filters can also be made into **dichroic mirrors**, which transmit long wavelengths and reflect short wavelengths. This allows **epifluorescence**.

Epifluorescence microscopy

Epifluorescence illuminates from above using a dichroic mirror and uses the objective as a condenser. As can be seen in *Fig. 17.16,* the light source is situated above and to one side of the specimen and the objective. A dichroic mirror is placed in the optical path and angled at 45°. The light from the lamp impinges on the dichroic filter. Short-wavelength

---

Box 17.7 // Fluorescence in everyday life

Fluorescence is widely used in areas other than microscopy. **Brighteners** are materials added to washing powders and detergents to make the washed clothes brighter. These compounds absorb short-wavelength light invisible to the human eye and convert it into visible light. The effect is that in sunlight the clothes glow and seem 'cleaner'. The effect is best seen in short-wavelength 'disco' lights, which have a violet colour that shows the white clothes literally glowing.

Fluorescent pens are used to identify valuables such as computers and DVD players by marking them invisible to the human eye but instantly visible under UV light.

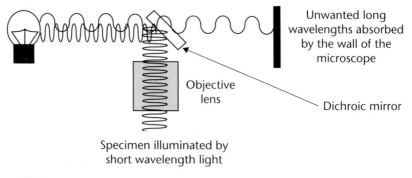

Objective
lens

Unwanted long
wavelengths absorbed
by the wall of the
microscope

Dichroic mirror

Specimen illuminated by
short wavelength light

Figure 17.16
Epifluorescence microscopy

light is reflected down the objective to illuminate the specimen. Any long-wavelength light is transmitted through the filter and absorbed by the black wall of the microscope. The short-wavelength light hits the object and elicits the fluorescence. The fluorescent light is emitted in all directions so some of the light will enter the objective and be magnified. This light reaches the dichroic filter but since the fluorescent light is a long-wavelength light, it will be transmitted by the angled filter and will reach the eyepiece. Most of the incident short-wavelength light will pass through the specimen and not be reflected back into the objective. Even if some short-wavelength light is reflected, it will not be able to pass through the dichroic filter and will instead be reflected back into the lamp house.

Epifluorescence has several advantages over transmission fluorescence

The epifluorescent microscope gives a much clearer and brighter image and is the main form of fluorescence microscopy used in laboratories as it has significant advantages compared with transmitted light fluorescence:

- The image is brighter, especially at high magnifications, since all of the incident light is concentrated on the area being viewed and not spread over the whole specimen as it would be in normal fluorescent microscopy. The intensity of the excitation is greater and this gives a brighter image.
- There is less fading since only the area being viewed is illuminated so only a small area fades at any one time.
- There is better resolution of detail since the light only has to pass through the coverslip and not through the slide. The brightest part of the image will be the part nearest to the objective and this reduces flare.
- There is less risk of eye damage. With a conventional brightfield microscope it is possible to omit the barrier filter so that short-wavelength light reaches the eye. UV light is dangerous to the eye and such a mistake could cause eye damage. The dichroic filter is both an excitation and a barrier filter. If it is left out, then no light will enter the microscope; if it is in place, then it acts as a barrier filter. The epifluorescent arrangement is much safer.
- Since there is only one filter, it is easier and quicker to change filter combinations.
- Filters are more accurately matched.
- It is possible to have conventional brightfield microscopy illumination set up to illuminate from below the specimen and have the incident light microscopy coming from above. It is quite simple to see the same field with both forms of illumination.

Simply switching the brightfield lamp off will allow the dimmer fluorescence to be examined and if the fluorescence is bright and the brightfield illumination is kept low, it can even be possible to view both forms of illumination at the same time.

Confocal microscopy

The latest addition to the range of fluorescence microscopes is the confocal microscope, which illuminates just a single very small point on the specimen. The fluorescence from this point is then measured accurately. To cut down the extraneous fluorescence and scattered light, the confocal microscope uses a very small aperture at the secondary focus of the objective. This aperture has the effect of restricting or eliminating any out-of-focus light. The confocal microscope produces a very clean signal from the spot examined, with virtually no noise from out-of-focus materials. Since only one spot can be examined at any one instant, it is necessary to scan the laser beam across the specimen and reconstruct the full image using a video display.

The confocal microscope also produces an image from a very thin plane within the specimen, which is often referred to as 'optical sectioning' as it can be used with relatively thick specimens but still produces an image of a very thin slice of the specimen (see *Fig. 17.17*).

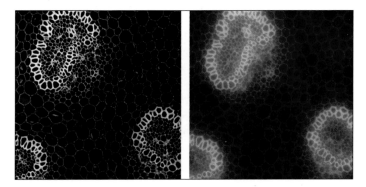

Figure 17.17
Two images of the same specimen of a cross-section of a lily stem. On the left is a confocal image which is crisp and clear whilst the image on the right is taken using conventional fluorescent microscopy and there is some loss of detail and distinct flare from out-of-focus light.

The most common form of confocal microscopy uses a laser which is scanned across the specimen to produce the image; this is the laser scanning confocal microscope (LSCM). The following account will concentrate on the LSCM, which is widely used in biology. Different designs of confocal microscope are available but are less widespread and often more difficult to operate.

Basic principles of confocal microscopy

With a conventional fluorescence microscope the whole of the specimen is illuminated with short wavelength light. The resulting longer wavelength light is captured by the objective and focused to give a visible image which can be viewed by the human eye or used in photography. The result is that fluorescent light from the upper surface and the lower surface and all the tissue in between is transmitted to the eyepiece. The optical system usually has a narrower depth of field than the thickness of the specimen so that only a very narrow slice is accurately focused. The fluorescence from the out-of-focus parts of the specimen are still transmitted to the eyepiece but the light is spread or smeared over the accurately focused specimen, resulting in flare and blurring of the image.

To avoid the flare the confocal microscope uses an opaque plate with a small pinhole to allow some light through. The plate and pinhole are placed at the conjugate plane of the focal point of the objective.

Figure 17.18 shows conjugate focal planes of a lens. On the left are two point sources of light (e.g. a fluorescent nucleus). The first (A) is further from the lens than the second (B). Both point sources can be focused by the lenses but at different points. The point further away will be in focus close to the lens (Af), whilst the object closer to the lens will be focused further away (Bf). Two planes that are mutually in focus, e.g. A and Af (also B and Bf), are referred to as conjugate planes. The light from point B when it reaches the conjugate focal plane of A (Af) will be spread over a large area, as A and Bf are not conjugate focal planes; however, the light from point A will form a focused point at which is its own conjugate focal plane. The light from A, when it goes beyond its conjugate focal plane, will form a focused point but beyond there it will diverge. Therefore, on reaching the conjugate focal plane of B (Bf) it will be spread over a large area. In both cases the out-of-focus light will still be seen in a magnified image but will be smeared over the whole image, thus blurring the detail and reducing contrast.

Having a pinhole at the conjugate plane of the point of focus cuts out most of the light coming from out-of-focus points (see *Fig. 17.19*). A plate with a pinhole at the focal point of A at the plane Af allows all the light from A to pass through, but the out-of-focus light from B is absorbed by the opaque plate.

Laser scanning confocal microscopy
Confocal microscopy is based on using such a pinhole at the rear focal point of the objective. Using brightfield microscopy, a pinhole is needed at the focal point of the objective. This produces a very clean signal but as only a single point is illuminated, it does not produce an image of the whole specimen. Viewing it by eye would give only a

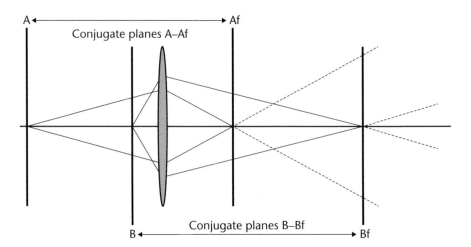

Figure 17.18
Conjugate planes of a simple lens

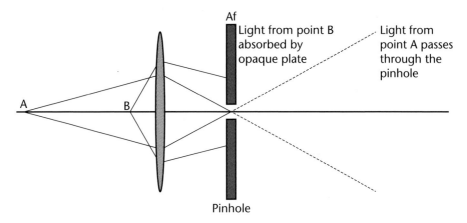

Light from point B absorbed by opaque plate

Light from point A passes through the pinhole

Figure 17.19
The effect of a pinhole stop at the focal point of a lens

Box 17.8 // Lasers

Lasers are a very precise form of light. They have a tight coherent beam which shows almost no spreading of light, as would occur with conventional light sources. It is difficult to use lasers as a light source for conventional widefield microscopy, as they will not illuminate the whole specimen, but they are well suited to scanning microscopy. Lasers are also virtually monochromatic so it is important to match the absorbance spectrum of the fluorescent dyes used with the spectrum of the laser.

Lasers also have their own safety issues, as the intense beam of light can damage the retina if shone directly into the eye. There are problems with ground-based lasers pointed skyward, as the intense light can temporarily blind aircraft pilots.

single spot of fluorescent light. The whole picture can be imaged by examining every spot on the specimen (i.e. scanning in a raster pattern), measuring the intensity of light with a photomultiplier and then using these data to reconstruct an image of the whole specimen and displaying it on an LCD screen such as is used in TVs and computer monitors (each pixel on the LCD being the equivalent of a measured spot on the specimen).

When transmitted light microscopy is used, both pinholes are on the central optical axis so the scanning can only be achieved by moving the specimen using stepper motors attached to the microscope stage. This is effective but slow and cumbersome, with a single scan taking minutes to assemble. Most modern confocal microscopes use a system based on epifluorescence and a laser as the light source to give an intense illumination. An arrangement of two mirrors in the optical path allows the laser light to be scanned across the specimen with only small movements in the angle of the mirrors. One mirror controls the X deflections (left–right) and the other the Y deflections (front–back). The laser is focused at a small point at the focal point of the objective and elicits the fluorescence.

The fluorescent light from the focal point of the objective will travel back along exactly the same optical path as the laser and since it is reflected by the mirrors it will be realigned with the optical axis ('de-scanning') and focused on the pinhole. Light from areas above or below the focal point will not be focused on the pinhole and will not be part of the signal. The fluorescent light will be quite weak, often a lot less than 100 photons, and needs to be measured with a very sensitive detector such as a photomultiplier.

With the confocal system, the specimen can remain stationary and the small adjustments in the mirror angles can be so rapid that a complete scan can be achieved in a fraction of a second, sufficient time to allow the investigation of active processes in living cells. 'Real time' imaging can also be facilitated, as the magnification can be increased without the delay and inconvenience of changing lenses. Magnification can be achieved very quickly by reducing the area being scanned, as the displayed image on the screen remains the same size.

There are compromises in the parameters used, as they affect the fluorescent light intensity. The size of the pinhole alters the thickness of the optical section; small pinholes give a thinner section but the light intensity is reduced. Faster scanning speeds allow rapid changes in metabolism to be imaged but at the cost of reduced fluorescence, as the

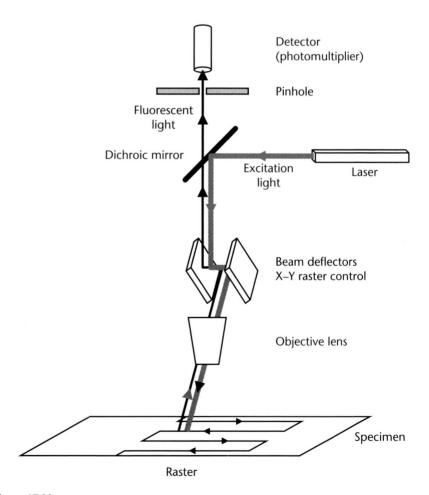

Figure 17.20
Outline of the arrangement of the confocal microscope

length of time the area is illuminated by the excitation light is reduced (shorter dwell time).

Figure 17.20 shows an outline block diagram of the component arrangement of an LSCM. Although the excitatory and emitted light are shown as solid and dotted lines running in parallel, the two beams of light would be completely coincident in the LSCM.

Three-dimensional confocal microscopy
If the specimen is relatively thick, it is possible to take several thin optical sections by readjusting the point of focus and rescanning. Calibrated stepper motors can adjust the focus by small regular steps, giving a series of aligned thin optical slices through the specimen. The slices can be viewed as a stack of images. This is called a Z stack; the x and y co-ordinates give length and breadth and Z is the conventional letter for height.

Interference microscopy
Interference is a phenomenon of the electromagnetic nature of light. As it travels along, the electromagnetic wave has an electric and a magnetic field that vibrate; hence it is called electromagnetic radiation. Since the vibration is a waveform, the peaks and troughs of the waves can be displaced. This produces waves that are said to be either in phase or out of phase (see **Fig. 17.21**).

Light from a single source is coherent when the propagating waves have the same wavelength and are in phase with each other. If two such coherent waves are combined, they will interact and produce a new wave with the same wavelength and phase but increased amplitude or brightness. This is **constructive interference**.

If two beams of the same wavelength but different phases are combined, then **destructive interference** occurs. The two beams interact but the combined wave has a lower amplitude and is dimmer than either of the two original beams. This is not as obvious as constructive interference. The beams effectively cancel each other out. If two beams have the same wavelength and amplitude but are completely out of phase, then the effects of their electric and magnetic fields will be equal but opposed and hence there will

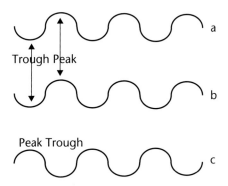

Figure 17.21
Light waves that are in and out of phase. The three waveforms above show the alteration of phase. All three waves have identical wavelength and amplitude and will appear identical to the human eye. They differ in phase but the human eye cannot directly see the phase of the light. Waves a and b are exactly in phase, with peaks and troughs coinciding exactly. Waves b and c are exactly out of phase (180° out of phase) with peaks coinciding with troughs

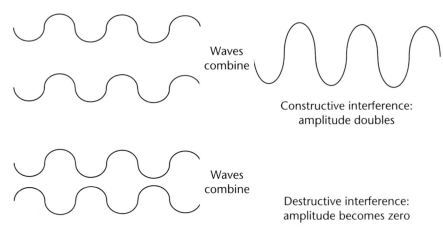

Waves
combine

Constructive interference:
amplitude doubles

Waves
combine

Destructive interference:
amplitude becomes zero

Figure 17.22
Interference. The two waves at the top combine with increased amplitude because they are
in phase, whilst the two equal but out-of-phase waves in the lower half cancel each other
out when they are combined

be no net magnetic or electric field; the light waves have cancelled each other out (see
Fig. 17.22).

Interference can be used as a method of converting changes in the phase of light,
which are invisible to the human eye, into intensity changes that are visible. **Interference
microscopy, differential interference contrast** and **phase-contrast microscopy** (see
below) make use of this phenomenon but are quite different in the final image produced.

Phase changes are not visible to the human eye
The human eye cannot detect phase changes in light so if we wish to observe whether a
specimen changes the phase of light, we need a method of converting the phase change
into an amplitude change. This is what interference microscopy achieves. By splitting a
coherent beam of light into two beams and passing one beam through the specimen, we
can detect whether the phase has changed by combining it with the other beam again. If
the phase is altered, then it will result in interference. The interference alters the amplitude
that the human eye can see.

The interference microscope (see *Fig. 17.23*) or interferometer microscope is a very
precise measuring instrument that can effectively measure the dry weight (mass) of
very small objects. It has been used to precisely map any small variations in section
thickness produced during sectioning and is an interesting way to assess the precision
and performance of any microtome. It is an unusual type of microscope and requires very
careful setting up and calibration. It uses two beams of light that are separated by a beam
splitter; one passes through the specimen, whilst the other passes through a blank area of
the slide. When the two beams are recombined, they interfere depending on the degree of
phase difference. The amount of phase difference is itself dependent on the dry mass of the
specimen. The two beams must have identical path lengths so when they are combined
again they would still be in phase. Only if the specimen alters the phase of the light does
the image appear.

Polarizers and analysers (as used in polarizing microscopy), a half-wave plate (which
effectively reverses the polarization of light) and a compensator are also needed. The

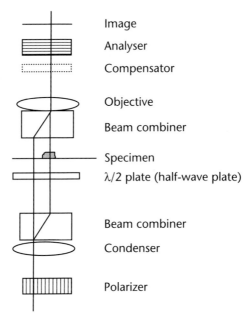

Image

Analyser

Compensator

Objective

Beam combiner

Specimen

λ/2 plate (half-wave plate)

Beam combiner

Condenser

Polarizer

Figure 17.23
Path of the light beams through an interference microscope. A single coherent light beam is split into two beams, which travel through different parts of the specimen before being recombined

adjustment of all of these components changes the image and it is this complexity that limits the use of the interference microscope.

The differential interference contrast microscope (e.g. Nomarski) uses a smaller separation of the light beams (comparable in size to the resolution of the microscope). These give a false three-dimensional image and have a much smaller halo than the phase-contrast microscope. This small offset halo gives the impression of a three-dimensional image, which is a characteristic of Nomarski microscopy and can be used to enhance the subtle localization of chromagen in immunohistochemical methods. These microscopes are generally used in a similar way to phase-contrast microscopes but are more expensive.

Phase-contrast microscopy

The phase-contrast microscope is in many ways related to the darkground microscope but the hollow cone of light produced by the annulus in the substage condenser is arranged to enter the objective, whereas in darkground microscopy it completely misses the front lens of the objective. The ring of light is arranged so that it is focused on a phase plate at the back focal plane of the objective. The phase plate is a simple circle of glass with a groove cut into it that alters the phase of light relative to the ungrooved region (see ***Box 17.9***). Light passing through the thicker region of glass will have its phase retarded relative to the light that passes through the grooved region (see ***Fig. 17.24***).

Light that passes straight through the specimen is focused on the groove, whilst any diffracted or refracted light will pass though the thicker areas. The two beams are recombined in the eyepiece and interfere. This produces interference and hence an image, even in unstained specimens (see ***Fig. 17.25***).

Box 17.9 // Negative phase-contrast

The phase ring can be made in the opposite way with a raised platform of glass instead of a groove. This means that undiffracted light is retarded rather than the diffracted light being retarded. This is sometimes referred to as **negative phase-contrast**.

The amount of retardation can also be varied from $1/4\lambda$ through $1/2\lambda$ to $3/4\lambda$. The most common application is for $1/4\lambda$ difference with positive phase-contrast.

The phase-contrast microscope was developed by Fritz Zernicke in 1932 at around the same time as the interference microscope was developed. In more recent years, the **modulation contrast microscope** has been developed. This uses a slit rather than an annulus and a different level of absorption rather than a phase delay. Phase objects are still visualized but the modulation contrast microscope is better for thick objects and the phase-contrast for thin flat objects.

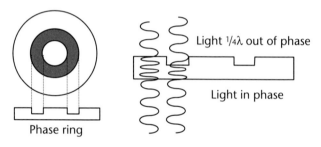

Figure 17.24

A phase ring as used in phase-contrast microscopy. The phase ring is situated in the back focal plane of the objective and consists of a grooved piece of glass. The groove is darker than the rest of the glass. As light passes through the glass, the wavelength alters. The light passing through the groove has its phase altered relative to the light passing through the full thickness of glass

---

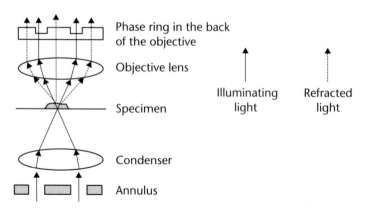

Figure 17.25

Phase-contrast microscopy. Light passing through the annulus forms a ring of light that will pass through the groove of the phase plate. If any light is diffracted or refracted, it will pass through the full thickness of the glass and have its phase retarded compared with the undeviated light

There is much less refracted light than direct light so the grooved area also has some light-absorbing material deposited so that the refracted light and undeviated light have similar intensities and the interference is then more apparent.

Phase-contrast microscopy is relatively cheap as phase plates can be built into objectives fairly simply and will not seriously impair the lens for normal microscopy when normal illumination is used. It then only requires an annulus to be placed in the substage condenser and phase-contrast illumination is achieved. Substage condensers with rotating filters and annuli are common so that switching from brightfield to phase-contrast or darkground microscopy is as simple as rotating a ring on the condenser. Phase-contrast does introduce a distinctive halo that reduces the resolution slightly, but the increased contrast it produces in living cells makes it a popular method in tissue culture work (see *Figs 17.11* and *17.26*).

The differential interference microscope detects phase shifts introduced by the specimen, whilst phase-contrast really introduces the phase changes to detect slight changes in RI. In practice, they tend to be used in similar ways to look at living cells without the need to fix and stain them.

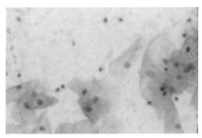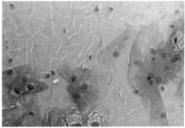

Figure 17.26
The photograph on the left shows a conventional brightfield image of a cervical smear; on the right is the same view using phase contrast microscopy. In the centre of the right hand field can be seen a y-shaped clump of *Candida* yeasts. These are just visible on the left as a faintly stained structure. The ability of phase contrast to pick out unstained or faintly stained material is clearly shown

17.3   Photomicrography

The ability to record what you see is just as important in histology as in any other walk of life, but to record a microscopic image the usual camera lens is replaced by the microscope and the camera body must be attached to the microscope.

Photomicrography has its own advantages and disadvantages compared with normal photography. The advantages are that for most histological specimens the specimen does not move so exposures can be quite long and this allows slow films to be used. Slow films often have better contrast, better colour rendition and finer grains so the quality of photographs taken with the microscope can be excellent. The disadvantages are that the microscope is not as versatile as most camera lenses (see *Box 17.10*). There is no equivalent of the aperture control, which is available on most cameras. This means that there is no control over the depth of field (the amount that is in focus) and exposure is entirely controlled by time. The microscope is also a fixed focus lens so the image cannot be framed differently by moving closer or farther away as it can with a normal camera.

> **Box 17.10 // Micrography**
>
> Micrography is the recording of microscope images and photomicrography is only one aspect. It is also possible to use drawings to record what is seen. Even in these days of cheap and easy photography, the use of drawings still has an important if lesser role.
>
> The recording of very thick specimens in which nerve fibres or similar structures take complex paths through the tissues are still best recorded by drawing. Photographs can only record a single focal plane, which can be a very thin slice of the section. The complexity will not be well recorded because of the narrow depth of field. A drawing can integrate several layers of focus into a single drawing and show the full extent of the complexity.
>
> Drawings can also exaggerate differences and become interpretations of the structure as well as a record. The human brain is a very good visual computer and can interpret much better than a camera can record. Drawings are still widely used in books for teaching where interpretation is important.

There are two main types of system used to produce photomicrographs: one is a dedicated system built by the manufacturer to match their microscopes, whilst the alternative is to attach a standard single-lens reflex camera to a microscope using a simple adapter. Alternatively, a solid state image grabber can be attached to the optical tube, replacing the eyepiece and connected to a computer through its USB port.

Use of digital photography

Photomicrography is increasingly using digital technology instead of conventional photographic film. The advantages of using digital imaging are:

1. Speed. Since there is no development needed, the image can be transferred to a computer and printed within a few minutes. Single images can be downloaded and printed and there is no need to use a whole film as with conventional films.
2. Image files can be transferred quickly. The transfer can be to another site using the Internet or into a computer application such as a word processor (e.g. Word) or an audiovisual system (e.g. PowerPoint).
3. The images are easily corrected or modified. This partially overcomes one limitation mentioned earlier as editing the image can mimic zooming and differential focusing. This type of modification needs to done responsibly. Overuse of modification of the image can come close to fabrication of data rather than just recording what is actually there.

> **Box 17.11 // Microphotography**
>
> Photomicrography is the taking of photographs of small objects and making them bigger. Microphotography is taking photographs of large things and making them smaller. This is done using very fine grain film and has its main use in taking photographs of documents. Many documents need to be stored in detail but are difficult and expensive to keep in their full format. These can be photographed on film and the film takes up much less space. This is done for newspaper archives in libraries where the information is held on microfilm rather than as the full newspaper.
>
> Microphotography is therefore quite different to photomicrography although the two words often get confused.

4. Ease of measurement using image analysis as the image can be imported directly into image analysis programs.

17.4 Physical arrangement of the microscope

As well as the optical arrangements of the microscope, the arrangement of the various mechanical parts can be quite important for the uses of the microscope. The standard laboratory microscope, as shown in *Fig. 17.4*, is the commonest, but other arrangements are possible.

Inverted microscopes

These have the light source and condenser above the stage and the objectives, microscope tube and eyepieces below the stage, which is 'upside down' to the usual arrangement. The stage still has its working surface uppermost but the rest of the microscope is upside down (hence the word 'inverted'). Inverted microscopes look upwards through the base of a specimen, which is useful in cell culture where the cells grow at the bottom of the culture flask. By looking through the base, the cells can be seen without loss of their covering of culture medium and without opening the culture vessel and compromising the sterile growth conditions. The main use of inverted microscopes is in cell culture laboratories and they are usually fitted with phase contrast, dark ground illumination or differential interference optics to visualize the unstained cells.

Incident light, reflected light or epi-illumination microscope

The light source and condenser are above the stage, lighting the surface of the specimen. The objectives capture light reflected from the surface instead of light transmitted through a translucent specimen. Although widely used in industry (microelectronics), metallurgy and materials science, they are not frequently used in biological applications.

Multihead and conference microscopes

Microscopes can also differ in how the image is presented. As well as the usual monocular or binocular heads, by using extra prisms the image can be channelled into two separate binocular heads, so the same image can be seen by two observers. Alternatively it can be split into four or more separate heads for teaching larger groups. Projection lenses can be used instead of eyepieces to allow projection of the image onto a screen, giving images up

---

Box 17.12 // Illumination loss

With both multihead microscopes and projection microscopes there is a problem with illumination. The light source needs to be much brighter for both systems. In a 20-head multihead system each of the viewing heads receives only 1/20th of the light, so a brighter light source is essential. In a projection microscope a field of 100 µm diameter can be enlarged to 2 metres diameter, thus light intensity is reduced by a factor of 20 000. One of the authors (DJC) has used a projection microscope which needed a carbon arc light source. The light was bright enough to give a reasonable image but if the slide was not kept moving it 'burnt' holes in the slide by photobleaching the dyes. Video cameras do not need such bright illumination yet can give hugely magnified images or multiple screens without loss of light intensity.

to 2 metres across. The image can also be projected onto the back of a ground glass screen to give a 'conference' head which can be viewed by several people at once. These are mainly intended for conferences or teaching, but with the availability of digital imaging, such microscopes either already have been or will be replaced by a microscope fitted with a video camera, so the image can be seen on TV-type videoscreens or transmitted over the internet (see **Chapter 18** for more details).

Comparison microscopes

The opposite process to splitting one microscopic image between several eyepieces is to take two microscopes and combine the images into a single view, using prisms and a bridge to link the two microscopes to a single binocular head. This is done in comparison microscopes and allows the images of two specimens to be directly compared. It is widely used in forensic science to compare bullets, paint chips or pieces of cloth to see if they match identically. Comparison microscopes are not used much in pathology or biological research.

Dissection microscopes and stereomicroscopes

These also use two microscopes (usually just at low magnifications) but the images are kept separate and are directed into an eyepiece for each eye. Human stereoscopic vision is retained, allowing good vision for manipulation such as dissection. They also have a longer 'working distance' (the distance between the specimen and the front of the objective when the microscope is in focus). This allows the dissector to get their hands easily under the microscope to carry out the dissection. They can be fitted to provide both transmitted and epi-illumination and are widely used in biological applications. Surgical microscopes and colposcopes are related to these laboratory instruments and can have very long working distances (up to 25 cm) for surgery; they are usually fitted with epi-illumination. If the correct orientation of small samples is difficult it is advisable to use a dissecting microscope to identify, for instance, the epithelial surface of skin or the digestive tract. A dissecting microscope is also essential in identifying glomeruli in fresh renal biopsies where they stand out clearly against a pale background, giving the classic 'strawberries in cream' appearance.

Suggested further reading

Bancroft, J.D. and Floyd, A.D. (2012) Chaper 3: Light microscopy. In: *Bancroft's Theory and Practice of Histological Techniques*, 7th edn. (eds K.S. Suvarna, C. Layton and J.D. Bancroft). Churchill Livingstone.

Bracegirdle, B. and Bradbury, S. (1995) *Modern PhotoMICROgraphy*. Garland Science.

Bradbury, S. and Bracegirdle, B. (1998) *Introduction to Light Microscopy*, 2nd edn. Garland Science.

James, J. and Tanke, H.J. (2012) *Biomedical Light Microscopy*. Springer.

Suggested websites

www.histology.leeds.ac.uk/ Leeds Histology Guide

www.microscopyu.com Nikon MicroscopyU (the source for microscopy education)

www.olympusmicro.com/index.html Olympus Microscopy Resource Centre
www.zeiss-campus.magnet.fsu.edu/index.html Zeiss Online Campus

Self-assessment questions

1. Why is the numerical aperture of a microscope important?
2. What is meant by empty magnification?
3. What is an achromatic lens?
4. Outline the principle of a darkground microscope.
5. Why do crystals of urea appear bright in the polarizing microscope?
6. How are the filters in a fluorescent microscope arranged so that the fluorescence can be observed?
7. Briefly outline the theory of confocal microscopy.

18 \ Digital microscopy

Learning objectives

After studying this chapter you should confidently be able to:

- **Understand the limitations of the human visual system**
  The human visual system is limited because of optical illusions, colour blindness and sensitivity. Objective measurement and computer recognition are often valuable adjuncts to human interpretation.

- **Describe the methods and use of simple measurement**
  Simple measurement techniques include methods of determining size (length, area and volume), number and optical density of microscopic features. Statistical methods can be adapted for computer and digital microscopy.

- **Outline the techniques of digital image capture**
  Microscopic images are converted to a grid of points (pixels), each represented by a numerical value for the intensity of the point. Colour can be captured by having three such grids; one for each of the three primary colours. Images are stored in standard file formats (e.g. JPEG and TIFF). Image processing involves manipulating the image to enhance certain aspects of the specimen and results in a new enhanced image. Image analysis uses measurements from the specimen to create numerical descriptions of the specimen. Image analysis can use the numerical data to interpret the image and is used in applications such as karyotyping and cytological smear screening.

The human eye is excellent for everyday living but some of its beneficial features are more problematic for precise measuring. One problem for accuracy is that intensity measurements are not constant. The human retina shows non-linear characteristics, with the perceived intensities being different to the actual brightness. Linearity of a detector means that doubling the intensity of light results in a doubling of the measurement. Doubling the intensity of the light entering the eye gives a less than doubling of the perceived brightness – the response is more logarithmic than linear. The eye also adapts to the ambient intensity so there is no standardization (see ***Box 18.1***).

The microscope is a magnifying system but it is not complete without something or someone to use its image. In most cases it is used by a human being for visual examination

---

Box 18.1 // Retinal adaptation

In very bright light the overall sensitivity of the retina reduces (bright adaption), so when a sudden change in brightness occurs (such as driving out of sunshine into a dark tunnel of trees), the eye becomes temporarily blinded. In very dark situations the eye becomes dark adapted and the reverse problem occurs. Going from a photographic darkroom into bright sun can be 'blinding'. In both cases the eye readapts to the prevailing ambient light after a few seconds and perception returns to normal.

of specimens. The human visual system is therefore part of the optical system. The human eye can introduce its own aberrations into the system, but providing any major aberrations are corrected by spectacles or contact lenses, they will not interfere with the optical quality of the final image. Most microscope manufacturers offer high eyepoint eyepieces that allow spectacle wearers to wear their spectacles whilst using the microscope. So it is to be hoped that the optics of the human visual system do not degrade the image.

18.1   Optical illusions and microscopy

One very neglected part of microscopy is the role of the human visual system in interpreting the image once it has been received by the retina. The retinal image is interpreted by the visual system and the visual system can be fooled. The human visual system is not a passive viewer but interprets what it sees and attempts to identify structures and patterns in any image projected on to the retina. The human visual system is extremely good at interpreting normal visual images and making assumptions about the nature of the real object that is producing that image – a constant game of joining the dots or filling in the gaps. It is not, however, perfect and psychologists have catalogued a wide variety of different **illusions** that fool or confuse the human eye. These illusions can sometimes occur in the microscopic image and be misinterpreted. The confusion is perhaps easier in microscopy than in everyday life because the microscopic image is unusual. The appearance of a section is only two-dimensional rather than having the usual three dimensions. The human eye unconsciously tries to interpret the flat, even plane of a section using the same techniques that it uses to interpret the three-dimensional world of normal life.

The eye–brain system can be fooled in a number of different ways, each of which causes significant illusions or mistakes.

Geometric illusions

These are illusions that occur when the size, shape or orientation of objects is wrongly perceived. Size distortions are the simplest and most extensively investigated illusions. The Müller–Lyer illusion is one of the best known of the geometric illusions (see *Fig. 18.1*).

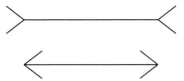

Figure 18.1
The Müller–Lyer illusion. Both horizontal lines are equal in length but the upper line appears longer than the lower line due to the presence of angled fins

In the Delboeuf illusion, the size of circles is misinterpreted when they are part of concentric pairs. This illusion does not rely on the circles being symmetrically arranged and the figures do resemble cells with nuclei so the appearance could affect interpretation of important parameters such as nucleocytoplasmic ratio (see *Fig. 18.2*).

Figure 18.2
The Delboeuf illusion. The inner circle on the left appears to be larger than the outer circle on the right due to the effect of the other circles present. An outer circle apparently enlarges the inner enclosed circle and an inner circle shrinks the outer circle

In the Titchener illusion, the presence of non-overlapping circles alters the apparent size of an inner circle (see *Fig. 18.3*).

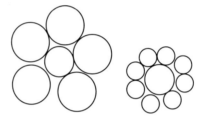

Figure 18.3
The Titchener illusion. The inner circles are equal but the one on the right surrounded with smaller circles seems larger than the one surrounded by larger circles

The Poggendorff illusion may lead to the wrong tracing of nerve or connective tissue fibres (see *Fig. 18.4*).

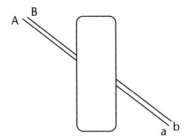

Figure 18.4
The Poggendorff illusion. Line A appears to connects with line a and line B with line b. However, the only straight connection is the line B–a

Contrast illusions

Contrast illusions occur when there is a false impression of the brightness of an area due to its surrounding areas. An example of this is the White illusion (see *Fig. 18.5*).

> **Box 18.2 // Optical illusions can kill**
>
> Optical illusions are not just trivial party games but have very real applications in life. The use of elongated letters that appear to stand upright is used as advertising at sporting events and as warning signs on roads. In at least one case, an optical illusion (the Poggendorff illusion) was given as a major factor in a fatal air crash. In other areas of science such as astronomy and radiography, the study of illusions is taken more seriously. In microscopy, there have been very few papers or articles directly related to the subject.
>
> Illusions are still used in microscopy, e.g. in the shadowing of objects where a false impression of depth is achieved by deliberately printing the negative of the photograph rather than the positive (see *Fig. 19.5* in *Chapter 19*).

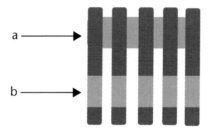

Figure 18.5
The White illusion. The extra bars (a) and (b) are the same brightness (20% grey) but (a) seem less bright than (b) because of their position between the darker lines, whilst (b) seem brighter as they are in the dark lines

Organizational illusions

The human brain always tries to organize images even when there is no real pattern. This gives rise to the children's game of seeing pictures in the fire or in cloud patterns. The use of ink blots by psychiatrists (the Rorschach inkblot test) to probe the human psyche is well known and the images seen are taken as an indicator of the person's mental state and not the physiology of the visual system. Different patients will see different images in the same pattern. Early microscopists often saw objects differently to how we would see them today (see *Box 18.3*).

The Kanizsa figure (see *Fig. 18.6*) consists of three dark circles with segments removed but there is a strong impression of a triangle that seems brighter than the surrounding paper. There is no triangle, yet the human visual system forces it on our consciousness. The non-existent triangle even seems brighter than the surrounding paper. The Kanizsa figure can be a very strong illusion and using stereo images the imaginary figure can appear as a three-dimensional tilted triangle.

Colour illusions

Colour-blindness is a well-known example of individual variation in colour perception and affects a substantial proportion (over 8%) of the male population. The affected person sees objects quite differently to the rest of the population. This can mean that a colour-blind person may miss a critical feature. For example, observing the standard Ziehl–Neelsen technique (red mycobacteria) is enhanced for a normal sighted person if the background is a contrasting green colour, but the red–green contrast is the worst possible option for a person with red–green colour blindness. It is usual to consider

Box 18.3 // Errors in early microscopy

Organizational illusions are one of the most insidious forms of illusion because they involve a definite interaction between the brain and visual system. We not only have the problem of seeing things in error but also of imposing our own structure on things that are unconnected.

An early microscopist named Hartsoeker examined human sperm and interpreted what he saw in the light of current theories. He drew a perfect image of a small embryo in the head of the sperm.

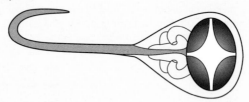

This fitted in with the **animist** idea of human conception. The theory proposed that the father's sperm provided the embryo, whilst the mother simply provided a suitable place for the sperm-borne embryo to grow. The microscope Hartsoeker was using was not up to the standards of a modern microscope so he 'organized' the vague pattern of light and dark into the shape he expected. I do not think he was being deliberately misleading; it was an honest mistake. But it was a mistake and he believed what he thought he saw.

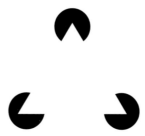

Figure 18.6
The Kanizsa figure. The three dark circles with missing segments give a strong impression of a triangle

that a person who cannot see a particular colour contrast is seeing things 'wrongly' but it is as well to remember that all colour vision is an illusion since humans only have receptors for three regions of the spectrum. It is quite possible to produce any colour with only three monochromatic lights, yet we consider colour to indicate the wavelength of light. Anyone with a known defect should take care to use the appropriate colour contrasts to avoid missing significant features. The use of special variants of staining techniques is often useful, e.g. replacement of fuchsin in the Ziehl–Neelsen technique with victoria blue makes it easier for a red–green colour-blind person to see the tuberculosis mycobacteria.

Although artefacts and technical mistakes are more common sources of error than the human visual system, it is as well to be aware of visual illusions. This is particularly true when trying to measure the size of objects. If we need to be sure that one object is larger than another, we should measure it objectively rather than relying on a purely subjective judgement.

18.2 Measurement and microscopy morphometry

One way of overcoming the problems with visual illusions is to accurately measure the features of interest within sections. Measurement in microscopy is not as easy as in normal life because the magnification alters the size and the specimens are too small to be measured directly using a ruler. It can, however, be even more important to measure objects since there is no obvious scale for humans to relate to the cell size.

Not only is the microscope itself a problem in measurement but also the preparation of tissue for microscopy alters the size of objects. The techniques for measuring in microscopy always suffer from the difficulties associated with preparation of sections, so that any measurement always has a degree of error and the size of this error is often unknown.

Tissue limitations

The preparation of the section introduces problems in measurement and these include:

- **Shrinkage/swelling.** These are commonly associated with processing. Shrinkage is the most likely result and tissues may lose 20% or more of their original volume in the preparation of paraffin wax sections. Most of the shrinkage is due to a combination of the fixative, dehydration and the heat from the wax bath. Although it is possible to measure the tissues before and after processing to gain some idea of the extent of the shrinkage, it still does not guarantee an accurate correction factor since different parts of the tissue may have shrunk to differing degrees.
- **Section thickness.** This is a problem since it is rarely known with any accuracy. It is also difficult to measure thickness once the section has been cut. Microtomes are not reliable enough to use the thickness settings on the microtome feed mechanisms. Firstly, there is no guarantee that the settings were ever anything more than a rough guide and secondly, any wear on the microtome or any slight slackness in any part can lead to alternate thick and thin sections. This is commonly found during cutting and it is likely that sections may vary by 20–30% between one section and the next, even when the ribbon seems quite even.

Measurements that can be made include size (length, area and volume), the number of objects in a given volume and optical density.

Direct measurement of structures

Size: length or linear measurement

The length or breadth of an object is probably the simplest microscopical measurement and can be done in a number of ways.

Estimating size by comparison with a known object. Biologically, exact size is often not critical and an approximate size is all that is needed. For example, the size of a superficial cell in exfoliative cytology is considered to be between 40 and 60 μm. It is not critical to know the size of a cell to fractions of a micrometre as long as the size is somewhere in that range. By comparing to a cell of known size, it is possible to get a sufficiently accurate idea of the size for many purposes. In most sections, the best cell for comparison is the

---

Box 18.4 // Estimating the size of a microscopic object

It is possible to measure using an ordinary ruler and the magnification. It is easiest to do with a monocular microscope. Look down the microscope with one eye and keep the other eye open. You will now see an image of the specimen with one eye and an image of the bench with the other. A ruler held 250 mm away from the eye will appear in focus at the same time as the specimen. Measure the specimen in millimetres by mentally merging the two images and then simply divide by the total magnification and you get a reasonable approximation of the size. This also means that if you have a good sense of scale, then a very approximate size can be guessed. Thus, if a cell looks to be 10 mm at a magnification of 1000, then the real size is 10 μm.

---

red blood cell, which is 7–8 μm. Thus, if a cell is about three times the diameter of a red cell, then its diameter is about 20–25 μm. This technique is widely used and is very easy with a little practice. It essentially gives an idea of the relative size of objects. A similar approximation can be done using an ordinary ruler (see ***Box 18.4***).

Using a graticule. Graticules are transparent discs of glass etched with a pattern. The pattern can be a graduated linear scale that can be used as a ruler, circles of different sizes, square grids or a multiplicity of other types. Linear-ruled graticules can be placed in the eyepiece so that the ruled patterns are visible at the same time as the magnified image of the slide. The eyepiece scale can be calibrated against a known size using a stage micrometer, which has an accurate 1 mm rule divided into 100×10 μm graduations. Once the eyepiece graticule has been calibrated, it is possible to measure lengths very accurately and yet quite simply. Graticules are comparatively cheap (only a few pounds) and, provided only a few measurements are needed, they are reasonably quick. The variety of styles of graticules make it possible to measure circles, distances, separation of lines, etc.

Taking a photograph and measuring the photograph. Provided you can calculate the magnification accurately (e.g. by photographing a stage micrometer at the same magnification), it is possible to measure the photograph using an ordinary ruler and then calculate the size of the original object. The measurement of photographs is widely used in electron microscopy, but can also be useful in light microscopy for moving objects since the photograph freezes the action and allows measurements to be performed.

Size: area

Area can be calculated if the object is a reasonably simple shape by measuring length and breadth and using geometry, but if the shape is complicated this can be difficult or impossible. The area can be calculated by using a photograph and then cutting out the shape and weighing the cut-out part. Provided it is calibrated and heavy paper is used, this is surprisingly accurate. (If the photograph itself is cut out then the paper should be bleached of silver since the silver deposit makes the weight irregular.) Generally, however, it is very tedious to measure areas by these direct methods.

Size: volume

Volume is even more difficult to measure by direct methods since volume needs to be measured in three dimensions so multiple sections are needed and the exact thickness of the sections needs to be known. By using thick board or plastic, an image can be built up from photographs. This is extremely tedious.

Number

This is at first sight apparently easy. For example, to count how many cells are dividing all we need to do is take a measured area and count how many cells are in that area. The normal field of view of a microscope is a standard area so it is not unusual to estimate the numbers of objects relative to one standard field. For example, you can count the number of mitotic figures in a high-power field to get an estimate of the numbers of dividing cells in a tissue. This is often done but it is only a relative count, not an absolute figure. Provided it is only used to see whether there are more or fewer mitoses than usual, then simple counting can be useful. If the exact number of mitoses is needed, then counting becomes more difficult. The actual number is not only related to the numbers seen in the measured area but also related to the size of the objects being measured, the thickness of the section and the distribution of the objects. These are related by the equation:

$$N = n \, \frac{t}{(D + t)}$$

where N = true count, n = observed count (cells are only counted if the centre of the cell is in the section), D = diameter of the cells and t = section thickness. However, exact calculation requires careful corrections.

Density

Density measurements are again apparently easy but in reality are more difficult. Simply shining a light through a section and reading the result in a colorimeter or spectrophotometer is inaccurate since the distribution of coloured objects is not uniform or even regular. It is the equivalent in a biochemical measurement of not mixing a tube of solution before trying to measure the concentration. The results are not consistent and can be very misleading.

The Beer–Lambert law relates the incident and transmitted light:

$$log \left(\frac{I}{I_0} \right) = -\varepsilon d c$$

where I = intensity, d = thickness, ε = absorption coefficient and c = concentration. If the thickness is constant and the material is the same then:

$$log \left(\frac{I}{I_0} \right) \infty \, c$$

Since the change in transmission is logarithmic, it follows that doubling the concentration does not double the absorption. Thus, if a material has an absorption A such that only one tenth (0.1) of the light gets through when it is evenly distributed, then when it is unevenly distributed the results can be quite different.

If a section has four separate areas:

A	A
A	A

there will be a quarter (0.25) of the incident light falling on each block. Of this light one tenth will be transmitted (0.025), so the total transmittance is 0.025 + 0.025 + 0.025 + 0.025 = 0.1:

0.025	0.025
0.025	0.025

and the overall transmission is **0.1**.

If the same quantity of material is concentrated in one sector then the total transmittance will be more (the absorbance will be less) than with an even distribution.

0	0
0	$4 \times A$

The transmittance will be 100% in the three empty sectors but 0.0001 in the sector with all of the material.

0.25	0.25
0.25	0.000025

Only the one area absorbs at all and since it is now four times the concentration virtually no light is transmitted through this area but the overall transmittance is **greater than 0.75**, much greater than that in the case where the same amount of material is evenly distributed. Thus, in this example the same amount of material ($4 \times A$) can give a transmittance of 0.1 or 0.75. The same amount of material can give transmittances as high as 99% depending on how small an area it occupies.

The only remedy is to measure large numbers of small areas and then integrate the result. This is more difficult and if the areas are small and highly concentrated, it may still be inaccurate.

Statistical micrometry

Statistical micrometry does not attempt to measure the actual size directly but attempts to get an estimate of the size and to **calculate how accurate that estimate is**. Absolute accuracy may not be possible but fortunately a high degree of accuracy is often not needed. As there is wide variation in the normal sizes of cells and structures, not only between normal individuals but also within the same specimen, and provided an estimate is within a few per cent of the true size, then it is usually good enough for even the most critical applications in biology. Statistical estimates are not only easier to apply but can give any required degree of accuracy. One example is point counting to measure cross-sectional area (see **Box 18.5**).

Box 18.5 // Statistical sampling

The idea of point counting is similar to that in opinion surveys. Rather than having to ask the whole population what they think about a topic, the pollsters ask a thousand people and then extrapolate to get an approximate answer, with about a 3% error. This is quicker, easier and cheaper than asking everybody and is reasonably accurate.

Pollsters do, however, sometimes get it wrong but they do have extra problems in that some people refuse to co-operate and distort the result.

Point counting

Imagine a carpet with a complex pattern of red and green. If you throw a coin randomly on to the carpet, the chances of it falling on a green part rather than a red part are directly related to the areas of both colours. If the red and green have equal areas, then the probability of the coin landing in a green area is the same as it landing in a red area, i.e. a half (0.5). If the red area covers three-quarters of the carpet, then it is more likely that the coin will fall on the red area than the green. The probability of landing on a red area is 0.75 and on a green area is 0.25, the same as their relative areas. If you throw the coin often enough, then the areas can be guessed by counting how many times the coin falls on a red area and how many on a green area. This is the principle of point counting.

If a graticule with a large number of dots on is superimposed on a section containing two components, A and B, then the chances of the point falling on A is A/(A + B) (i.e. A/ total area). If the number of dots counted is large enough, then it is possible to estimate A and B as a ratio and this is the **area fraction** or **area proportion**.

A typical graticule will have 25 or 100 such points. Crosses are often used instead of simple dots as they are easier to locate when seen against a section. Only the centre of the cross is counted; the arms just help to identify the centre point. Using a 100-point eyepiece graticule and a section of kidney, the calculation might be something like the example shown in *Fig. 18.7*.

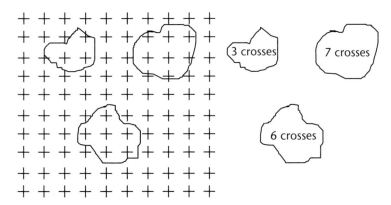

Figure 18.7
The graticule is used with a section of kidney and in the cortex 16 of the crosses of the graticule are superimposed on top of glomeruli. The area of glomeruli is therefore 16/100 of the area of the graticule. Thus, in a cross-section of kidney, the cross-sectional area of the glomeruli is 0.16 or 16%

Whether the dots fall on an object or not is a random statistical event so it is also possible to get an estimate of the accuracy of the measurement. The accuracy is measured by calculating the relative standard error (RSE). Thus, if an RSE of 1% is achieved, we can be reasonably sure that the true measurement is within 1% of our estimate. The RSE can be calculated from the following formula:

$$\text{RSE} = \sqrt{\frac{1 - \text{area proportion}}{\text{hits}}}$$

In the example used in *Fig. 18.7*, the RSE would be:

$$RSE = \sqrt{\frac{1 - 0.16}{16}} = 0.23$$

This means the estimate is only 23% accurate. If we wanted 5% accuracy, we would need to count more grids. A single count is not accurate enough. By assuming that the proportion will not change greatly, we can tell how many grids are needed. Thus, taking an RSE of 5% (0.05), we can use the equation:

$$0.05 = \sqrt{\frac{1 - 0.16}{\text{hits}}}$$

Solving this suggests that we need to count about 20 grids to get an estimate accurate to 5%. If we need to know the actual areas rather than just the area proportion, we need to know the area the graticule covers on the section and then by proportion we can calculate the actual area in square metres. This area of the graticule can be measured using a stage micrometer.

Volume measurement

Unlike the direct methods, measuring volume using point counting is no more difficult than measuring area, provided the following conditions are met:

1. The objects are homogeneous (evenly arranged) in the tissues.
2. The sections are thin compared with the objects.
3. More than one level (i.e. section) is measured.

Then the following applies:

Area proportion = volume proportion   (theorem of Delesse)

Thus, in the example of the kidney above, we could estimate that the volume proportion of glomeruli is 16% of the total volume of the cortex. To get the total volume of glomeruli, we only need to know the volume of the renal cortex and it can be calculated by simple ratios. In this case we could *not* use the volume of the kidney as a whole, since glomeruli are not evenly distributed within the kidney; there are no glomeruli in the renal medulla, although they may be reasonably regular within the cortex.

Many of the measurements of microscope images that are made by computers rely on the micrometry techniques outlined above. A digital image of tissue consists of a large number of individual pixels and can be considered as a grid of points like the cross-hairs on a graticule, with each pixel being equivalent to one point on the graticule. The digitized image does have many more points than would be used in a visual count. Visual graticules rarely exceed 100 points (10 × 10 grid), whilst a digital image will usually be in at least 1024 × 1024 pixels. The accuracy is correspondingly better, as the more points that are counted, the more accurate the estimate.

18.3 Image digitization

The commonest way of digitizing images is to use a digital camera. Almost any digital camera can be used with a microscope but many have fixed lenses and will require extra optics to bring the microscopic image into focus. The extra optics will disturb the carefully matched optical system of the microscope; focusing becomes difficult as the

LCD screens are small and exposure calculation can be difficult. For simple microscopy it may be adequate, but it is far from ideal and image quality suffers. Single lens reflex (SLR) cameras have interchangeable lenses, so using the microscope can be considered as just another lens in the SLR range. They will generally work much better than a simple fixed lens camera, as all the optical components belong to the microscope. Focusing is usually easier, as the image can be seen through the camera viewfinder and focusing is similar to ordinary visual microscopy.

Microscope manufacturers produce dedicated photomicroscopes, usually with an extra 'eyepiece' tube for the camera. These trinocular systems are ideal for photomicrography, as all the optics are matched. Focusing is achieved using the binocular eyepieces as normal and the light can then be redirected to the camera, usually using a simple sliding arrangement, and the image photographed. The switch from visual to photographic viewing is needed, because if the visual eyepieces were open at the same time as the camera was in use, there could be stray light entering via the eyepieces which would degrade the image.

The image is captured by the camera as a series of dots called pixels ('picture cell' or 'picture element') and the image produced in this way is called a 'bit-mapped image'. Bit mapping needs three pieces of data for each pixel:

1. A number representing how far across the screen the pixel is (x position).
2. A number representing how far down the screen the pixel is (y position).
3. The light intensity of the pixel. A total black has a value of 0 and a pure white a value of 255 (detector totally saturated with light).

The more pixels used in the recording of an image, the greater the resolution and detail in the image but the more data that needs to be stored. The intensity data for each pixel may be a single byte or several bytes. Monochrome images typically use only 1 byte (equivalent to 256 shades of grey from black to pure white), whilst colour images typically use more than 1 byte. For colour, 3 bytes is common, 1 byte for each of the three primary colours (red, green and blue; the RGB system), and this gives 16 777 216 possible colours ($256 \times 256 \times 256$). The use of 3 bytes for colour is usually referred to as 24 bit colour, as a digital byte has 8 bits and $3 \times 8 = 24$. Larger numbers of bytes can be used for colour; for example, 2 bytes per primary colour (48 bit colour) is possible but rarely used.

The x and y positions are not usually stored as separate numbers. Since the pixel values are stored in order as a list, the x and y position of each pixel can be calculated from its position in the list. Thus, the position of the pixel is implied rather than stored in memory.

A 1024×1024 monochrome image will therefore need 1 megabyte (Mb) of memory and a colour image will need 3 Mb. Many applications do not need colour and monochrome image processing is still used. As the number of pixels rises, so does the amount of storage needed; a 2048×2048 image requires 4 Mb of memory for monochrome and 12 Mb for colour.

Box 18.6 // Human vision and digital images

Monochrome digital images show 256 shades of grey, whereas the human eye can accurately distinguish about 30–40 shades of grey, depending on the lighting conditions. This allows digital images to have a finer perception for measurement than a human ever could. Similar limitations apply to colour perception, so again digital images have more distinctions than the human eye.

Colour capture

Digital image capture is mostly achieved using semiconductor technology which can be fabricated as single chips, with many thousands of individual semiconductors on one chip arranged in a grid. Each sensor produces one pixel on the final image. The commonest microscopic digitizers use CCD (charge-coupled device) detectors rather than CMOS detectors (complementary metal–oxide–semiconductor). At the time of writing, CCD detectors give better quality images but are more expensive and use more power than CMOS detectors, so CMOS is more common in simple photographic cameras, where cost and battery life are more important than outright quality. However, CMOS technology is improving and may replace CCD as the sensor of choice. The number of sensors on a chip is also increasing. At the time of writing over 20 000 000 sensors (20 megapixels) are to be found in cameras and 13 megapixels in mobile phone cameras.

Each sensor is essentially monochrome, as it will output a single electrical charge indicating light intensity. By using a coloured filter over the sensor it can be made to respond to specific colours of light. If full colour RGB data are needed, then it requires one sensor to detect red light, and two other sensors to detect green and blue. In some systems the light from the microscope is split by a prism into three beams directed to three separate chips, with each chip detecting one colour. In an alternative arrangement each sensor is individually covered with a filter, so a single chip detects red, green and blue using only one camera. The filters are arranged in a pattern to allow colour to be regenerated by interpolation. The commonest arrangement is the Bayer grid (see **Fig. 18.8**) with two green and one red and one blue receptor in a 2 × 2 arrangement which is repeated over the whole area of the chip.

It often comes as a surprise that many 'colour' images (e.g. from confocal microscopy or FISH analysis) are actually captured as greyscale values and will then have a colour added to help discrimination. This is not sleight of hand as each greyscale value corresponds to

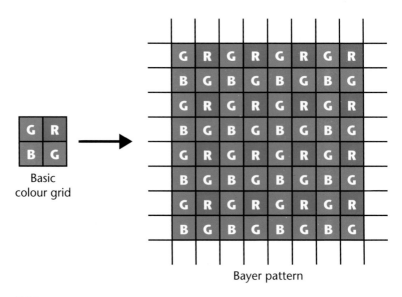

Basic colour grid

Bayer pattern

Figure 18.8
The Bayer pattern of colour pixels, as used in most digital cameras

a given colour of fluorescence and the greyscale is then allocated a colour that is a (fairly) true representation. This is similar to the various deep space telescope surveys that can analyse in X-ray, UV, near or far IR wavelengths or even detect polarization of light. The telescope images are allocated colours and images can be superimposed for comparison purposes. The main aim of the added colour is to help investigators to visualize the images that have been created from types of light (or electromagnetic radiation) that is not detectable by the human eye.

As all four pixels are used to generate the colour, the pixel dimensions in terms of colour are less than the stated pixel count for the chip. This method of taking colour images is found in most digital cameras for general use.

The RGB system of colour is the most widely used but is not the only one. The RGB system of using three complementary colours is similar to the human visual system and photographic films. The retina detects colours with the cone cells; there are three types of cone which are roughly sensitive to red, green and blue, although the exact sensitivity is not quite the same as with digital RGB. However, human perception and ideas about colour are more in line with HSL (hue, saturation and luminance) which is a different system for colour information. The HSL system uses hue (equivalent to the colours of the rainbow), saturation (indicating depth of colour) and luminance (how bright the colour appears). HSL can easily be converted to RGB and vice versa by image processing software. HSL is mainly used in computer graphics and similar applications. A third system, CMYK, uses three colours: cyan, magenta and yellow, and a further attribute K (for **K**ey or blac**K**). The CMYK system is used mainly in printing, as it corresponds to the ink colours used.

File types and digital images

Because digital images can generate very large computer files, it is often necessary to reduce the size for storage or to send over the internet. Files can be compressed in a variety of ways, some of which degrade the image and some of which do not. For routine photographs or snapshots, a slight degradation is often not visible, but for scientific purposes the files must not be degraded. Formats that degrade the image are referred to as **lossy compression** because some information is lost in the process. Formats that retain all the information are referred to as **lossless compression**. Sometimes some formats are not recognized by individual computer programs so the choice of image format is often crucial. The different file formats can be recognized as they use a file suffix to aid recognition, e.g. image.jpg is a JPEG file.

Box 18.7 // Applications of digital imaging

Digital imaging is widely used in the film industry to merge computer-generated material with live action shooting, to add or remove objects and to quickly change aspects of a scene without needing to completely reshoot it. It is also becoming increasingly used in business to read documents automatically (using character-recognition programs), for speed cameras and other types of law enforcement (reading number plates) and for quality control.

The leaders in the field of videodigitization are NASA who use digitization to send pictures back from space. The Hubble space telescope is a great example of the data collection that can now be achieved with high-resolution cameras.

File formats for digital photographs include:

- **Bitmap** (suffix .bmp). This is an uncompressed image file and so contains all of the information (lossless) but can result in a very large file. It is used only if no better file format is available.
- **GIF** (suffix .gif). This was originally produced by CompuServe as a method of transferring graphics over the internet. It gives a good compression but it is a lossy compression. It is limited to 256 colours and so will not give a good record of a colour photograph. It is excellent for graphics.
- **JPEG** (suffix .jpg). This is a lossy compression but gives a good saving in file size. It produces good results from routine photographs but is not suitable for scientific images that are to be analysed. It is very widely recognized.
- **TIFF** (suiffix .tif). This is a lossless compression but does not compress files as much as JPEG. This is the preferred format for most scientific images.

In addition, there are many less-common formats such as RAW, PNG, PSD, PCT and TARGA, which have specialist or restricted applications.

Image processing and image analysis

There are two distinct ways in which computers transform digital microscopic images. Image processing involves using an algorithm or program to convert the image to make details or structures easier to recognize and interpret. The result of image processing is a new altered image. Image processing is where data are extracted from the image which can be used to analyse the structure.

The result of image analysis in not an image, but a series of numbers accurately representing some aspect of the specimen. Often both images and numbers are used – the image is first processed to clarify and simplify it prior to the image analysis step.

Image processing

The image can easily be modified in the following ways to allow the detail in the image to be more easily seen by the human eye:

1. **Intensity shifting.** If the image is too dark or too light overall, the intensity can be shifted by adding or subtracting a constant from the intensity value for each pixel.
2. **Contrast stretching.** The difference between grey shades can be exaggerated by using a mathematical operation (e.g. multiplying by a factor or taking the logarithm of the number).
3. **False colouring.** Grey shades are converted into colour differences rather than shades of grey. This uses a **look-up table** to convert each pixel. Each possible shade of grey is simply assigned a particular colour and this is stored as a table of colours in the computer. As each pixel is being displayed, the computer 'looks up' the colour corresponding to the value of the pixel in the table and displays that colour instead of a grey shade. There does not need to be any regularity in a look-up table and colours are often chosen simply to look attractive or to enhance contrast. False colouring is quite effective if a particular set of objects can be shown to stand out from the background. If a histochemical test has been done and the colour is fairly difficult for the human eye to see, computer enhancement will allow the colour to be boosted by using a look-up table.
4. **Inverting.** This alters the image from a positive to a negative form.

5. **Find edges.** Edges are where there is a sudden change in density or colour. Intensity is stored as a single number for each pixel, and edges can be detected by simply subtracting the numbers stored for adjacent pixels and then storing the new number as the value of the pixel.

| 4 | 4 | 4 | 100 | 100 | 100 | 4 | 4 | 4 |

This grid shows a series of 9 pixels from an image. The pixels cross an object with a density of 100 in a background of density 4 (a bright object on a dark background). Subtracting the neighbour to the left and then storing it as the new pixel value is shown below:

| x | 4−4=0 | 4−4=0 | 100−4=96 | 100−100=0 | 100−100=0 | 4−100=−96 | 4−4=0 | x |

which results in a new set of values:

| x | 0 | 0 | 96 | 0 | 0 | 96 | 0 | x |

Pixels in the evenly coloured background and in the evenly coloured object become zero (black) whilst the pixels in the line become non-zero and show white/grey. For line detecting in two dimensions, each pixel is recalculated from its eight neighbours using a grid with weightings. The full explanation is beyond the scope of this book but the use of 3×3 (or larger, such as 5×5, 7×7) matrices, referred to as kernels, is a powerful and versatile way of processing digital images.

In general there are different categories of transformation. Pixel transformation alters isolated or separated pixels; the use of look-up tables such as in false colouring is an example of this. Area or spatial transformation involves groups of cells, such as in line detection or area measurements. Frame transformations, such as intensity or contrast changing and merging of images, produce Z stacks in confocal microscopy. Geometric transformations are ones in which the relationships between pixels, shape or orientation of the image are affected. Geometric transformations include rotation of the whole frame, lateral or vertical movements to align images for Z stacks.

Image analysis

The computer can also be used for measurement. This is often complex but commercial programs are available and are becoming more common. Image analysis is now one of

---

Box 18.8 // Image analysis expertise

It is difficult to find out exactly how good current technology has become, because the most advanced systems are those used by military forces. These are used for automatic recognition of enemy aircraft, tanks and soldiers. The details are classified as secret but it seems likely that they are better than the current commercial recognition systems. However, some transfer is occurring. In 2001 QinetiQ was formed as a private company from parts of the British Ministry of Defence research sections. QinetiQ is using its expertise to develop medical image interpretation software. As this technology filters out of the secret military area, it is revolutionizing microscopy. The prediction of crowd behaviour in queues, populating high capacity venues and general pedestrian flow in thoroughfares are other important applications.

the most rapidly expanding areas of computer technology, driven by advances in optical and IT technology. This type of analysis is not restricted to histology but is being used in industry, commerce and everyday life (see ***Box 18.7***).

In computer recognition, objects are often classified differently to the way a human would normally describe them. Several different measurements can be used that are relatively easy for a computer system to measure but would be less easy for the unaided human eye. With complex shapes there are a variety of parameters that can be measured by simple formulae including:

- area;
- Feret diameters (or 'calliper diameters'), where the largest Feret diameter is the length and the smallest is the breadth (see ***Fig. 18.9***);
- length/breadth (maximum Feret/minimum Feret);
- convex perimeter (like pulling a string around the shape);
- roundness [= perimeter^2($4\pi \times$ area)];
- radius of an equivalent circle;
- convex perimeter [= $2\tan(\pi/2n) \times \Sigma n$Feret];
- roughness (= perimeter ÷ convex perimeter).

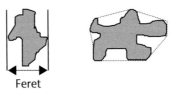

Feret

Figure 18.9
Diagram illustrating a Feret diameter and convex perimeter

The application of this type of technology may well revolutionize the whole of cellular pathology with automatic analysis of sections and computer diagnosis. The question is not so much whether this will happen as when. Automated computer analysis of isolated cells in smears or films is possible, but tissue sections are more difficult as they have incomplete cells where the plane of the sectioning cuts through a cell. There are also issues with overlapping cells, and cell borders can be quite vague. Some of the difficulties can be alleviated by changing the way the tissues are prepared, to suit the automated system rather than the human eye, for example by intensifying the nuclear stain. The choice of counterstain or even complete avoidance of a counterstain may suit the automation better, whereas human perception would have different requirements.

Computer-assisted karyotyping

Slides of metaphase spreads (see ***Chapter 15***) are good subjects for computer image analysis, as the background is fairly clear of confusing materials, the chromosomes are reasonably well delineated and recognition is relatively straightforward. Slides can be automatically scanned and mitotic figures identified and classified. This may involve three steps of computer-controlled image processing:

1. Image enhancement using algorithms or human-assisted thresholding, enhancement filters, contrast adjustment and others, e.g. 'straightening'. This can give well-banded chromosomes from even quite poor images or specimens.

2. Identification of mitotic spreads and classification of spreads for quality. This involves image analysis using algorithms such as 'find edges'.
3. Separation of touching chromosomes, counting of chromosome numbers, preliminary identification of individual chromosomes and banding patterns including abnormalities. Rearrangement of chromosomes into a karyotype.

These steps may require human interaction with the computer software. The systems can often be trained to improve or customize the recognition. The systems can be very flexible and use different staining methods such as Q-banding (quinacrine), G-banding (Giemsa), GTG banding (G-bands after trypsin and Giemsa), FISH and others.

The system can scan much faster than a human, thus reducing the time needed for diagnosis. The results, including images, can then be assessed by a trained cytogeneticist for final diagnosis. The images and associated information can then be filed and stored.

Computer-assisted cytological screening

Cytological smears are not quite as easily assessed as karyotypes, because the slide background is more cluttered, cell outlines are less well delineated than stained chromosomes, and the criteria for recognition of abnormal cells are more difficult to programme into the system.

The systems currently available are used to 'triage' the smears in screening into the categories of normal, suspect (or borderline) and abnormal cells. The latter are identified for later assessment by a trained cytologist.

Most of the abnormalities associated with malignancy are nuclear changes. Thus, the first part of the analysis is to recognize the nucleus with such algorithms as 'find edges' and then measure nuclear parameters such as size, regularity of shape and nuclear density. On the basis of these measurements the cell can be classed as normal or abnormal. This is not entirely straightforward; as with any parameter, there will be a spread of values for normal cells and a different spread of values for abnormal malignant cells. In *Fig. 18.10* the spreads are shown as a Gaussian bell-shaped distribution, but this is not necessarily the case in practice.

The distribution of normal specimens is shown on the left and that of abnormal specimens on the right, but the two curves overlap. Any specimen showing a score between A and C could be either normal or abnormal. If the criterion for abnormal is

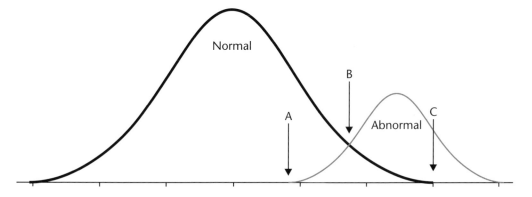

Figure 18.10
Distribution of normal and abnormal cells

taken as A, then all specimens with a score less than A will be diagnosed as normal and all to the right, abnormal. Setting the cut-off point at or below A will result in no false negatives, as all the specimens to the left of A are normal, so all the true positives are included in the abnormal group. However, this cut-off will result in significant numbers of false positives, as the normal curve extends to the right of A.

If the cut-off point is set at C or higher then there will be no false positives, as all the samples to the right are abnormal and all the true negatives are in the normal group. However, there will be significant numbers of false negatives, as the abnormal curve extends to the left of C.

It would be reasonable to use computer screening with a primary threshold sufficiently less than A to eliminate normal specimens and these would not need further screening as there will be no false negatives in this group. Specimens with scores above C need to be examined carefully and a definitive diagnosis given. Between the primary cut-off and point C the specimens should be at least rescreened to confirm their status and the true positives then carefully examined and given a definitive diagnosis.

If the cut-off point is set anywhere between A and C (e.g. at B) then there will be both false positives and false negatives. The FocalPoint screening system curves are shown in *Fig. 18.11* and the Food and Drugs Administration (FDA) in the USA allows the automated rejection to be used, provided there are no more than 25% rejected in this way.

Improving disease detection through intelligent Pap imaging

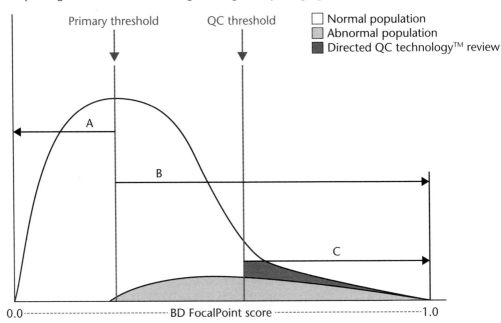

A. Slides with scores below the primary threshold can be archived with no further review
B. Slides with scores above the primary threshold are reviewed by cytotechnologists
C. Slides with scores above the QC threshold are re-screened by cytotechnologists

Figure 18.11
The FocalPoint triage system. Image reproduced courtesy and © Becton, Dickinson and Company.

> **Box 18.9 // Availability and use of computer recognition technology**
>
> Recognition systems are not only being used in laboratories for screening, but are widely used in other areas. An example of this would be in subatomic physics with, for example, the Large Hadron Collider, where protons are smashed together to try to find evidence of the Higgs boson. Sorting through the products from millions of disintegrations in each experiment, most of which are of little interest, would otherwise be impossible. Computer screening eliminates all of these 'trivial' results and presents only the unusual results for further expert analysis, thus solving the same problem as cytological screening.
>
> In border controls when aiming to detect known terrorists, drug smugglers, etc., face recognition systems are often used. This will pick out only a few 'suspects' from the thousands of people passing through an airport, for further investigation by staff. This is again a familiar problem and a similar solution to another screening problem.
>
> The technology is also used in advertising. A person standing in front of an advertising videoscreen is scanned, and characteristics such as sex and age are estimated by automated recognition technology; the adverts are then targeted more precisely to that person.
>
> The cost of some such systems of digital recognition is falling to the point where they are within the reach of the general public. For example, Amazon released a smartphone in 2014 that can recognise not just faces but also 100 000 000 real world objects from a snapshot taken with its inbuilt camera.

In both karyotyping and cytology screening the computer is an aid to screening, rather than being a replacement for human interpretation. The most tedious and time-consuming task is scanning every cell on every slide to find the occasional aberrant cell. The automated screening system scans a slide, eliminates cells which get a low score and then identifies the cells most likely to be aberrant. The laboratory staff only need to examine these already identified suspicious cells to make the final diagnosis. Only a single cell on a slide may be genuinely abnormal, yet the screening needs to check every cell. A screener could spend a whole working day examining thousands of cells without any of them being abnormal – prescreening by an automated method would eliminate a lot of this tedious work.

Some things about this arrangement are worth commenting on. The retention of human beings as part of the system is partly an emotional response, as many people are wary of computers taking over such a sensitive (literally 'life and death') decision, but it is also a logical decision, as trained human staff are still better at correctly diagnosing abnormalities than automated systems. There is a fear that laboratory staff could be 'deskilled' by not examining so many slides, but in practice the skill level is likely to rise as the workers are exposed to the more critical decisions much more frequently. Computers are much faster and more reliable at simpler tasks such as primary screening and can relieve the pressure on laboratory staff, thus releasing them to attend to the more critical aspects of diagnosis. Computers can speed up screening and eliminate backlogs so that reporting is more rapid; this is important in the case of malignancy where any delay is a disadvantage. Computers can also be cost-effective and can release funds for other purposes. Finally the different qualities of computers and humans beings can be complementary and they can act as part of a mutual quality control process for the other.

> **Box 18.10 // Consequences of telepathology**
> The ability to remotely access images and data may have profound effects on laboratory working. Working from home becomes possible, as does interrupting a person's day off to ask them to assess a specimen remotely. The use of computer tablets and wifi could potentially mean microscopy being available on a warm sunny beach. Whether the availability of telepathology would mean staff 'skiving off' but still being available for consultation, or being pestered whilst on holiday, remains to be seen.

Digital microscopy and telepathology

The ability to digitize images is not only useful for computer recognition but has other applications. Integrated systems are now available to automatically scan slides and the images can then be archived and attached to the patient's notes in the health databases so that they can be accessed easily if the patient moves to a distant location. Computer storage is now so inexpensive that large numbers of high quality images can be stored quite cheaply. This aspect of storing images is not limited to pathological images but is also used for such images as X-rays and ultrasound scans.

Digital images can also be transmitted over the internet so that colleagues in different geographical locations can share and discuss the same image. This facility can be used between two hospitals within the same group, with experts in other parts of the country or even internationally. It may also be used to access images from home, thus allowing working from home. It is also possible using digital video cameras to remotely control a microscope for focusing and stage movements, allowing interactive diagnosis rather than just a static image. There is a very practical application for pathology at a distance where skin lesions can be viewed superficially at low magnification and the live image sent for an opinion to a centrally located dermatopathology expert. This approach has proved to be accurate and efficient in less populated areas where travel becomes an issue, such as Australia.

Although it is possible to store images of every case, it may not be appropriate to use this to completely replace retention of the specimens themselves. The image is a fixed record and cannot be further investigated so there could be loss of valuable information with the loss of a specimen. In forensic pathology, original specimens from the archive have been used to solve crimes and gain convictions decades after the crime was committed because of advances in science (such as DNA analysis) which were not available to the forensic investigation at the time of the original crime. With such advances in technology, an archive of original specimens becomes a valuable resource to retrospectively research diseases. It is not possible to predict what might be able to be discovered, so original specimens can still be more useful than simple images. Much has been learned from the investigation of Egyptian mummies which could never have been found if only the statues and hieroglyphics were available.

Intelligent systems – evaluating the degree of abnormality

Ultimately the ability to interpret images relies on learning and recognizing what is abnormal. Humans are particularly good at facial feature recognition, and similar abilities are needed to recognize abnormal and normal cells. Computers can either be programmed to use known differences between normal and abnormal cells or can be programmed to 'learn' the differences. The first step of the process is to determine a set of key features and their normal range of appearances. In cellular pathology these features usually comprise size, number, shape and staining characteristics. Once a sufficient number of representative cases have been reviewed according to these criteria, the distribution of each feature will have been established. Using this distribution, new cases can be judged to be normal or abnormal. If some form of scoring can then be applied for each feature, it may indicate the degree of overall abnormality. This logical approach is being developed to evaluate both breast and prostate cancers and, if successful, will be extended further. The aim is to render decision-making less subjective and introduce some degree of rigour by including objective measurements.

Computing power has increased dramatically and has approximately doubled every two years (this was predicted by George Moore and is therefore referred to as Moore's Law). At the same time the cost of digital data storage has also fallen dramatically. This means advanced computers can 'learn' by storing large banks of images, making the identification of the abnormal a relatively straightforward task. The aim now is to implement a Bayesian belief network (BBN) to analyse the accumulated data. Bayesian networks have proved particularly useful in battlefield surveillance for the identification of friend or foe (IFF) and this is very similar to the task of analysing cellular pathology samples. All that is required is to establish a number of key features that relate to a disease and then add a probability weighting factor to each that will adjust the significance of their contribution; essentially a form of ranking. These features, as well as including the histological findings, may include clinical observations, for example family history, genetic susceptibility or a mammogram.

Suggested further reading

Baak, J.P.A. and Oort, J. (2011) *A Manual of Morphometry in Diagnostic Pathology*, 2nd edn. Berlin: Springer-Verlag.

Bancroft, J.D. and Floyd, A.D. (2012) Chaper 3: Light microscopy. In: *Bancroft's Theory and Practice of Histological Techniques*, 7th edn (eds K.S. Suvarna, C. Layton and J.D. Bancroft). Churchill Livingstone.

Cook, D.J. (1983) Optical illusions in microscopy. *Microscopy*, 34: 533–552.

Crocker, J. and Murray, P.G. (2003) *Molecular Biology in Cellular Pathology*. Wiley.

Carroboles, M. *et al.* (2013) A morphometric tool applied to angiogenesis research based on vessel segmentation. *Diag. Pathol.* 8 (Suppl 1):S20.

Floyd, A.D. (2012) Chapter 23: Quantitative data from microscopic specimens. In: *Bancroft's Theory and Practice of Histological Techniques*, 7th edn (eds K.S. Suvarna, C. Layton and J.D. Bancroft). Churchill Livingstone.

Hamilton, P. and Allen, D. (1995) Morphometry in histopathology (Review). *J. Pathol.* **175**: 369–379.

Montori, R. *et al.* (1996) How to develop and use a Bayesian Belief Network. *J. Clin. Pathol.* 49: 194–201.

Suggested websites

www.histology.leeds.ac.uk/ Leeds Histology Guide

www.microscopyu.com Nikon MicroscopyU (the source for microscopy education)

www.olympusmicro.com/index.html Olympus Microscopy Resource Centre
www.virtualpathology.leeds.ac.uk Virtual Pathology at Leeds
www.zeiss-campus.magnet.fsu.edu/index.html Zeiss Online Campus

Self-assessment questions

1. Why is it unreliable to estimate the size of an object 'by eye'?
2. In a section across the bronchus of a person who had chronic bronchitis there are some mucous glands. A point count is made of 200 points: 110 points fall on the glandular tissue and 90 fall on other tissues in the wall. What is the proportion of mucous glands in the tissues?
3. What are the problems with measuring density in a section of stained tissue?
4. Briefly outline how microscopic images can be captured as digital information which can then be used by computing technology.
5. Distinguish between image processing and image analysis.
6. Briefly explain how digitally based microscopy is assisting in cytological screening.
7. Briefly outline how telepathology is affecting pathology.

Electron microscopy

Learning objectives

After studying this chapter you should confidently be able to:

- **Describe the principle of the transmission electron microscope**
  The transmission electron microscope (TEM) uses a beam of electrons rather than visible light to examine and magnify the specimen. The electron beam has a much shorter wavelength resulting in a much greater resolution than the light microscope. The electron beam can be focused to form a highly magnified image by using magnetic fields instead of solid glass lenses. Specimens for electron microscopy need to be very thin and dry and the TEM needs a very high vacuum to work. This means that the cells are always dead.

- **Describe how specimens can be prepared and stained for the conventional transmission electron microscope**
  Specimens need to be fixed very promptly (typically with osmium tetroxide or glutaraldehyde) to preserve the cell structure. The specimens are then dehydrated and impregnated with a plastic monomer, which is then hardened by polymerization. The plastic allows much thinner sections as it gives firmer support to the tissues. The sections can be cut on an ultramicrotome using glass or diamond knives. The sections are finally stained using heavy metal salts to increase the contrast.

- **Describe the principles of the scanning electron microscope**
  The scanning electron microscope uses a thin beam of electrons, which scans across the specimen in a raster. The beam illuminates one spot on the specimen and all of the electrons that bounce back from the specimen are collected and counted. The number of electrons is used to alter the intensity of the beam in a cathode ray tube, which is scanning in synchrony with the illuminating beam. This produces a three-dimensional image of the surface of the specimen.

- **Outline the methods of preparing specimens for the scanning electron microscope**
  The specimens do not need to be embedded and sectioned as they are for TEM. They do need to be dried and coated to make them electrically conductive. Drying must not distort the specimen so critical point drying is preferred to simple desiccation.

- **Describe the principles of X-ray analysis in electron microscopy**
  When an atom is struck by a high-energy electron, it will dislodge an electron from the outer electron orbit of the atom. The dislodged electron may then fall back into the outer orbit and loses energy as X-rays in the process. The amount of energy is related to the orbit and is characteristic of each element. By measuring the energy of the X-ray it is possible to identify the nature of the element undergoing the transition. By scanning the beam across the specimen, it is possible to map the distribution of elements in the specimen.

There are two quite different forms of electron microscopy. The first is transmission electron microscopy (TEM), which was the first type to be produced, with the first commercial models appearing in the 1930s, so when people refer to the electron microscope they are usually referring to this type of microscope. It should be noted that all electron microscopes can only produce greyscale (black and white) images. These are sometimes digitally enhanced with a false colouration to highlight specific features.

Since the 1960s, another form of electron microscopy has become commercially available; this is scanning electron microscopy (SEM), and subsequently scanning transmission electron microscopy (STEM) has become available. Since TEM is in many ways similar to conventional light microscopy it is also sometimes called conventional transmission electron microscopy (CTEM) to distinguish it from STEM. In biology and medicine, CTEM is still probably the most common and most useful form of microscope, but SEM is becoming ever more useful and certainly gives very striking images for illustrations. In the 1990s an environmental version of the electron microscope emerged. This operates under a minimal 'soft' vacuum and can image living insects and plants that have not been fixed or freeze dried, for example aphids on leaves.

19.1   Conventional transmission electron microscopy

Light microscopy has always been limited since it cannot resolve features of less than about 200 nm apart. Attempts to overcome this resolution limit continued for many years with people trying UV light (see ***Box 19.1***), X-rays and similar short wavelengths related to light. These were generally either very little improvement on the ordinary light microscope or difficult to use, and often both. The electron microscope was developed using electrons rather than electromagnetic radiation and has proved to be much more successful. It can be used with only a little more difficulty than the light microscope.

Electron microscopes are much more expensive than light microscopes and need a special electrical supply because of the high voltages and currents used. They need a large water supply to remove the excess heat and a very strong and rigid floor to enable them to be used since even slight vibrations can seem huge when highly magnified. They are generally much larger and require a room of their own and basically are not able to be moved. Throughput of specimens is much slower and running costs are much higher and because of this electron microscopes are much less common than light microscopes. Versions that plug into the mains have been developed and compact bench top models are now available.

> **Box 19.1 // Ultraviolet microscopy**
>
> The UV microscope is different to the fluorescent microscope, which also uses UV light. In the fluorescent microscope, UV light is used to make the specimen glow visibly (fluoresce). The microscopist looks at the visible light and the UV light is removed by a barrier filter and never forms part of the final image. A true UV microscope uses UV light as the illuminating light and it is the UV light that forms the image. Using UV light is difficult since UV light is absorbed by glass so all of the lenses need to be made of quartz instead of glass and all of the slides need to be quartz. This is expensive. The image cannot be viewed directly since UV light is not visible and is also dangerous to the human eye, so all examinations must be done on photographs. This is expensive. The relatively small increase in resolution afforded by UV light is simply not worth the trouble.

Electron microscopy is made possible since electrons, although often considered as particles, can also have wave-like properties. The wavelength of an electron beam is related to the energy used in the production of the electron beam. Very short-wavelength electron beams can be produced using very high voltages. The same underlying laws govern the light and electron microscopes. Thus, as with the light microscope, good resolution is achieved by using short wavelengths. The extremely short wavelengths of the electron beams used in the electron microscope make it possible to achieve very high resolution. The modern electron microscope has a resolution of around 2 nm for biological specimens and is almost to the point of resolving individual atoms with crystalline materials. In practice, the resolving power of the modern electron microscope is so great that in biology the limiting factor is not the wavelength or the resolution of the microscope but the tissue preparation. Biological preparations cannot be prepared of sufficiently high quality to make use of the available resolution and, in any case, the resolving power is greater than any biologist would reasonably need.

Components of the electron microscope

The electron microscope is quite similar in principle to the light microscope but with the EM, instead of a lamp to produce a beam of light there is an **electron gun** producing a beam of electrons. The electron gun consists of a tungsten filament, which is heated to emit electrons and surrounded by a negatively charged shield. The shield has an aperture so as the electrons are repelled from the shield they can only escape through the aperture. The electrons are then accelerated using an anode, which produces a strong electric

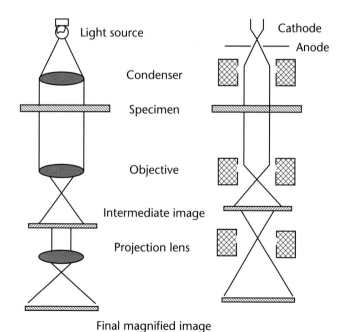

Figure 19.1
Comparison of the light microscope (left) and electron microscope (right). The light microscope is shown inverted for easier comparison. This schematic diagram shows that the two forms of microscopy share many similarities

field between the electron source and the anode itself. This accelerating voltage is what determines the wavelength of the electrons and is usually between 20 and 120 kV. The electrons are then focused using magnetic lenses.

The general layout is similar to the light microscope (see *Fig. 19.1*) with a **condenser** focusing the electron beam on to a specimen and an **objective lens** focusing the electrons that have passed through the specimen and magnifying the image. The image is then further magnified by a **projection lens** (equivalent to an eyepiece in light microscopy) and finally viewed on a fluorescent screen or photographed using standard photographic techniques. It is now far more common to capture the black and white images digitally with a solid state camera and display them on a screen. Although the principle is similar, there are often modifications in commercial instruments. In most electron microscopes, there are extra lenses in the system, such as a second condenser focusing the electron beam on to the specimen and extra diffraction and intermediate lenses magnifying and focusing the light after it has passed through the specimen.

Several significant differences exist between electron and light microscopy

Although there are similarities to the light microscope, there are also several differences. Some are advantages but some make electron microscopy more difficult and more expensive:

- Electrons have a much shorter wavelength so resolution is improved. This allows high magnification without it becoming empty magnification. Factors other than simple resolving power limit the magnifications used. For biological specimens, it is not normally possible to achieve resolutions better than about 10 nm.
- The lenses are magnetic and not glass. Each lens is a massive coil of thick copper wire that draws large electric currents to generate a strong magnetic field. Electrons are charged particles so are deflected by passing through a magnetic field. This is similar in some ways to the refraction of light when passing from air to glass. The field can be shaped using **pole pieces** to produce the same magnifying effect on electron beams that a convex glass lens produces with a light beam.
- The geometry of the magnetic field cannot be as complex as that of a glass lens and this limits the ability of magnetic lens designers to totally control lens aberrations. To overcome this limitation, the electron microscope usually works at much narrower

---

Box 19.2 // Apertures in electron microscopy

The apertures in the electron microscope are pieces of molybdenum or platinum with a small hole (200–300 μm) in the centre. This limits the size of the electron beam. The small apertures can be easily contaminated and so become asymmetric. This results in **astigmatism**. This must be compensated for, especially at high magnifications. The compensation is carried out with a **stigmator**, which is a special multipole magnetic lens that can produce a correcting magnetic field. The control of this correction is one of the key skills for an electron microscopist. Regular cleaning and polishing of these apertures is an important part of regular maintenance.

The small apertures do increase the depth of field so the electron microscope is actually able to focus a greater thickness of a specimen than an equivalent light microscope lens. The greater depth of field and the very thin specimens used in electron microscopy mean that the whole of the specimen can be in focus at once so there is not the constant readjustment of focus that is needed in light microscopy.

apertures (see ***Box 19.2***) than the light microscope since most aberrations are worse when a large aperture is used. Since the wavelength of the beam is so small, the reduced apertures do not seriously diminish the resolution and aperture size is less critical than in the light microscope where aperture size (numerical aperture) is the limiting factor in resolution.

- The magnification is altered by changing the electrical current through one or more of the lenses rather than changing lenses, which is the usual way of increasing the magnification in light microscopy. Electrons spiral in an electric field so altering the magnification results in the image rotating as well as getting larger.

- Electrons are easily absorbed or deflected by many substances including air and water. To avoid this absorption, the whole microscope tube must be kept in a very high vacuum. The electron microscope requires complex pumps to remove all of the air to keep the operating column and specimen chamber under a high vacuum, as if it is returned to atmospheric pressure the microscope will need to be 'outgassed' for several hours to remove all traces of air.

- Special air locks are needed to put specimens into the microscope without losing the vacuum in the main microscope tube. This slows down specimen changes.

- The specimens must be very thin and dry. This invariably means that electron microscopy specimens are dead. Until recently there were no electron microscopy equivalents to techniques that allow living cells to be examined by light microscopy.

- The intense energy of the electron beam in the high vacuum of the electron microscope can be quite destructive to specimens. Even dead cells need to be well protected by embedding in plastic to survive being examined in the electron microscope and unlike light microscopy the plastic is left in place and is not removed for staining or viewing. Ultimately, however, the heat generated by the beam will vaporize both plastic and specimen, leaving neatly drilled holes and, eventually, loss of the entire section.

- Electrons are deflected by any material with a static electrical charge. The specimen is constantly being irradiated with a beam of electrons during viewing and without special precautions this would very rapidly charge the specimen and make viewing impossible. For this reason, the specimen must be put on to an electrically conducting copper grid (see ***Box 19.3***) to disperse electric charge, which may be coated with carbon to help conduct the charge. The spaces in the grid allow electrons to pass through while the copper bars act as a mechanical support and electrical earth to disperse any static electricity.

Box 19.3 // Copper grids for electron microscopy

The copper grids used in electron microscopy have grid bars that are more intrusive than the glass slides used in light microscopy. It is impossible to arrange the specimens on the grid so it is pure chance where the individual cells lie. The grids bars all too often interfere with visualization of the area of the tissue that is most interesting. They are quite small but have an apparently large area when highly magnified and searching across a grid can seem very tedious. There are no real shortcuts since the position is controlled solely by a very fine mechanical x–y movement. The bars on some grids are made slightly different to allow the operator to know roughly which part of the grid they are observing by providing some 'landmarks' to navigate by.

- Electrons are not visible to the human eye and must be viewed using a phosphorescent screen. This phosphorescent screen uses a similar material to the phosphor used in television screens, which also use electrons to produce an image. The alternative is to produce a permanent record using photography. Electrons will directly affect photographic emulsion so placing a sheet of film in the plane of the image will produce a negative. More recently, digital photography technology has been adapted to allow it to be used in the electron microscope. It is more common these days to view a real-time image on a monitor screen. The electron microscope screen is visible immediately but is mainly used for identifying which area of the specimen is to be photographed. Photographs are used as the main image in electron microscopy as they can be further enlarged and are more convenient to view and handle (e.g. to measure sizes). The capture of digital images permits the use of sophisticated software packages to measure dimensions and calculate areas and perimeters.
- The image is always a monochrome image. Coloured electron microscope images are artificially coloured, often using computer techniques.
- The contrast between the nucleus, cytoplasm and other structures is enhanced using heavy metals rather than dyes. Most dyes will not deflect or absorb electrons sufficiently to act as 'stains'. There are only a few heavy metal salts that are suitable for electron microscopy so the range of reagents is more restricted. The fact that only shades of grey are possible also limits the amount of contrast that can be achieved.

Preparation of tissues for the electron microscope

The general principles of histological preparation are similar to light microscopy, but the need for very thin sections and the greater detail required in the final preparation mean that there are significant differences in the actual procedures.

Fixation

The general principles of fixation that apply to light microscopy are still valid for electron microscopy but fixation is even more critical since subcellular changes, especially in phospholipid membranes, can occur within seconds in some conditions. Usually the tissue must be fixed within minutes and becomes almost useless for critical work if left for a few hours. By contrast, post-mortem specimens for light microscopy are often left intact within the body for more than 24 h before the specimen is placed in fixative.

To speed up fixation for the electron microscopy, the specimen is either perfused *in vivo* or if this is not possible it is cut into very small pieces by chopping with a razor blade. Each fragment may be less than 0.5 mm in size. This allows rapid penetration of fixative. Fixation is often carried out at low temperature to slow down cellular degeneration. Fixatives that are used in electron microscopy include:

- **Osmium tetroxide.** This preserves lipids very well and is the main fixative for membranes and membranous organelles.
- **Glutaraldehyde.** This preserves proteins well and is the main method of preserving the cytoskeletal materials.
- **Formaldehyde.** Formaldehyde for electron microscopy use needs to be much purer than for simple light microscopy. It is usually prepared by depolymerizing pure paraformaldehyde rather than using the impure formalin solutions that are usual in light microscopy. Since formaldehyde results in fewer cross-links than glutaraldehyde, it is often useful in histochemical and immunological techniques. Stronger cross-linking would chemically alter the tissues making these techniques useless.

Fixatives can also be mixed to give a compound fixative (e.g. Karnovsky's fixative uses formaldehyde and glutaraldehyde) or the fixatives can be used in sequence to give a secondary fixation (e.g. glutaraldehyde followed by osmium tetroxide). The fixative must be buffered, usually to pH 7.2–7.4, but occasionally a higher or lower pH is used. The buffers used include phosphate buffers, cacodylate buffers and collidine buffers. Some fixatives such as glutaraldehyde may react with some buffers. The fixative solutions must be carefully controlled to give optimum results. Many buffers are highly toxic; cacodylate contains arsenic and veronal contains sodium barbitone.

Osmolality is also important. Often a slightly hypotonic solvent is used (290–320 mOsm). Calcium chloride is often added to fixative solutions to help stabilize membranes.

The fixation time for electron microscopy is usually much shorter than for light microscopy, often just 1–4 h at 4°C. This is partly a reflection of the much smaller block size, which means that diffusion is not a major limiting factor.

Although fixation is important, most laboratories have slightly different techniques that work well in that laboratory so there is no single recognized technique used by everyone.

Processing

The waxes used in light microscopy are useless for electron microscopy as they would rapidly vaporize in the electron microscope due to the high vacuum and energy of the electron beam. Instead, plastics are used to embed and support the tissues. These plastics can be acrylics or epoxy resins. Acrylics were used in early electron microscopy techniques but have largely been replaced by epoxy resins. The use of these materials has already been covered in ***Chapter 5***.

Sectioning

Sectioning of the plastic-embedded tissue is done using a glass knife (see ***Box 19.4***) and an **ultramicrotome**. The glass knives are made by breaking a sheet of glass to give a fresh and extremely sharp edge (see ***Fig. 19.2***). The knives are disposable and have a very short but useful life. Two or more knives may be needed to cut a single block depending on the material being cut.

The basic components of the ultramicrotome are the same as for a normal microtome but the feed mechanism is much finer. The feed may be mechanical or by thermal expansion of a metal rod attached to the microtome chuck. An electric heating coil surrounds a metal rod holding the chuck and as the rod expands it pushes the chuck forward. The thermal expansion is easily altered by varying the current through a heating coil and can give very fine advancement of the chuck. The degree of advancement is, however, less predictable than a mechanical feed so that section thickness cannot accurately be set

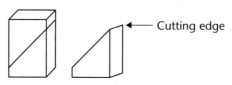

Figure 19.2
Glass knives for electron microscopy. A small rectangle of glass is scored and then fractured to give a triangular knife

> **Box 19.4 // Microtome knives for electron microscopy**
>
> Originally glass knives were made by hand and this required great skill to obtain a good edge. Nowadays, knife-making machines are available. These machines take 25 or 38 mm pre-cut strips of 6 mm thick plate glass and lightly score them. By applying an even pressure, the glass can be fractured. This produces squares of glass, which are then induced to free-break diagonally to produce triangular knives with the fine cutting edge across the thickness of the glass. The mounted angle of the cutting edge can be varied: 45° is suitable for most materials, 35° is better for soft blocks and 55° is better for hard blocks.
>
> Producing disposable glass knives is the commonest and cheapest way to cut sections but diamond and sapphire knives can also be bought. These are harder and sharper than glass knives but considerably more expensive. They come with special holders that act as the floating-out bath.
>
> Diamond knives are made from industrial diamonds and polished to give a fine cutting edge. The edge is fragile so care is needed to avoid damage but if used carefully they are very long lasting and can give months of use. As they are used repeatedly the edge will build up a layer of debris if not cleaned regularly. This can be done with the pith from an elder plant, which is rigid but soft and will gently scrape away the dirt without damaging the edge. Diamond knives can be resharpened when they become blunt but this is expensive.

in advance. Instead of relying on the settings on the microtome, the section thickness is measured by the interference colours seen when the section is viewed at an angle. Sections may appear pale gold (approx. 90–120 nm), silver (approx. 60–90 nm) or grey (less than 60 nm); silver sections are ideal. Other sections are too thick to be useful.

As sections come off the edge of the knife they are floated on to a small waterbath attached directly on to the knife (see *Fig. 19.3*). The sections can be flattened on the waterbath by allowing some chloroform or xylene vapour to soften the plastic and then gently teasing the section with a hair probe.

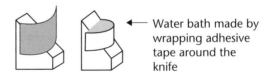

Water bath made by wrapping adhesive tape around the knife

Figure 19.3
Water bath attached to a glass knife

Semi-thin pilot sections

It is rarely the case that the block dimensions and surface area will be of an appropriate size to allow ultra-thin (90 nm) sections to be picked up directly onto a copper grid for high resolution work. Similarly, it may be that a significant area of interest within the specimen (e.g. a glomerulus in a kidney sample) will not be represented in every section. It is therefore vital to produce semi-thin sections that can be quickly reviewed on a glass slide in the first instance. At a thickness of 1–2 microns, these sections are about 10–20 times thicker than those ultra-thin sections destined for heavy metal staining and viewing in the electron beam. Semi-thin sections are cut on the same ultramicrotome, with the same style of glass knife that is used to produce ultra-thin sections. The section can be carefully lifted from the knife edge with curved forceps, placed on a single drop of water on a clean microscope slide and allowed to dry on a hotplate. Once dry, differential staining is achieved by adding a few drops of 1% toluidine blue in 1% borax to the slide, which is heated to 50°C on the hotplate. Staining takes about 30 seconds. The slide can be coverslipped in DPX after first rinsing in water and drying thoroughly on the hotplate.

Semi-thin sections are reviewed using a light microscope to determine the particular area of interest. This is then identified on the block surface, and the surface area of the block reduced to frame this area alone. Careful and accurate work is needed and can only be achieved by viewing the block surface through a binocular eyepiece at a magnification between ×40 and ×60 with angled illumination. The reduced block area will help to minimize any mechanical strain or vibration against the knife and helps in the production of the highest quality ultra-thin sections.

Sections are picked up on a fine copper grid (2.3 or 3 mm diameter), which supports them in the electron microscope. The grids have one side shiny and one side dull. Although it probably makes little difference to the result, it is probably best to standardize on one orientation (e.g. shiny side up) so that you always know on which side of the grid the section has been placed. Grids come in a variety of different grid patterns and hole sizes that are a matter of personal preference rather than being critical in use. Grids may need to be coated for some applications. The coating is a thin film of plastic, usually formvar or carbon or even both. Films are easily made by allowing a formvar solution in acetone to evaporate on the surface of purified dust-free water. Grids are then scattered shiny side down and film plus grid picked up on parafilm. This gives extra support but the coating film can sometimes interfere with the clarity of the final preparation. After being picked up on the grid, the section is then dried by placing in a closed container on top of filter paper. Plastic petri dishes are a cheap and convenient container for grids but commercially produced holders are available.

Staining

Unstained plastic sections are usually low in contrast, so no detail can be seen with the electron microscope. Contrast can be introduced into the specimen by staining with heavy metal salts that can absorb or deflect the electrons. The two most commonly used are uranyl acetate and lead citrate; both are highly toxic and the uranium salt is also radioactive.

Staining can be done at any stage in the preparation. Osmium tetroxide is a stain as well as a fixative so staining of lipid structures often starts during fixation. Other stains can be added during the processing of the block, e.g. the use of ethanolic uranyl acetate during dehydration. Both of these techniques stain all of the tissue in the block and are referred to as block staining. The difficulty with block staining is firstly that the penetration of the reagents may be a problem since the large ionic heavy metal stains used in electron microscopy do not penetrate easily through intact cell membranes. Secondly, in addition, the reagent may interact with one or more reagents in the processing (e.g. phosphotungstic acid reacts with epoxy resin).

The finished section on the grid can be stained by direct immersion in a single drop of the staining reagent and this is probably the commonest method for staining. Penetration is very poor since the section is embedded in plastic. Only the surface really absorbs the stain but this is usually sufficient to give a good contrast. Automated grid staining machines are advantageous in high throughput laboratories. Stains must be filtered just

Table 19.1   Selectivity of various metal salts used in electron microscopy

Metal salt	Selectivity
Lead salts	General contrast enhancement
Uranyl salts	General contrast enhancement but more selective for DNA
Phosphotungstic acid	Polysaccharides
Osmium tetroxide	Lipids including membranes

before use since heavy metals tend to precipitate easily, especially lead citrate which reacts with exhaled carbon dioxide to produce lead carbonate. For a number of reasons, electron microscopists often hold their breath, particularly so in this case as even tiny precipitates of metal salts appear as the size of boulders when magnified by a factor of 100,000×. Filtration of the small volumes needed for staining is conveniently achieved by the use of a Millipore filter attached to a 1 ml syringe. The addition of drops of sodium hydroxide around the periphery of a covered Petri dish will absorb CO_2 from the local atmosphere above the sections.

Different metals have slightly different affinities for tissue components so that the final image can be altered to some extent (see ***Table 19.1***). Their specificities, however, are probably even less reliable than with dyeing so interpretation is not straightforward. It is probably better to think of electron microscope stains as making materials more readily visible rather than identifying materials by their staining reaction. The metals can also be used in combination to increase contrast but, unlike dyes in light microscopy, which have different colours, they all just alter the tone of grey so they cannot be easily distinguished.

Specialized electron microscopy techniques

Shadowing

Shadowing can give an impression of three dimensions and is used in measuring height. TEM is only capable of giving a two-dimensional image, much like an X-ray or photograph, but by choice of preparative technique it is possible to give an impression of three dimensions (see ***Box 19.5***) and to measure the height as well as the length and breadth of an object. This is usually used with discrete objects deposited on the grid by drying from a suspension (e.g. bacteria, viruses or crystals) rather than being used on sections. The object is coated with a heavy metal but instead of it being evenly coated from directly above the grid, it is coated at an angle. Any object that rises above the grid will result in a 'shadow' area (see ***Fig. 19.4***). By knowing the angle and the length of the shadow the height of the object can be calculated.

> **Box 19.5 // Shadowing makes use of illusions**
>
> Shadowing is the technique of deliberately using an optical illusion to show the structure in three dimensions. The only problem is that some structures are ambiguous. The human brain always assumes that light comes from above and interprets shadows in that way. If a picture is viewed upside down, then the result is that grooves become ridges to correlate with the shadows and ridges become grooves. Thus, the whole relief becomes inverted and can be misinterpreted.

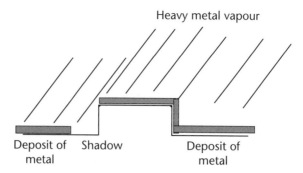

Figure 19.4
Shadowing with heavy metal vapour. The metal is deposited at an angle so that a shadow occurs

The metal vapour is generated by using a low voltage arc between two metal electrodes in a vacuum. The metal condenses on the cold surface of the specimen and leaves a deposit.

When a normal electron microscope photomicrograph of a shadowed specimen is viewed, the shadowed area is bright because it is electron transmitting, whilst the coated areas are black (electron dense). This looks odd to the human eye so the final photograph is printed in negative form so that the shadow is dark and the rest of the image is light (see *Fig. 19.5*). This gives an impression of three dimensions. Shadowing is also used to give this effect of three-dimensional relief even when direct measurements of height are not needed.

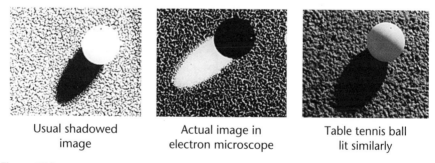

Figure 19.5
Shadowing. The first image is the one that is usually used and shows a small polystyrene ball shadowed with metal from the top right. This is a negative image. The positive image, which would be seen in the electron microscope, is seen in the centre with an electron-transparent white shadow. The image on the right is a table tennis ball lit in the same way from the top right. Although the left-hand image looks superficially like a three-dimensional photograph it is a false impression since the true three-dimensional image shows a shadow (half-tone shadow) across the ball. This is missing from the electron microscope-shadowed image

Replica making
The surface texture of specimens is usually not seen with TEM as it usually uses sectioned material but it can be seen by making a mould of the surface using a heavy metal coating.

The replica matches the surface exactly and can be removed and examined; this gives a good representation of the surface structures. Coating needs to be very even, unlike the shadowing technique, so the specimen is either placed directly under the electrodes in an evaporation coating chamber or is coated by **sputter coating.** Sputter coating uses a stub of metal (e.g. gold) that is connected to a high-voltage supply (1.25 kV) and the chamber is then filled with argon. Gold ions are discharged from the surface of the stub by the high voltage and are focused and accelerated by a magnetic field. This gives a plasma beam of ions, which coats the specimen evenly. (This technique is widely used as part of the freeze–fracture technique (see *Box 19.6*), which examines the inside of membranes.) Even specimens that are too thick or too bulky to examine in the electron microscope can be examined by making replicas of their surface.

Box 19.6 // Freeze–fracture technique

The freeze–fracture technique allows investigation of the inside of membranes. The membranes of cells are double layers of lipid held together by hydrophobic interactions. The centre of the membrane contains no water and will not freeze and the hydrophobic bonds are quite weak. The tissue is first frozen at –196°C and then cut with a razor blade or scalpel. The tissue breaks (fractures) rather than cutting cleanly and the tissue is likely to cleave along the weaker lipid region in the centre of the membrane. A replica of the surface is made and the tissue is dissolved away leaving the replica, which can be examined in the electron microscope. The inside of the membrane shows the presence of structures that span the membrane such as receptors and intercellular junctions.

Negative staining

Staining is usually done by staining the solid material with heavy metals but in negative staining it is the spaces in the specimen that are filled with an electron-dense material giving a silhouette of the objects (see *Fig. 19.6*). It gives information about size, shape, surface structure and groupings of particles but no information about internal detail of the specimen. This is useful with suspensions of viruses, bacteria, etc., but less useful with sections of embedded material that are solid throughout. Typical stains include potassium phosphotungstate, cadmium iodide, sodium tungstate and uranium nitrate.

A mixed suspension of particles (bacteria, viruses, etc.) and the electron-dense stain (e.g. phosphotungstate) is made and thinly spread on a carbon-coated copper grid and

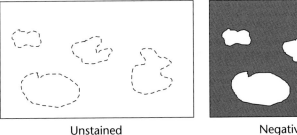

Unstained   Negatively stained

Figure 19.6
Negative staining. The untreated specimens on the left have no contrast but by filling in the spaces in between, the outline of the objects becomes apparent

dried. The suspension can be sprayed on or a drop can be placed on the grid and excess fluid removed after a few moments using a piece of filter paper.

Electron microscope histochemistry

Histochemical techniques can be applied to electron microscope specimens but usually involve **block staining** of fresh tissue followed by embedding and sectioning rather than staining of frozen ultrathin sections on electron microscope grids. 'Blocks' is a very flexible term in electron microscopy. A 20 μm frozen section cut in a conventional cryostat can be an electron microscope block since it is large enough to be a complete block for electron microscopy. A cryostat section can be cut and stained by a suitable histochemical technique and then processed to produce an electron microscope block.

The final product of the histochemical technique must be electron dense. This occurs with metal salt methods, e.g. Gomori phosphatase methods, and some tetrazolium salts are electron dense (e.g. the osmate of TNST, 2,2′,5,5′-tetra-*p*-nitrophenyl-3,3′-stilbene ditetrazolium chloride). Some reaction products (e.g. diaminobenzidine product) are **osmiophilic** (i.e. absorb osmium salts) and so will become stained when the block is fixed with osmium tetroxide.

Electron microscopy immunotechniques

These are similar to the immunotechniques discussed in ***Chapter 12*** but use electron-dense labelling instead of fluorescent dyes. Antibodies can be labelled directly with ferritin (an iron-containing protein) or colloidal gold spheres that are clearly visible with the electron microscope beam. Colloidal gold is probably the most popular since the colloidal gold particles are perfect spheres and can be manufactured chemically to precise diameters. Perfectly spherical objects are rare in sections so positive identification and localization of the spheres is comparatively easy. By using particles of two different sizes (diameters) to label different antibodies, it is possible to identify the sites of binding of more than one antibody in a single section.

Immunoenzyme techniques are also applicable, e.g. immunoperoxidase, provided the final product can be made electron dense, as is the case with diaminobenzidine used in peroxidase methods, which is osmiophilic as mentioned above. The technique can be done either before or after embedding in plastic.

Paraffin-embedded material in electron microscopy

Although it is usually better to process material directly for electron microscopy (see ***Box 19.7***), it is possible to use paraffin-processed material as electron microscope blocks. This can be done either from paraffin blocks or from sections.

Box 19.7 // Perfect fixation is not always possible

The preservation of material for the electron microscope is critical to get the best results; however there are occasions when the preservation is poor but it is still worthwhile looking at the material in the electron microscope. In this way, the electron microscope is a useful tool for archaeologists looking at mummified remains or human materials recovered from bogs. The tissue preservation will be extremely poor but since it is not possible to go back thousands of years to get a better specimen you have to do the best you can with what you have. Ensuring that fixation and preservation are done properly will always give better results but we cannot ignore the benefits of any technique just because it cannot be done in a perfect way.

One advantage of this is that areas of interest can be identified using light microscopy and only relevant areas are then examined by electron microscopy. Since the size of a typical electron microscope section is <1 mm in diameter, this can greatly reduce the number of blocks needing to be examined, saving much time and work. Wax blocks simply need dewaxing in xylene, transferring to alcohol and treating with osmium tetroxide to increase their contrast. They can then be processed as if they were normal tissues.

Using paraffin sections works best with thicker sections than is usual for light microscopy (>7 μm). The best way is to dehydrate the section (to alcohol or acetone) and then invert an embedding capsule containing resin over the area of interest in the section. The resin can permeate the tissue and is then polymerized. The whole thing (glass slide, section and resin capsule) is then cooled in liquid nitrogen and the capsule snapped from the slide. The glass and tissue usually separate very cleanly. The resin block is then trimmed and sectioned as for a standard resin-embedded sample.

Preservation of material previously sectioned in wax is adequate rather than good but may be better than tissue stored for several days in routine fixatives. The ability to use sections or blocks is useful in **retrospective studies**. This is where the laboratory has a specimen that could be useful in research in its archives as a paraffin block or section but not as an electron microscope block. Thus, it is possible to do electron microscopy on blocks from many years ago.

19.2 Scanning electron microscopy

SEM is quite different to TEM and has no direct counterpart in conventional light microscopy. SEM is totally reliant on video technology and the user never directly sees the specimen, as only a video reconstruction of the specimen is seen on the monitor screen. SEM is used mainly for examining the surface of structures, although by fracturing a block of tissue it is possible to study an exposed surface that was originally an internal structure.

SEM can examine relatively massive and dense structures since the electron beam does not need to penetrate right through the specimen. Unlike TEM, SEM is not often used at very high magnifications; indeed, it is often used at magnifications lower than would be possible with light microscopy. The main benefits of SEM are in the way that it visualizes objects as three-dimensional photographic images rather than giving better resolution (see *Box 19.8*).

Box 19.8 // SEM is best for illustrations

The main benefit of SEM is to give quite graphic images of the structures. It is an ideal tool for illustrations in books, images for education and similar visual applications. It is less useful scientifically than CTEM, which shows greater detail and more of the internal structure. The applications of SEM in diagnosis are quite small compared with the use of CTEM, but there are a few times when it is useful, for example in hairy cell leukaemia where the cellular projections are seen much better in surface views than in transverse sections.

There is very little that can be done with SEM that cannot be done, albeit with more difficulty and effort, by CTEM. Images of the surface and three-dimensional shape can be obtained using replicas and sections taken at different levels. Thus, SEM images are prettier but CTEM is more useful.

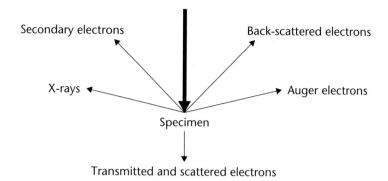

Figure 19.7
Diagram of scattering of the electron beam

The examination of the surface by SEM relies on the electrons that are reflected back by the specimen or are emitted from the surface of the specimen when it is irradiated by a beam of electrons. The interaction of electrons with the specimen is illustrated in *Fig. 19.7.*

Electrons produced from the incident beam

The incident electron beam from the microscope produces several types of electrons and also X-rays. The electrons have different energies and properties and will give different appearances to the specimen if they are used to produce the final image.

Back-scattered electrons have a high energy and are produced by electrons in the beam being deflected when they 'hit' an atom in the specimen (like a ball bouncing back from a wall). The back-scattered electrons are produced within a pear-shaped region (see *Fig. 19.8*) underneath the incident beam and can emerge at quite wide angles. The degree of scattering is partly dependent on the nature of the specimen, i.e. the elements present, so back-scattered electrons give information about the nature of the specimen and not just surface shape.

Secondary electrons have lower energy and are produced by electrons being dislodged by the passage of the main electron beam (like a ball dislodging a brick out of the wall). These low-energy electrons can only escape from the specimen from a relatively thin

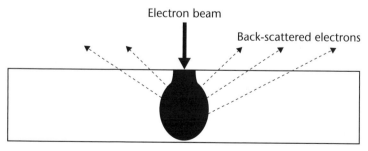

Figure 19.8
Back-scattered electron production. Electrons are dislodged from a pear-shaped area underneath the electron beam and can escape from quite deep in the specimen as they have considerable energy

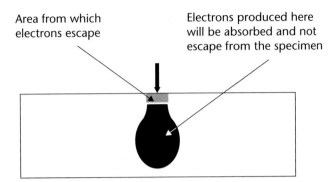

Area from which electrons escape

Electrons produced here will be absorbed and not escape from the specimen

Figure 19.9
Secondary electron production. Electrons are dislodged from atoms quite deep in the object but their low energy prevents them escaping from the specimen

'skin' (about 2 nm thick) on the surface of the specimen. These electrons therefore give a good representation of the surface shape (see *Fig. 19.9*). Secondary electrons are less affected by the nature of the material.

Auger electrons are produced by the X-rays interacting with the specimen and have relatively low energy. They are generally not important in SEM.

X-rays and other electromagnetic emissions (cathode luminescence and bremsstrahlung) are also emitted but play little role in SEM. They may be used to identify elements present in the specimen.

Detection of electrons coming back from the specimen

Detection of the electrons in SEM is completely different to TEM. There is no attempt to focus the electrons to form an image and all of the electrons are collected and counted. The image is made by scanning a very narrow illuminating beam of electrons across the specimen. At each point on the specimen as the incident beam of electrons hits atoms in the specimen, some electrons will be emitted. *All* of the electrons are then collected and used to generate an electrical signal. The greater the number of electrons emitted, the greater the electrical signal generated. The electrical signal is then used to feed into a video monitor (essentially a high-resolution TV system), which is scanned in parallel with the electron beam in the electron microscope.

The strength of the signal alters the brightness in the monitor and so an image is built up on the screen, dot by dot, as the SEM beam scans across the specimen and the monitor scans across its screen (see *Fig. 19.10*).

Higher magnifications are achieved not by increasing the power of the lenses but by the electron beam scanning smaller areas of the specimen. The monitor still displays the same sized picture but it now represents a smaller area and this gives greater magnification.

Electron collection system

The electrons are collected by a specialized system such as an Everhart–Thornley collector (see *Fig. 19.11*). Other detectors are also available but the Everhart–Thornley collector is one of the most frequently used.

The actual detector consists of a **scintillating system** that converts the electrons' energy into flashes of light, which are measured by a **photomultiplier** (see *Box 19.9*) and this produces an electrical signal equivalent to the number of electrons entering the

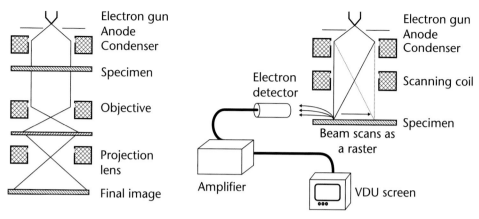

Figure 19.10
Schematic diagram comparing CTEM (left) and SEM (right)

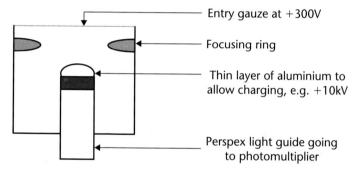

Figure 19.11
SEM electron collector (Everhart–Thornley version)

Box 19.9 // Use of photomultipliers in SEM

Scintillation is where electrons emit a pulse of light as they enter certain materials such as Perspex. The light is weak but can be amplified using a photomultiplier. This instrument converts the light back into an electrical signal. It is ironic that the detectors used in electron microscopy are still using optical methods and are not purely electronic. Photomultipliers are one of the most sensitive ways of detecting very weak signals and can detect single photons that are equivalent to single electrons.

---

detector. The scintillator is kept at a high potential (+10 to +12 kV) so all of the electrons are attracted and accelerated into the scintillant. This accelerating voltage means that electrons are collected from a wide area and have sufficient energy to generate a strong flash of light in the scintillant.

An entry gauze is used to shield the system and this is also held at a voltage; by varying this voltage the type of electrons collected can be controlled. If it is held at +300 V, then both back-scattered and secondary electrons will be detected. By altering the voltage from +300 to −100 V, or by switching off the 10 kV high tension, the secondary electrons will no longer be attracted but the higher-energy back-scattered electrons will still be collected.

The back-scattered image is weaker but gives information about the subsurface

The back-scattered image penetrates further into the specimen and is more dependent on the chemical nature of the specimen, but it is also a much weaker signal as fewer electrons are reflected in this way. As a result there is a lot of noise in the image. For most purposes, SEM is only used to detect the secondary or emitted electrons ('emissive mode'), which gives surface structure rather than penetrating into the specimen. Tilting the specimen can reduce the effect of back-scattered electrons. The incident electron beam glances across the specimen rather than hitting it full on and this minimizes the number of back-scattered electrons.

Secondary or emitted electrons give a surface view

The incident electrons penetrate quite large distances (up to 100 μm) into the specimen but the relatively low energy of the secondary electrons makes them unable to escape from deep inside the specimen. These deep electrons are absorbed and lose energy and so are trapped deep in the specimen. The electrons that can escape are those within 2 nm or so of the surface. If there is an edge or a hole in the specimen, however, then more secondary electrons can escape (see *Fig. 19.12*) and the image is brighter at these points. This gives a 'backlighting' effect that is aesthetically pleasant.

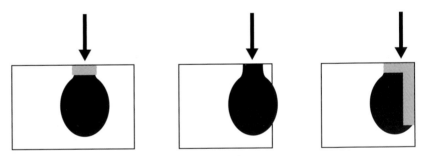

Figure 19.12
Edge effect in SEM. The secondary electrons can only escape from the specimen from a shallow region close to the surface but are produced from a larger pear-shaped region below the point where the electron beam hits the specimen. As the beam scans across, the pear-shaped region overlaps the edge of the object (centre). The region from which electrons can escape is now much larger (right) so there are more electrons escaping and this makes the edges appear brighter in the final image

Advantages and disadvantages of SEM

Unlike CTEM, SEM does not focus the electrons coming from the specimen and there are no lenses involved in producing the image so there are no lens aberrations with SEM. Despite the relatively high magnifications used in SEM, there is tremendous depth of field (i.e. the image is in focus at both the front and the back at the same time) and this makes for a very pleasing image (see *Box 19.10*).

Like TEM, the image is only monochrome but many workers colour the image artificially in order to bring out particular features.

Box 19.10 // Depth of field

The depth of field is impressive when compared with light microscopy. SEM can examine quite large specimens such as complete insects or plants or blocks of wood. SEM can show the front and back as acceptably in focus at the same time, even when they are several millimetres apart. The use of a light microscope results in a narrow focus that may be only a few micrometres deep. The only advantage of the light microscope is the use of colour but for monochrome pictures, SEM is better than any of the alternative methods.

Photographs are taken using a high-resolution monitor

Unlike CTEM, it is not possible to use a sheet of film to record the electron image directly and with SEM, photographs are made by photographing the VDU screen. The normal monitor image is relatively coarse so a second high-definition screen is used for making photographs. In this case, the scan is performed very slowly, which gives better definition, but this mode is not useful for direct viewing since only a single dot is visible on the screen and it moves too slowly to allow the human eye to see a complete image. A camera pointed at the screen records all of the dots by using a long exposure and so the photograph appears as a complete image.

As SEM is used mainly for its three-dimensional aspects, it uses whole, intact specimens rather than thin slices or sections of material. Thick sections treated by freeze-etching or freeze–fracture are sometimes used to see the internal detail of specimens but otherwise sections are not used for SEM.

Specimen preparation for SEM

The usual way of preparing specimens is to dry them completely and then coat them with a thin metal or carbon film to protect them and make them electrically conductive. Electrical conductivity is essential to allow static electrical charges to be earthed. If the specimen is not earthed in this way, then as electrons are absorbed by the specimen it becomes highly charged and repels all of the electrons before they can reach the surface of the specimen. In consequence, with an unearthed or uncoated specimen there is no image to be seen.

Specimens are dried and coated for use in SEM

Simple air-drying is not recommended since it is uneven and as the water evaporates there are surface tension effects, which may damage delicate objects or distort the surface. Instead **critical point drying** is used. Critical point drying relies on phase transitions from a fluid to a vapour without evaporation or boiling. The gas occupies the same space

---

Box 19.11 // Scanning microscopy

Although there is no direct equivalent of SEM in light microscopy the confocal microscope uses scanning technology to build up an image. However, confocal microscopy works by limiting the light captured whilst SEM tries to capture all of the electrons coming from the specimen. This results in confocal microscopy having a very small (approaching zero) depth of field whilst SEM has a very large (effectively approaching infinite) depth of field so the two scanning forms of microscopy are not equivalent.

---

as the liquid so there is no dramatic change in pressure. The phase change can be achieved using CO_2. The CO_2 changes from a liquid to a gas by a simple rise in temperature. The critical temperature for CO_2 is 31°C; above this temperature CO_2 cannot be a liquid even at very high pressure and will become a gas.

The specimen can thus be dried without evaporation by putting it into liquid CO_2. Liquid CO_2 is not miscible with water so the specimen is first dehydrated in alcohol or acetone and then transferred into liquid CO_2 in a pressure chamber (commonly called a 'bomb'). The liquid CO_2 is under pressure and kept below its critical temperature (Tc) until all of the dehydrating agent has been removed. The liquid CO_2 can then be converted to a gas simply by slowly warming it up above 31°C. There is no evaporation – it just turns into a gas instantaneously without alteration in pressure or volume. Significantly, there is no surface tension effect that will damage delicate structures, as the densities of gas and liquid are identical at the Tc. The gas can then be slowly released from the chamber of the drying apparatus, ensuring that the pressure is dropped gradually to allow the gas to escape from the specimen in a slow and controlled manner. Rapid drops in pressure will result in explosive escapes of gas from pockets trapped in the specimen and are likely to damage the specimen. The specimen is now dry and can be sputter coated with a film of metal to confer good conductive properties. It is then ready to be viewed by SEM.

Box 19.12 // Other forms of scanning microscopy

Other forms of electron microscopy and scanning microscopy have also been developed. For example, the scanning transmission electron microscope (STEM), which detects transmitted electrons passing through sections rather than the various types of 'reflected' electrons of SEM but is otherwise similar in its nature to SEM.

The atomic force microscope and scanning tunnelling microscope scan the specimen surface using an extremely fine electrical probe. Height contours are mapped by recording the current required to maintain the probe tip at a constant distance above the surface. These machines are currently of much more use as research tools for particular applications and are often more useful in non-biological fields such as physics and materials science.

Other SEM techniques

Electron microprobe analysis

The final type of emission from scanning microscopes that can provide information is the X-rays that are released when electrons are dislodged from their atomic orbits. As electrons move back into the hole created by these dislodged electrons they emit X-rays. The electron shells of atoms are characteristic of each element so the X-rays produced by the electrons 'dropping' into empty orbits are of different wavelengths with each element having certain characteristic wavelengths. By irradiating one spot with a thin electron beam, as is done in SEM, and then measuring the wavelength and intensity of the X-rays produced, it is possible to detect any elements present and even to estimate how much of each element is present. This can be done using SEM, STEM or TEM, although the transmission techniques probably give better resolution. Possible applications in tissue samples include the identification of toxic metals, the micro-fragmentation of hip joint prostheses, and the role of trace metals (e.g. aluminium and iron) in senile plaque formation in Alzheimer's disease.

Freeze-etching

In this technique the surface of the tissue is first frozen solid and then dried by sublimation (see **Chapter 6**). This exposes the structure of the dry material in three dimensions. The surface is very delicate and must be protected by gold coating. The sample is still easily damaged and will not keep. Replicas of the surface can also be prepared for use in CTEM but this technique is mostly used with SEM.

Suggested further reading

Griffin, R.L. (1990) *Using the Transmission Electron Microscope in the Biological Sciences*. London: Ellis Horwood.

Hand, N.M. (2012) Chapter 8: Plastic embedding. In: *Bancroft's Theory and Practice of Histological Techniques*, 7th edn (eds K.S. Suvarna, C. Layton and J.D. Bancroft). Churchill Livingstone.

Slayter, E.M. and Slayter, H.S. (1992) *Light and Electron Microscopy*. Cambridge: Cambridge University Press.

Suvarna, K.S., Layton, C. and Bancroft, J.D. (2012) *Bancroft's Theory and Practice of Histological Techniques*, 7th edn. Edinburgh: Churchill Livingstone.

Woods, A.E. and Stirling, A.W. (2012) Chapter 22: Transmission electron microscopy. In: *Bancroft's Theory and Practice of Histological Techniques*, 7th edn (eds K.S. Suvarna, C. Layton and J.D. Bancroft). Churchill Livingstone.

Self-assessment questions

1. How does CTEM differ from light microscopy?
2. Why do electron microscopy sections need to be embedded in resin rather than wax?
3. Why does SEM not suffer from lens aberrations?
4. What is critical point drying and why is it used in SEM?
5. Why is it necessary to coat SEM specimens with gold?
6. Why is it more difficult in CTEM specimens to identify cellular materials using staining than it is in light microscopy?
7. How can the X-rays produced by the electron beam striking a specimen be used in electron microscopy?

Some aspects of the organization of a histology laboratory

There are probably as many different ways of organizing a laboratory to produce histological sections and cellular preparations as there are actual laboratories. This is in part a reflection of the differing needs and different services. The organization needed to produce a rapid and reliable service for routine diagnosis is not the same as would be needed to produce large numbers of almost identical sections for teaching. The specific needs of a research laboratory are to be able to solve problems that may not have been attempted before and are quite different to the problems encountered by diagnostic and contract laboratories. These, by contrast, offer a standard level of consistency, ensuring that cellular preparations will always have the same treatment and represent a predictable level of quality.

20.1 Organization of a routine diagnostic laboratory

Whilst no two laboratories may be identical in their activities there are, however, several features of their organization that are essential and that are common to most, if not all, laboratories. The organization of a routine diagnostic laboratory will now be considered in more detail to illustrate the way in which the specimen flow within laboratories and the reporting of results can be organized effectively.

Specimen types and fixation

Specimens can be of several types and widely different sizes. Specimens that are taken from a living patient are **biopsy** specimens. The term is usually used as an indication that the sample is removed when a diagnostic investigation is being carried out. In some cases, the whole lesion is removed as an **excisional biopsy** (for example, a piece of skin with a mole that may have become cancerous), so the distinctions between the two are not always clear. The nature of the sampling device and procedure is, however, a good guide, for example punch biopsy of skin, trephine biopsy of bone marrow and Tru-Cut needle biopsy of kidney. Large pieces of tissue may be referred to as **resections** or **amputations**.

Surgical specimens are usually fixed immediately so that post-mortem changes can be avoided. However, they are taken by surgical staff and the staff responsible for fixation are often relatively junior nursing staff. This can mean that tissues may be less well treated after removal (see ***Box 20.1***).

Fixation is possibly the most important single part of the processing, yet for most surgical specimens the fixation is done by surgical nursing staff rather than histologists. A badly organized system will cause many problems. Many laboratories simplify the fixation of samples by always fixing everything in formaldehyde. If these specimens are then delivered fairly quickly, the laboratory can usually produce good results.

The specimen may still be ruined by insufficient fixative. Very large specimens may be placed in barely adequate containers and then 'topped up' with fixative. Thus, a specimen with a total volume of 700 cm^3 (700 ml), for instance, may be placed in a 1 litre container and then 250 ml of fixative added. This would not be enough to fix the specimen completely, but provided the specimen is taken to the laboratory quickly it may be recoverable.

Specimens removed after death are **post-mortem** or **autopsy** specimens. Post-mortem specimens are usually taken several hours after death and so often show artefacts of cellular degeneration (called, not surprisingly, post-mortem artefacts). As they are taken by pathologists and fixed by post-mortem technicians, who are closely associated with the laboratory and its specific requirements, post-mortem samples are usually fixed correctly.

Box 20.1 // Tissues are removed for therapy as well as for diagnosis

The reasons for removing specimens are often quite varied and only occasionally is the main reason concerned with pathology. Many specimens are removed to treat the patient and a pathology report is often a secondary consideration used to confirm a clinical opinion. This means that on some occasions the received specimen is less than ideal for histology. I came across a comment about a specimen in which the report read, 'It is impossible to tell whether the specimen was ruined by a pair of large Spencer Wells or a pair of small Spencer Wells'. Spencer Wells forceps are a type of surgical forceps with serrated jaws. The surgeon had crushed the jaws on to the specimen making it useless for diagnosis. The surgeons providing specimens need to be aware that if they mistreat specimens, then the pathologist may be unable to give an adequate report. Inappropriate fixation (or lack of any) can be equally damaging.

If damage is inevitable during surgery to save the patient's life or health, then it may be justified, but education of surgical staff is needed if they are to provide good-quality specimens for diagnosis.

Labelling

The lack of proper forms and mislabelling is very dangerous. A couple of small specimens without labels can all too easily be confused. Once tissue is placed in fixative it changes in both colour and shape so deciding which unlabelled specimen is which can be a nightmare. Specimens arriving badly documented should not be accepted. The only thing that should be checked is that the specimen is being fixed correctly, but otherwise it should not be processed until it is labelled correctly. It is good practice to ensure that specimen types are alternated as they accumulate. A batch or sequence of similar sample types is more susceptible to mislabelling.

How the specimen is labelled and the form used will vary with the local system, but for diagnostic hospital specimens it must positively identify the patient and the specimen and should include:

- patient's full name;
- gender;
- patient's hospital number;
- date of birth;
- relevant medical history;
- possible diagnosis;
- examination requested;
- person and department submitting the request;
- date of the request.

On acceptance, the specimen should be logged and, if relevant, given a laboratory number. Laboratories with a high turnover often have automated or semi-automated systems that will print labels and processing cassettes; these may also include a bar code. Low-throughput laboratories may make do with a hand-written system. Hand-written systems are probably less reliable and need more care but if care is taken they can be satisfactory.

Specimen treatment

Specimens may be of a specialized type, such as a needle biopsy, which may need to be split into separate samples for different treatments (see *Fig. 20.1*). Every laboratory will have slightly different requirements and every specimen type will have different demands so there is no universal system.

Figure 20.1
Needle biopsy. The specimen is split, with the light microscopy (LM) part being fixed in formalin, the piece for frozen sections being left unfixed and the electron microscopy (EM) resin-embedded specimens being fixed in glutaraldehyde

The specimen is allowed to fix for an adequate length of time (12 h or more, depending on size). The specimen is then examined and cut into suitable blocks for processing (sometimes called 'cutting up' or 'grossing') and an accurate description of the whole specimen is made, together with a description of where the individual blocks are taken from the specimen. This will include number, size, shape, colour and texture of the original specimen and constitutes the official macroscopic report, which may be very simple or quite complex. A simple specimen would be a routine vasectomy where only a single block may be needed, whilst a complex or very large specimen may need many blocks and their exact location and orientation may be crucial. Accurate recording of where a specimen was removed from a large piece of tissue may be crucial. If there is an invasive tumour that is spreading through the tissues, the surgeon needs to know whether the tumour has spread beyond the edge of the removed tissue. This means he needs to know exactly where the borders have been crossed. Marking of the borders can be done in several ways (e.g. sewing on of tags, writing with indelible ink) and some laboratories are now using digital cameras, which record the specimen exactly. Since images are captured electronically as data files in electronic format, they can then be readily inserted into the 'macro' part of the final histopathology report.

The tissue samples (blocks) are then placed into individual cassettes, which may be colour coded to identify particular types of specimen; for instance, a biopsy sample for rapid processing or a post-mortem sample for less urgent processing. The blocks are then processed, usually on an automatic tissue processor. The following day, the blocks are placed into moulds and oriented. At this stage a simple check should be made to confirm that each block corresponds to the description given during the 'grossing'. The blocks are cooled and then sectioned. Some specimens may require more than one section to be taken at differing planes in the block ('levels'). This is a standard approach designed to locate abnormalities that may be present within the sample but within an unknown area. The sections are stained and coverslipped. Once again, larger laboratories may have a degree of automation to all of this, whilst smaller laboratories will be more reliant on manual techniques. Some special stains may be done automatically. Each laboratory, and indeed an individual pathologist within a laboratory, may have different requirements for the number of levels in a block and the type and number of special stains done routinely so it is difficult to give a single description of the usual procedures.

The stained sections should be checked for quality. It is difficult to diagnose accurately from a poor-quality section. If the sections are of an acceptable standard, then they are passed to the pathologist for diagnosis.

Diagnosis

The diagnosis may be straightforward, but if there is an outstanding problem the pathologist may need to have further special stains performed: histochemistry, immunohistochemistry or electron microscopy investigations to help inform the final diagnosis. It may even be necessary to consult with other pathologists for a second opinion in a few difficult or unusual cases. Sections may need to be prepared and sent to a distant laboratory

---

Box 20.2 // Work rates

The workload in laboratories is quite variable but a rough estimate of the typical rates of work is given here.

For a person working effectively with no distractions, then blocking out into a mould takes about 30–90 s. To section the embedded sample block takes about 1–4 min assuming there are no significant problems. Staining is usually quite variable as it depends on whether it is automated, what stain is being done and how many slides are having the same treatment. A single slide needing a special stain might take a couple of hours, but the biomedical scientist would probably do other jobs or several other stains at the same time.

Many laboratories aim to have the routine tasks of cutting and staining done by lunch time and use one or more staff on an early start to get the embedding done ready for the main start of the day so that cutting can begin immediately. The afternoon is then used for urgent rapid samples, routine maintenance, preparing reagents, grossing for the following day and doing any special stains requested by the pathologist.

The typical output from a single competent histologist can be up to about 50–80 slides per day (depending on automation) assuming that they have to do everything concerned with the preparation. This corresponds to about 12,000–20,000 slides per year.

where there is an expert in some particular field of pathology. Rapid broadband transfer of digitized microscope images via the internet/intranet supports local and national/international conferencing for diagnostic and teaching purposes (see *Chapter 18*). These facilities are currently more widely used in radiography than in pathology, but are likely to become more widespread as the technology becomes more accessible and familiar.

The final diagnosis is then entered into the computer records and also sent to the clinician requesting the result. Some computerized systems do this automatically, but other places still rely on paper copies to be sent to the ward or GP.

Record keeping

The final stages in the laboratory are to file the slides and blocks and dispose of tissues that are no longer needed. Tissues, blocks and sections can be retained for several purposes. The retention of tissues/slides is more useful than keeping only the report as it is sometimes possible at a later date to check on the tissue in a way that was not possible at the time of the original examination. For example, it is now possible to perform DNA analysis on 50-year-old blocks. At the time such analysis was impossible as the detailed genetic structure of DNA was not known. Paper records can only reflect what was known at the time.

Archived specimens of tissue and blocks may be needed for:

- Further diagnosis and clinical management of the patient at a later date. This allows the progress of the patient to be monitored. If the patient's condition changes, it may be possible with hindsight to re-examine the specimen and find early signs of the change that it was not possible to recognize at the time.
- Teaching purposes. Good examples of particular diseases are often used in teaching and observation of the actual tissues is the best way to illustrate the condition.
- Use in research. A good archive of tissues can be an invaluable resource. Some diseases are very rare and the accumulated total of many years of cases is often the only way of collecting enough samples to give a reasonable coverage of the disease. Also, some diseases have only become recognized recently, so it seems likely that there are further diseases of which we are currently unaware but which may become important later. The availability of historical samples allows researchers to find out when the disease first appeared, even though it was not recognized or even known at the time. This has occurred with AIDS, where evidence of human immunodeficiency virus infection can be found in samples taken before the disease was recognized. The situation is similar for bovine spongiform encephalopathy (BSE) and Creutzfeldt–Jakob disease.
- Evidence in legal cases. This occurs where the patient alleges incorrect diagnosis or treatment.
- Control material. Most, if not all, special stains need to be validated by running a known positive control slide alongside the test slide. It is always necessary to be on the lookout for reported cases that contain evidence of essential pathologies. These are varied and may include amyloid deposits, acid-fast bacteria or Gram +/− bacteria, *Pneumocystis*, glycogen, varied acid mucins and fibrin, to name but a few examples.

Storage and disposal of records and specimens

Paper or computer records that relate directly to the patient's diagnosis are normally kept for at least the lifetime of the patient and in most cases are kept permanently. Paper records of laboratory procedures (servicing records, quality control records, day books) may be kept for between 2 and 10 years but often are retained permanently. Specific recommendations have been made by the Royal College of Pathologists and these give details of how long to store each individual type of record. All patient details are subject to confidentiality and should be stored with limited access, whether electronic or manual.

Paraffin blocks and sections are stored permanently whenever possible and as a minimum for the lifetime of the patient. Wet tissue is normally kept for at least 4 weeks after the final report. The Royal College of Pathologists again has guidelines that give details for different specimen types.

Disposal of unwanted material is a problem, as it is personal to the patient and socially sensitive. Strictly speaking, following surgery the tissue still belongs to the patient and they may ask for the material to be disposed of in a particular way or even ask for the return of the material. It is normally considered reasonable to comply with such a request provided there is no obvious reason to refuse (a serious risk of infection, for example; see *Box 20.3*). In the absence of a direct request, the material should be disposed of carefully and respectfully, e.g. by incineration, taking care to avoid any specimen being identifiable.

Box 20.3 // Return of tissues may be hazardous

The return of specimens may involve dangers to the recipients if there is a risk of infection. Even fixation in formaldehyde does not guarantee safety as the infective agent in Creutzfeldt–Jakob disease and BSE ('mad cow disease') can still be infective. Also, the materials used to preserve the specimens may be a hazard. Nineteen objects from the Indian Arts Research Centre due to be returned to the Hopi Indians who considered them sacred were found to be contaminated with arsenic.

Post-mortem tissue is also a problem but is slightly more complicated. The deceased person may have made known their wishes, in which case they should be respected, as for a surgical case, but a surviving spouse or relative may also make a similar request, even though there is no record of the deceased person themselves having made such a request. Normally such requests are followed. The exception may be a coroner's post-mortem where foul play may be a possibility. In such cases, requests to destroy the tissue rather than storing it could be an attempt to destroy evidence so the coroner's permission is needed for any disposal.

In the past, small amounts of unwanted tissue left from clinical diagnosis were used for medical research, teaching, therapy (see *Box 20.4*) and for quality control within the laboratory, often without any specific consent. This had been held to be ethical and justified by long-standing consensus and a useful way of dealing with surplus tissue. This has now changed and any use not directly related to the patient's diagnosis must be specifically sanctioned. Most patients would not want their tissue returned to them and the very idea of being presented with the remnants of, say, their appendix would seem bizarre to many patients. Having said that, the authors know of patients who asked for the return of gallstones to be polished and used in jewellery! The respectful disposal of surplus tissues is often the preferred option. Surveys have indicated that more than two thirds of patients would be willing to give permission for their tissues to be used for acceptable and reasonable purposes, provided they were fully informed of the use.

Products of conception, including all samples from later term foetuses to early ectopic pregnancies, must be disposed of with respect, according to a set procedure and the religious preference of parents, who may choose to be in attendance.

Box 20.4 // Harvesting of pituitary gland for treatment

From the 1960s to the mid-1980s the treatment for a congenital lack of growth hormone and resultant very short stature was human growth hormone (GH). Unlike many other hormones, GH is tissue specific, and animal-derived GH was ineffective. The only source of human GH was pituitaries from cadavers. The pituitaries were harvested at post-mortems and preserved in acetone. The pituitary is small, about the size of a pea, so extraction used many pituitaries in a batch. As a rough approximation, the hormone extracted from a single pituitary was sufficient for just one week of treatment for one patient. There were insufficient pituitaries being harvested to supply the demand, so treatment was rationed and aimed to bring the patients to an acceptable, but still very short, stature. To encourage maximum harvesting there was a small payment for each pituitary. Estimates of the number of pituitaries removed in Britain over the two decades range up to one million. There was no consent sought for the vast majority of these removals, but at the time most people would have allowed the removal of such a small tissue for the benefit of severely affected children, had they been asked. Over 1800 patients in Britain and over 25 000 worldwide were treated with human cadaver growth hormone. In the mid-1980s it became apparent that some of these patients were developing Creutzfeldt–Jakob disease (CJD; see *Chapter 11*) and eventually over 60 British patients were identified as having caught CJD from cadaver GH. The use of cadaver GH was stopped and replaced with recombinant GH which is safer and is available in larger quantities, so rationing is no longer needed. Although the incidence of CJD in the general population was low, the pooling of pituitaries for extraction and the number of treatments needed would mean that an individual patient could be exposed to GH extracted from several hundreds or thousands of pituitaries. Although precautions were taken against contamination by microorganisms, the cause of CJD and the very nature of prions was unknown so no precautions could have been taken.

It is easy with hindsight to see this as a sorry episode but the use of cadaver GH was done in an altruistic way to help the children and even the payments were often donated to charity. It is wrong to judge the past by today's standards. To quote the opening lines of *The Go-Between* by L.P. Hartley, 'The past is a foreign country: they do things differently there'.

Human Tissue Act 2004

The use of tissue samples has become more complex in recent years and is regulated by the Human Tissue Act 2004. The change in law was initiated by the events at Bristol Royal Infirmary and Alder Hey Children's Hospital in Liverpool in 1999–2000, where large amounts of tissue from post-mortems were held without relatives knowing that the tissue had been retained. These regulations are monitored by the Human Tissue Authority (HTA).

The main requirements of the law are that retention and use of tissues requires consent and information. The use of tissue samples for research, teaching and other reasonable purposes is not specifically excluded but does require informed consent. Institutions retaining human tissue must be licensed and have a Designated Person who will supervise the activities. The institutions will have regular inspections by the HTA to ensure compliance with the Act. Most of the inspections will be 'desk inspection' of documentary evidence but site inspections can also be carried out. Certain breaches of the regulation are considered as offences under the Act and may lead to legal proceedings and prosecution (see *Box 20.5*).

Tissue obtained prior to September 2006, when the Act came into force, are not subject to these laws, so there is no requirement to seek out the relevant people for permission to use archival tissue. It should, however, be used in a responsible way which should include anonymization of the samples.

> **Box 20.5 // Offences under the Human Tissue Act 2004**
>
> - Removing, storing or using human tissue for Scheduled Purposes without appropriate consent.
> - Storing or using human tissue donated for a Scheduled Purpose for another purpose.
> - Trafficking in human tissue for transplantation purposes.
> - Carrying out licensable activities without holding a licence from the Human Tissue Authority (HTA) (with lower penalties for related lesser offences such as failing to produce records or obstructing the HTA from exercising its power or carrying out its responsibilities).
> - Being in possession of human tissue, including hair, nail, and gametes (i.e. cells connected with sexual reproduction), with the intention of its DNA being analysed without the consent of the person from whom the tissue came or of those close to them if they have died (medical diagnosis and treatment, criminal investigations, etc. are excluded).
>
> The first four offences only apply in England, Wales and Northern Ireland, although the Human Tissue (Scotland) Act 2006 has similar offences and penalties. The offence of DNA theft applies throughout the UK.
>
> *N.B. All of the above apply to tissue used for both diagnostic and research purposes.*

The requirements for tissue samples can be simplified as follows, but this is not definitive. For full details refer to the HTA website (www.hta.gov.uk/).

If obtained before September 2006: There is no legal requirement for permission, but good practice of anonymity and ethical approval should apply.

If obtained after September 2006: Where surgical tissue has been obtained from a living patient, consent to use tissue is not a legal obligation but anonymity and ethical committee approval are required. Post-mortem tissue always requires permission.

Electron microscopy laboratories

Electron microscopy laboratories tend to be a much smaller set-up. There are far fewer electron microscopy laboratories and often only the very large hospitals have a histological electron microscopy unit and these are more heavily involved in research than a typical histology laboratory. They are more often organized with a particular person following a specimen from start to finish rather than specimens being passed from one person to another at each step.

Research laboratories

These are organized for a different purpose, so the organization tends to be slightly different. The need to investigate one particular problem in depth often means a small and coordinated team including senior academic staff, post-graduate students and technical support staff who will follow an organized programme of research and may well need to call in expertise and facilities not available 'in house'. This means that collaboration is crucial. In research, the reproducibility of results is the main criterion of a method's reliability, meaning quality is essential. This means choosing the most appropriate methods and applying them skilfully. Most techniques are non-routine and speed is secondary to quality. There is likely to be a much greater variety of techniques used, a greater diversity of methods and a less predictable working schedule.

> **Box 20.6 // Professional responsibility**
>
> At whatever staff grade a person is employed, they are expected to behave responsibly towards both patients and information about patients.
>
> The Institute of Biomedical Science suggests that professional staff should:
>
> 1. Exercise their professional judgement, skill and care to the best of their ability.
> 2. Fulfil their professional role with integrity, refraining from its misuse to the detriment of patients, employers or professional colleagues.
> 3. Seek to safeguard patients and others, particularly in relation to health and safety.
> 4. Treat with discretion all confidential and other information requiring protection and avoid disclosing to any unauthorized person the result of any investigation or other information of a personal or confidential nature gained in the practice of their profession.
> 5. Act in good faith towards those with whom they stand in professional relationship and conduct themselves so as to uphold the reputation of their profession.
> 6. Strive to maintain, improve and update their professional knowledge and skill.
> 7. Promote the study and development of biomedical science and the education and training of biomedical scientists.
>
> The Health and Care Professions Council (HCPC) has an important regulatory role on professional behaviour.

Every histology laboratory I have ever walked into has been organized slightly differently. To make an analogy with kitchens, some are large with lots of staff, plenty of specialist equipment and can produce hundreds of meals at a time, whilst others are small, fairly basic in their equipment and only produce a few meals. Provided the food is good, then who really cares how the kitchen is organized? The same applies to histology laboratories. Provided they work effectively and produce good, well-stained sections reliably, then the details of the laboratory are of less importance.

20.2   Quality control

It is now expected that every diagnostic pathology laboratory will maintain at least a minimum level of quality that is judged to be satisfactory. If this standard is reached in the UK, then the laboratory can become accredited by Clinical Pathology Accreditation Ltd. In the USA, there is a similar system operated by the College of American Pathologists. This includes several aspects of quality. Many of these relate to management including aspects such as administration, staffing, facilities (buildings and equipment), staff development, record keeping, ethics and confidentiality. Detailed consideration of these aspects is beyond the scope of this book, but it is worth noting that the key to a reliable and reproducible test lies in the strength and efficient organization of the entire management chain in the laboratory (including quality control). Accreditation is essential for those laboratories which wish to be recognized as training laboratories.

One aspect particularly relevant to the science of cellular pathology is the quality of the final stained preparations. In the past, quality was maintained internally by direct visual assessment by senior staff. This, however, can be subject to local 'drift' that may not be detected. The UK has introduced an external quality audit under the UK National External Quality Assurance Scheme. In this scheme, each laboratory submits a series of stained preparations, which are assessed by a panel of competent, independent judges against a list of predefined standards. Each of the assessors assigns a mark for each preparation and the scores given are aggregated into a single final figure for each preparation. The laboratory receives a set of marks (out of 10) for each of the submitted preparations plus a bar chart indicating the performance of all of the other laboratories. This allows the laboratory to take remedial action if a low standard of quality is achieved. A mark below 6 is considered unsatisfactory. Consistently low marks will be noted by regional organizers of the scheme who will then make assistance available to improve performance.

Outside the realms of cellular science there are similar EQA schemes available for clinical chemistry, immunology, haematology and microbiology. Together these cover the blood and infection sciences.

20.3 Safety in the histology laboratory

The safety of staff has become a much greater issue in the management of UK laboratories since the introduction of the Control of Substances Hazardous to Health regulations in 2004. Safety is an area that involves not only management but every person in the workplace and has implications for the wider public and the environment. This area comes within the remit of the Occupational Health department.

Help can be obtained from the Health and Safety Executive who maintain a useful website and produce numerous publications to help staff comply with current UK regulations.

Every aspect of the running of a laboratory should undergo a risk assessment but particularly any procedures that involve an obvious hazard. The assessment will indicate how risks can be reduced but it is probably impossible to eliminate every hazard completely, so the regulations say that risks should be reduced 'so far as is reasonably practicable'.

Hazard assessment

The term 'hazard' includes anything capable of causing harm such as chemicals, electrical equipment and machinery, whilst 'risk' is used in the sense of probability or chance that somebody will be harmed by that hazard.

It is suggested that the assessment should include five steps:

1. Look for possible hazards.
2. Identify the people at risk from each hazard and how the harm might occur. This may involve suggesting that a particular group of people should not work on particular tasks (e.g. pregnant women need to avoid some chemicals and sources of ionizing radiation).
3. Determine the risk, whether the current working practices are safe and how frequently an accident or incident may occur. Consider whether better precautions can be implemented to reduce the risk further. For any particular hazard, this may involve considering whether the hazard can be avoided completely, whether the hazard and risk can be safely controlled or whether serious incidents can be prevented. Care should be taken to select the least hazardous option.
4. Keep a clear and accessible written account of the assessment so that anyone can refer easily to the conclusions.
5. Review the assessment periodically and revise it when necessary.

The severity of the effects of a hazard can be markedly different, so it is usual to put them into classes which are given a numeric value; the more severe the effects, the higher the risk number (see *Table 20.1*). Severity can range from an insignificant injury that is easily dealt with and treated within the laboratory to disasters which can result in multiple deaths. The likelihood of an event occurring can vary from those which are unlikely ever to occur in practice, to common events which may occur several times a month. It is possible to collate both of these variables into a grid, so that during a risk assessment each event can be ranked and measures implemented to deal with the risk in an appropriate way. Low frequency events which will only cause insignificant injury will require little effort, monetary outlay and planning to keep the staff 'reasonably' safe. As the severity rises or the frequency increases, then the following must become more of a priority: the attention to planning to minimize the risk, the effort in training, the ensuring of compliance with the planning and the cost of improving the working environment.

A score of below 8 (e.g. a moderate severity event that occurs infrequently would be $3 \times 2 = 6$) would be considered low risk and can be controlled by the usual precautions and adherence to the standard operating procedures.

A score of above 8 but below 15 constitutes a moderate risk. Careful consideration and the involvement of managers in the decision-making process are needed to minimize risks.

A score of above 15 indicates that the risks are high and likely to be unacceptable – immediate action is needed to stop the activity until the risk is reduced. Consultation

Table 20.1   Risk assessment grid based on severity and frequency of events

Severity ↓	← Likely frequency →				
	Rare (most unlikely to ever occur)	Unlikely (less common than once every 5 years)	Possible (about one event in a given year)	Probable (may occur 2–4 times in a year)	Expected (may occur more often than every month)
	1	2	3	4	5
1 Low (insignificant injury)	1	2	3	4	5
2 Slight (minor treatment required, short absence from work may be needed)	2	4	6	8	10
3 Moderate medical treatment needed, may need several days off work or hospital admission)	3	6	9	12	15
4 Severe (may be multiple injuries, permanent disability or fatality)	4	8	12	16	20
5 Disastrous (multiple fatalities)	5	10	15	20	25

with national authorites may be required in order to make appropriate policy decisions. An example of such a catastrophic risk might be Lassa fever or an Ebola virus outbreak. Investigation of cases in inappropriate facilities could lead to a major epidemic and many fatalities. Such high-risk investigations should only be carried out in specialist laboratories away from major population centres; Lassa fever and Ebola material would be investigated at highly specialized centres such as Porton Down on Salisbury Plain.

Side-effects of safety failures

The harm caused to an individual is the most important effect of the failure of safety and the resulting accidents, but there are other important considerations which need to be addressed as part of the response to an accident. There may be reduced performance because of loss of members of staff, either temporarily or permanently. The loss of crucial members of staff may mean that some critical services may not be able to continue. Contingency plans are needed to ensure that there is sufficient cover within the laboratory to prevent the loss of a few workers resulting in an inability to carry out critical work.

Adverse publicity needs to be considered and contained to prevent overreaction to the event. Public concern and loss of confidence can have an adverse effect on performance. Public concern and responsible reporting is to be welcomed rather than ignored, but overreaction is a serious concern.

With very serious accidents, some outside agencies and regulatory authorities will become involved, and with catastrophic events there may be parliamentary involvement such as questions in Parliament and even full public enquiries. These are distressing for the laboratory staff and very expensive.

If serious negligence or recklessness is proved, it can result in litigation and individual or multiple claims for compensation for thousands or potentially possibly millions of pounds.

If there are a small number of people who can be proved to be directly responsible, criminal charges are a possibility, and if those accused are found guilty, these charges could result in prison sentences.

Thus, although the harm to people is the first and most important aspect of safety, it is not the only consideration. Equally important are the acquisition, use and disposal of reagents, ensuring that they have a minimal impact on safety both in the workplace and the wider environment.

Consideration of safety failures

Whenever a failure of safety occurs and there is a significant event, there needs to be a review to identify the cause of the failure. Clearly, more serious problems need more thorough investigation. The purpose of these reviews is not to apportion blame but to identify weaknesses in the current working practices to avoid a repetition of the event. Simply to blame individuals is counterproductive. Individuals may be directly involved in the failure but the review should try to help the individual by means of better training, supervision or clearer safety protocols, rather than using them as a scapegoat for the failure.

Hazardous chemicals

Many of the risks in a laboratory will relate to chemical hazards. Each chemical needs to be considered carefully but this is relatively straightforward since the manufacturers will provide data hazard sheets with their products. In the risk assessment, it should be standard to identify how the material should be stored safely and how it can be used safely, including such precautions as using fume hoods and protective clothing and the safe disposal of the materials when they are no longer required. It should also include the steps to be taken if the precautions fail or there is an accident resulting in an emergency situation, including the medical treatment of the affected person.

Chemical exposure has two dimensions, both of which need to be considered. One is concentration – the more concentrated a toxic material is, the more likely it is to cause damage. This is particularly the case with volatile materials that can be breathed in. The use of effective fume hoods is one way of limiting exposure to such air-borne risks. The second aspect of respiratory toxicity is the length of exposure. The hazards are usually specified by manufacturers on the packaging or in their literature but can be in a variety of forms. They may be relatively non-specific with such terms as **toxic** or **irritant** (see *Box 20.7*) without giving precise details of the dosages causing these effects. They may be in coded form with a defined list of meanings; thus, R45 means the compound 'may cause cancer' and S20 indicates the precaution 'When using, do not eat or drink.' More precise information is given in occupational exposure levels, which are often legally mandatory. The exposure is related to time; thus, **time-weighted average** (TWA) levels indicate the maximum concentration averaged over a complete working day (this may allow higher levels for short periods of time provided they are balanced by lower levels at other times); the **short-term exposure limit** (STEL) indicates the time-weighted average for a 15 min period, which controls the levels during the peak times; and the **ceiling limit** or **ceiling exposure value** gives a value that must never be exceeded, even for a brief period of time. These phrases and codes are often specific to one country or even one regulatory authority within a country, so other terms may be found such as **threshold limit value**, **permissible exposure limits** and **maximum allowable concentration**.

Box 20.7 // Categories of hazardous chemicals

Irritants cause inflammation and this is usually reversible. **Corrosive chemicals** cause irreversible destruction of living tissues. **Sensitizers** cause allergic reactions. This is usually only specified if a significant proportion of the population would develop an allergy if they were exposed to the agent. A much wider range of chemicals may cause allergies in a small proportion of people exposed. **Carcinogens** cause tumours in animals but the dosage is often critical. **Toxic** implies that the chemical can result in death if ingested, inhaled or absorbed through the skin in a large enough dose.

Safety personnel

Safety needs to have someone who is identifiably responsible. This may be a supervisor or a designated safety officer. This person should not only prepare the assessment but should monitor its implementation – a good safety protocol is useless unless everyone adheres to it. There should also be appropriate monitoring of the health of the people performing the task. Monitoring may be relatively simple involving noting whether a member of staff suffers from obvious effects such as recurrent dermatitis or respiratory problems, but if appropriate may mean specific tests, for example, for liver function or early signs of chronic effects. This type of monitoring is often undertaken on an annual basis by the Occupational Health Department and in Histology would relate to exposure

---

> **Box 20.8 // Fire hazards**
>
> Fire hazards have their own nomenclature problems. The simplest description is anything that will burn or cause a fire, but things are not quite that simple. Materials may be described as explosive, an oxidizer, combustible, flammable or even inflammable. They also have a defined flashpoint, which indicates the ease with which they ignite.
>
> **Explosive** materials may be caused to explode by a variety of mechanisms. Some may explode if simply shaken (which need not necessarily be vigorous shaking), or if they are exposed to electrical currents or raised temperatures. The explosion may also cause a fire. **Oxidizers** do not burn directly but may promote or initiate fires in other chemicals. A **combustible** material is defined as having a flash point above a specified level. The specified level will vary depending on the regulatory authority (in the USA it is set at 100°F/37.8°C by the Occupational Safety and Health Administration and 141°F/60.5°C by the Department of Transportation). A **flammable** material is defined as having a flash point below the specified threshold. Despite the apparent negative prefix, the term inflammable means exactly the same as flammable and is now discouraged as a safety term.

to formaldehyde or glutaraldehyde fixatives, either as vapour or liquid. Many laboratories provide an assessment of respiratory function.

Safety is a state of mind

Although safe working requires clear specific precautions such as wearing rubber gloves and protective goggles, such precautions are best considered as an additional means of protection and not treated as a replacement for other control measures. The best protection is probably an awareness of danger and a personal commitment to safety. If the laboratory can instil an ethos of caution and care, then it will be better protected than relying on a few prescriptive safety measures. Ignoring or disregarding safety is obviously detrimental but so too are overreaction and outright fear of dangers. These may lead to mistakes being made because of anxiety causing loss of concentration. The best attitude to risk is a healthy unease, so that care is taken without outright worry or tension.

A concept known as Murphy's or Sod's Law states that 'If anything can go wrong, it will,' with an addendum that 'If something can go wrong in more than one way, it will go wrong in the way that will cause the most damage.' This seems at first to be a cynical and pessimistic outlook but despite the jokiness it is a good reminder to laboratory workers of the need to make appropriate and wide-ranging safety considerations. When assessing all hazards and risk, the worst-case scenario should always be considered. Safety considerations and an awareness of potential hazards (together with maintaining a high-quality output) must remain of paramount importance at all times when working in a laboratory environment.

The assessment of risk is not always just about considering the obvious. For example, the use of liquid nitrogen, e.g. in frozen sections, may require the material to be transported between floors of a building. With liquid nitrogen, its very low temperature is the main hazard. Thus, carrying it by hand in a Dewar flask, especially in large quantities, is a risk if it is dropped or spilt on to clothing or living flesh. A safer alternative is a specially designed wheeled trolley incorporating a secure damage-resistant flask. Unfortunately, wheeled trolleys are difficult to manoeuvre on stairs and require a lift. However, although liquid nitrogen is not directly toxic (80% of the air we breathe is nitrogen), it is suffocating if it evaporates and replaces the oxygen in air. Thus, when liquid nitrogen is transported

> **Box 20.9 // Be aware of the less obvious hazards**
>
> Hazards are not always obvious. In microtomy, for example, the dangers of the ultrasharp knives are usually well recognized. I have personally worked in histology for many years and have suffered only three minor cuts from microtome knives. However, in the same period I had many minor cuts from slides, coverslips and even paper – these seem less dangerous and threatening so fewer precautions are taken.

in a lift it should not be accompanied in person in case the lift breaks down. This would mean the person would be in an enclosed space with the consequent risk of suffocation, as the cold nitrogen gas evaporating would replace the air in the lift. In 1999, a technician died in this way resulting in the prosecution of the Medical Research Council who employed him. However, putting liquid nitrogen in a lift unsupervised would be a risk to unaware passengers boarding at intermediate floors, so further precautions are needed, such as supervision of intermediate floors during the transport. Consideration of worst-case scenarios and Murphy's Law is needed to identify and control such risks.

RIDDOR

Should an unforeseen incident occur, then it must be reported under the terms of RIDDOR (The Reporting of Injuries, Diseases and Dangerous Occurrences Regulations 2013). This is done in order to ensure that in future the safety of personnel engaging with work in that area can be guaranteed.

Equally important are the reporting and recording of diseases and physical conditions that may have resulted from working with reagents and/or equipment according to the method specified, i.e. the Standard Operating Procedure/Policy (SOP). This has particular relevance to low-level exposure to reagents that may have a cumulative effect over many years, and also to repetitive physical actions that have not been ergonomically assessed.

At the other end of the spectrum a record must be kept of 'near misses' or dangerous occurrences which may not have resulted in injury on that particular occasion, but which may well do so if the situation is repeated. Often this will indicate faulty equipment (electrical or mechanical) that needs to be withdrawn from use and either repaired and safety tested, or condemned as unsuitable for use.

Suggested further reading

Bancroft, J.D. (2012) Chapter 2: Safety and ergonomics in the laboratory. In: *Bancroft's Theory and Practice of Histological Techniques*, 7th edn (eds K.S. Suvarna, C. Layton and J.D. Bancroft). Churchill Livingstone.

Dunk, L. (2012) Chapter 1: Managing the laboratory. In: *Bancroft's Theory and Practice of Histological Techniques*, 7th edn (eds K.S. Suvarna, C. Layton and J.D. Bancroft). Churchill Livingstone.

Glencross, H., Wang, Q., Ahmed, N., and Smith, C. (eds) (2010) *Biomedical Science Practice: Experimental and Professional Skills*. Oxford University Press.

Pitt, S. and Cunningham, J. (2009) *An Introduction to Biomedical Science in Professional and Clinical Practice*. Wiley–Blackwell.

Suggested websites

www.legislation.gov.uk/uk/pga/ Human Tissue Act (2004)
www.cpa-uk.co.uk Clinical Pathology Accreditation
www.hse.gov.uk/pubns/leaflets.htm

Answers to self-assessment questions

Chapter 1

1. An artefact is an alteration in tissues brought about by the treatment of the tissue following its removal from the body.
2. (a) Materials normally present in cells can be lost from tissues during processing, e.g. glucose. (b) Materials that were not present in life can be added to the tissues, e.g. fixation pigments. (c) Tissues can be distorted by processing, e.g. shrinkage of cells.
3. Glycogen is probably present in the cells because the two different preparative techniques show the presence of glycogen. The distribution of glycogen within the cells cannot be determined with any certainty from these sections because the two techniques give conflicting results. One, or both, techniques have introduced an artefact and further techniques are needed to resolve the problem.
4. No single technique is perfect for every specimen and every investigation. Different people have different requirements and each person will choose the method that best matches their needs. Thus, frozen sections are good for demonstrating lipids but are not ideal for most routine applications as the tissue cannot be stored easily.

Chapter 2

1. Health is an abstract concept and is defined differently by different authoritative sources. For many people it is simply the absence of significant disease and means that there is no need to go to the doctor. A better concept is to define health as wellbeing and to specify the different areas of physical, mental and social wellbeing which all contribute to a healthy state.
2. The major signs of inflammation are pain, redness, heat and swelling in the inflamed area. The symptoms of redness, heat and swelling are produced by alterations in the blood vessels and blood flow through the tissues. The enlargement of the blood vessels brings increased blood into the area, giving rise to the redness and heat. Escape of fluids from the blood vessels results in increased volume of tissue fluids and swelling. Pain has several different aspects; it is partly due to tissue distension by oedema, partly to the direct effects of the extraneous cause of the inflammation (e.g. bacterial toxins). Chemical mediators such as histamine and bradykinin may also contribute to the pain.
3. In histological sections only the dilation of the blood vessels and the direct effects of the alterations of vascular activity can be observed. The slowing of the speed of blood through the affected vessels is the cause of haemoconcentration. Fluid escaping into the tissues expands tissue spaces and the increased number of white cells emigrating from the vessels can be seen.
4. Chronic inflammation shows fewer changes in blood flow, with the result that the classic symptoms are diminished or absent. The white blood cells in the area show a predominance of macrophages and lymphocytes rather than the predominance of neutrophils characteristic of acute inflammation. Acute inflammation is short-lived but chronic inflammation can last for many years if not successfully treated.
5. In healing, recovery occurs with very minor forms of damage where there is no loss of

cells and the tissue returns to normal with no formation of new tissues or scarring. Regeneration occurs in actively mitotic tissues and any lost cells are simply replaced as part of the usual replacement of worn-out tissue. No scar tissue is formed and the tissue regains its full physiological capacity. Repair occurs when there is significant loss of a tissue which is not actively mitotic. The tissue can repair by stimulating cells into mitotic growth, but this results in the production of scar tissue rather than tissue identical to that which has been lost. There is some loss of normal physiological capacity.

6. Atrophy is shrinkage of an organ, either by a reduction in cell number or a decrease in cell size. Causes of atrophy include reduced blood flow to a tissue (ischaemic atrophy), failure to regularly use or stimulate a tissue (disuse atrophy), ageing (senile atrophy) and starvation.

7. Hypertrophy is a growth abnormality where there is an increase in the volume of tissues caused by an increase in the volume of individual cells, and is typically associated with increased work demand. Hyperplasia is an increase in tissue volume caused by an increase in the number of cells and is typically due to irritation of the tissue. Metaplasia is where one tissue is replaced with a different type of tissue, often due to excessive stimulation. Dystrophy is an alteration in maturation of cells and may predispose to full malignant change.

8. Benign tumours have relatively normal cell morphology and tissue architecture. Mitotic activity is approximately the same as in normal tissues. Malignant tumours show differences in cell morphology particularly in the nucleus, which is often enlarged and irregularly shaped. Nuclear borders are less well defined, nucleoli are often prominent and the nucleus stains abnormally. There is also pleomorphism with variation not only from normal but also between adjacent tumour cells.

9. Malignant tumour cells can grow erratically, giving an irregular border to the tumour. They can be invasive and penetrate through tissue barriers such as basement membranes and so penetrate into adjacent organs. Malignant cells can also become detached into fluids such as blood and lymph and be carried to distant sites where they may establish secondary or metastatic tumours. The ill-defined edge of the tumour makes surgical removal more difficult and often requires the sacrifice of significant amounts of healthy normal tissue to ensure complete removal. Invasiveness and metastasis may result in several organs being affected, making surgery inappropriate. Benign tumours are usually well-defined and slow-growing. They do not spread by invasion or metastasis and so are readily treated by surgical removal.

10. Grading is a histological assessment of the nature of a tumour on the basis of the degree of nuclear abnormality, the indications of invasion and metastatic threat and estimation of the growth rate and mitotic activity. Staging is a clinical assessment based on size, how far the tumour has spread and whether invasion and metastasis have occurred.

11. Cysts are enlargements of tissues with a central fluid-filled cavity. Cysts can be tumorous, with the tumour cells secreting fluid into the central hollow cavity (e.g. ovarian cysts). Alternatively they can be retention cysts, where normal secretions cannot escape and cause distension (e.g. sebaceous cysts), infective cysts (e.g. parasites such as the canine hydatid worm which can infect humans as a secondary host and form cysts in the liver) and congenital cysts from remnants of embryonic tissues remaining into postnatal life (e.g. thyroglossal cysts).

12. Necrosis is cell death due to an external influence, such as a noxious substance, or interference with blood supply. Apoptosis is simply the catabolic destruction of cells following loss of energy regulation, resulting in loss of metabolic control and rupture of lysosomes. Nuclear changes include pyknosis, karyorrhexis and karyolysis. Apoptosis is much more rapid and the cell actively co-ordinates its own destruction. Catabolism is specific and controlled. The nucleus becomes condensed and appears crescentic in tissue sections. The cell actively fragments into membrane-bound apoptotic bodies which are rapidly engulfed by other cells. Necrosis can effectively be thought of as cell murder and apoptosis as cell suicide.

13. Gangrene is infection of necrotic tissues, particularly by anaerobic bacteria. This results in tissue becoming black because haemoglobin becomes degraded and smelly because of the anaerobic conditions.

Chapter 3

1. Tissues removed from the body may change by autolysis, where the cell's lysosomal enzymes break down the structure of the cells. Tissues may also change by putrefaction, where bacterial or fungal contaminants break down the tissue structure. Tissues may change if the cells remain alive after removal from the body as they will respire anaerobically. Cells will deplete their nutrient reserves and the accumulation of waste products may alter the tissue's structure.

2. Post-mortem changes can be prevented by stopping enzyme activity. Enzyme activity can be halted by chemicals or heat. Chemical fixation is most common and requires the cells to be treated with a large excess of fixative as quickly as possible. Perfusion of the fixative through the blood vessels or slicing of the tissue may be needed to ensure rapid penetration to all cells.

3. A good fixative has the following properties:
    * Penetrates tissues quickly and evenly.
    * Kills cells quickly and evenly; the killing stops abnormal metabolism.
    * Prevents autolysis.
    * Prevents putrefaction.
    * Does not add any extraneous material to the tissue.
    * Does not swell or shrink the tissue.
    * Prepares the tissue for later treatments such as staining and does not prevent any later investigation that might be needed.
    * Prevents desiccation and drying of tissue, which would cause shrinkage and distortion.
    * Safe to use (non-toxic, non-flammable).
    * Reasonably priced.
    * Convenient to use (shelf life, storage, etc.).

4. Dettol is not a good fixative, despite killing bacteria and thus stopping putrefaction. The prevention of putrefaction is only one of the three types of change that need to be inhibited. Only if the chemical also stops autolysis and cellular metabolism without damaging the tissues will it be a useful fixative. Dettol is an antiseptic and can be used on open wounds without killing the tissues. It will not stop normal tissue enzymes so it is not a good fixative.

5. Formaldehyde is readily washed out of tissues and leaves no residual aldehyde groups. Glutaraldehyde is a bifunctional aldehyde and is not easily washed out of tissues so there will be aldehydes from the glutaraldehyde left in the tissues following fixation and these will give a non-specific background reaction with Schiff's method for aldehydes.

6. Fixatives are intended to kill tissues so they are all toxic chemicals. Some are extremely nasty substances and may cause specific problems such as inducing allergies or predisposing to cancer.

Chapter 4

1. Wax embedding is used to support the tissues during the cutting of sections. In the absence of a support medium, the sections will collapse and the structure of the cells and tissues will be lost. Wax is the commonest embedding medium as it is cheap and allows good sections to be prepared.

2. Fixation (to preserve the tissues), dehydration (to remove the water and replace it with alcohol), clearing (to remove the alcohol and replace it with a solvent miscible with wax), wax impregnation (to replace the clearing agent with molten wax), blocking out (to set the tissues in a mould with fresh wax) and finally solidification of the wax to form the wax block. Other steps such as decalcification (needed with mineralized tissues) and vacuum impregnation (not absolutely essential but useful) could also be included in the list.

3. A graded series of any reagent, including alcohol, is a series of baths of gradually increasing strength. The purpose is to remove the water more gently to minimize tissue damage. If tissues are placed directly into 100% alcohol, it can cause more hardening and distortion of the tissues than a slower removal

using baths of increasing strength. An example of a graded series would be 50%, 70%, 90% and 100% alcohol. Alternatives to alcohol might be acetone, propylene oxide or Cellosolve.

4. Clearing agents were so called because they rendered the tissue transparent. This is not true of all reagents but the name has been retained.

5. (a) Benzene gives good results but is carcinogenic and is no longer used routinely. (b) Xylene is less of a health problem than benzene; it clears tissues rapidly but hardens the tissues and is more useful for clearing sections after cutting than for preparing blocks of tissues. (c) Cedarwood oil is slow and expensive but gives good results. (d) Petrol hydrocarbons are considered safer than many other materials and give acceptable results.

6. Vacuum embedding is the reduction of pressure in the wax bath during wax impregnation and is used to remove air bubbles and help to remove the clearing agent.

7. Mineralized bone can be softened using acids or chelating agents to remove the calcium salts.

8. Undecalcified sections can be prepared by embedding the tissue in a hard embedding medium such as plastic and then sectioning with a hardened tungsten knife or by preparing a ground section in which most of the bone is ground away leaving a single thin section.

9. The bone marrow fragments would have lost any mineralized material during fixation, as acetic acid is also an effective decalcifying agent. Thus, there is no way to be sure whether the patient's bones were fully mineralized or whether the patient suffered from osteomalacia. (If osteomalacia is suspected, then the bone must be protected from acids during processing.)

10. Automated processing can process a large number of blocks simultaneously. The processor can be entirely contained and sealed, which minimizes the risks to staff. Automated processing allows the use of microwaves, ultrasonic agitation and precise timing, which can significantly reduce processing times. Machines can be programmed to have tissues ready to be blocked out at predetermined and convenient times. Automation is also more consistent.

Chapter 5

1. Plastics are harder than most other embedding media and so can be used to cut thinner sections (including sections for electron microscopy). They can also be used to cut sections from hard tissues like bone. Preservation of plastic-embedded tissue can be better than with wax embedding since plastic embedding involves fewer reagents and does not overheat the tissue.

2. The epoxy resins are preferred for electron microscopy since they are stable in the electron beam of an electron microscope unlike wax and methacrylates.

3. Epoxy resins can form bonds with active groups in the tissues (e.g. amines). This prevents dyes binding to the groups and reduces staining. Also, the resins cannot easily be removed from the tissues without destroying the tissues, so dyes cannot penetrate into the sections and this reduces staining intensity. Neither wax nor methacrylates directly react with tissue groups and can easily be removed, allowing dyes to access all stainable groups.

4. The usual way to flatten sections and remove wrinkles from sections is to float them on warm water. Water-soluble waxes cannot be floated on to water as they will dissolve, allowing the section to disintegrate.

Chapter 6

1. Slow freezing allows large ice crystals to grow and these can mechanically damage the tissues. Slow freezing also causes concentration of tissue fluids, as it is pure water that freezes first. The hypertonic tissue fluids may cause cell shrinkage by osmosis.

2. (a) Frozen sections are used to diagnose malignancy because of the need for an urgent diagnosis. Stained frozen sections can be available within a few minutes of the tissue being removed from the patient. (b) Frozen sections are used for lipid demonstration, since the tissues are cut without needing to be embedded or processed. Most lipids are lost if the tissues are processed using organic solvents. (c) Frozen sections are used for

enzyme techniques, as they can be cut without the need to fix or process sections. Fixation will destroy most enzyme activity.

3. In the cryostat sectioning technique, everything (tissue, knife and microtome) is kept below freezing so neither the block nor the section thaws during sectioning. In the freezing microtome technique, it is only the block that is kept frozen and sections usually thaw during sectioning. The cryostat is better for most purposes but the freezing microtome has some advantages for large blocks or when thick sections are needed.

4. The boiling of liquid nitrogen when it contacts warm tissue may slow down the rate of cooling by producing a layer of insulating gas around the block. The use of isopentane cooled with liquid nitrogen avoids the boiling problem. The boiling effect does help to prevent minor splashes causing freezing burns to the histologist using the liquid nitrogen.

5. Ice is not completely stable at −20°C and will gradually recrystallize and cause tissue damage.

Chapter 7

1. Chromophores are groups that bring colour to a dye structure. Without a chromophore group, organic compounds are colourless. Auxochrome groups allow dyes to bind to the tissues. Auxochromes are usually ionizable groups that allow the dye to dissociate into charged ions.

2. Basic and acidic dyes bind to ionized groups in the tissues. Basic (cationic) dyes will bind to acid groups in the tissues such as nucleic acids. Acid (anionic) dyes will bind to basic proteins in the cytoplasm and connective tissues. If the two dyes have contrasting colours, they will show the nucleus in one colour and the cytoplasm and connective tissues in the contrasting colour. For example, methylene blue is a basic dye and will stain the nucleus blue, whilst eosin is an acid dye and will stain the cytoplasm pink.

3. Altering the pH of a dye solution will alter the ionization of the tissue. If the tissue groups are not ionized, then no staining will occur. A high pH will favour staining by basic dyes and a low pH will favour staining by acid dyes. Salts also affect the staining by altering the ionization and charge on the tissues. At low salt concentrations, the ions may help dyes to bind by masking surface charges that would otherwise repel the charged dye molecule. High concentrations of salt, however, are inhibitory to staining, as the salt ions will compete with the dye ions for the tissue-binding sites. Thus, low concentrations increase dye binding ('salting on'), whilst high concentrations decrease staining ('salting off').

4. With mordanted dyes, the mordant acts as a bridge between the tissue and the dye, allowing binding. In the absence of the mordant, staining will not occur. If a small amount of mordant is added, it will bind to the dye forming a dye lake and this dye lake will then bind to the tissues colouring them. If, however, there is more mordant than dye, there will be a lot of mordant without any dye attached. This unattached mordant will compete for the tissue-binding sites and inhibit binding of the dye lake (mordant with attached dye) and so reduce staining.

5. It is the mordant that binds to the tissue and so it is the mordant that determines where the dye will bind. Aluminium salts bind to the nucleus most strongly, whilst other metals will bind to elastic fibres or muscle proteins.

6. The name of the phenomenon is metachromasia. The colour change is believed to be due to polymerization of the dye on the surface of the tissue component.

7. An example of a red nuclear stain is neutral red and an example of a blue nuclear stain is methylene blue. A red nuclear stain is used when the main staining reaction produces a green or blue colour, and a blue nuclear stain would be used to contrast with a method producing a red primary reaction.

8. Haematoxylin itself is not a dye and must be oxidized to haematein to stain effectively. The dye solution will improve as atmospheric oxygen converts the haematoxylin to haematein; this is called ripening. The haematein can be further oxidized to a colourless product, which cannot act as a dye. Thus, a haematoxylin solution will begin to deteriorate when there is more

haematein being overoxidized than there is haematoxylin being oxidized to haematein.

9. Dyes differ in their molecular size and their ability to penetrate is dependent on this size. Small molecules will penetrate more readily than larger dyes. When in direct competition, the larger dyes will tend to displace or mask the smaller, paler dyes. Tissues differ in their permeability, with red blood cells being very dense, most cytoplasm being intermediate and collagen being very permeable. Using a small yellow or orange dye, an intermediate-sized red dye and a large blue dye, it is possible to have conditions such that only the smallest dye will stain the dense red cells and they will stain yellow or orange. The medium dye and small dye will both be able to penetrate and stain the general cell cytoplasm, but the larger red dye will displace or mask the yellow dye and so the cytoplasm will appear red. Although all three dyes will be able to penetrate the permeable collagen, the largest and densest dye will prevail, so the collagen will stain blue.

10. Argentaffin reactions occur when there is sufficient reducing strength in the tissues to give a visible silver deposit without using any extra reducing agent. The argyrophil reaction produces submicroscopic silver deposits, which can be enlarged and made visible using extra reducing agents such as quinol or formaldehyde.

11. Silver is the best material for metallic impregnations as its salts are easily reduced and small deposits of silver catalyse the reduction of more silver. The deposit is also stable and is a very dense black colour, which gives a good contrast.

12. DPX mounting medium requires tissues to be dehydrated with alcohol and cleared in xylene. These reagents will dissolve most lipids, so the staining would be lost in the processing. Water-based mounting media do not require this processing and tissues are mounted directly from water.

13. Resinous mounting media give greater clarity to the specimen, as they have a higher refractive index than most water-based mounting media. Resinous mounting media also usually set or dry firmly and remain clear for longer. This means that most resinous mounting media can be considered as a permanent mount and stored indefinitely, whilst water-based mounting media are usually best considered as temporary mounts.

Chapter 8

1. Histochemical methods must produce a final product that is insoluble and highly coloured to ensure that the location of the reaction is accurate and visible. Biochemical techniques are less restricted and can use reaction products that absorb in the UV region and can employ products that are soluble. Histochemical tests need to ensure that the material does not relocate but remains *in situ*.

2. The PAS reaction detects glycols (hydroxyls on adjacent carbon atoms, R-COH-COH-R) or some derivatives of this group such as the amino and alkyl amino derivatives. The main materials found in tissues that have this arrangement are carbohydrates and proteins that have carbohydrate attached (glycoproteins and proteoglycans).

3. Periodic acid readily oxidizes the glycol groups to aldehydes, which can be detected with Schiff's reagent, but has less tendency than potassium permanganate to oxidize the aldehydes to acids.

4. The biggest division is into neutral mucins and acidic mucins. The acidic mucins can be subdivided further into sulphated and non-sulphated mucins. Sulphated mucins are subdivided again into strongly and weakly sulphated mucins. The non-sulphated or carboxylated mucins can be subdivided into sialic-containing and uronic-containing mucins.

5. Alcian blue will stain acidic mucins blue, but this is sensitive to pH and salt concentration. By using several strengths of salt (e.g. magnesium chloride), it is possible to selectively stain and therefore identify different groups of mucins. The critical concentration of the salt is the one that will completely inhibit one group of mucins but allow more strongly acidic mucins to stain.

6. Hyaluronic acid- and sialic acid-containing mucins can be distinguished using enzyme digestion with hyaluronidase or sialidase. The enzymes will selectively remove the reactive groups and so prevent them from staining with alcian blue.

7. Lipid techniques are often unpopular, since lipids are difficult to fix and are easily dissolved or relocated. This means that the more usual preparative techniques are not appropriate, so lipids require special techniques, which are more trouble to the histologist.

8. The dye will dissolve more readily in the lipid than in the solvent, so the lipid becomes highly coloured whilst the remainder of the section remains uncoloured. This is an application of the partition effect.

9. Only osmium tetroxide fixes lipids, although potassium dichromate may have some fixative action on phospholipids.

10. During the degeneration process, the phospholipid of the myelin becomes degraded and forms neutral lipids. This is a change from an amphipathic lipid to a hydrophobic lipid and this alters its staining properties.

Chapter 9

1. Basophilic staining for DNA can be controlled using enzyme digestion with deoxyribonuclease. The enzyme will destroy the DNA but leave RNA and basophilic mucins unaffected. By using two adjacent sections, one of which is treated with deoxyribonuclease before staining, it is possible to identify the basophilic materials that have been destroyed by the enzyme and these will be the sites of DNA.

2. Intercalation is the insertion of a molecule into the slot between adjacent pairs of bases in the DNA helix.

3. The Feulgen technique uses acid hydrolysis to produce aldehydes from DNA and these aldehydes are then detected using Schiff's reagent.

4. Proteins can be stained using simple dyeing with acid dyes such as eosin. This is not very selective or specific. Most, but not all, proteins will stain to some extent but some other materials may also stain. Proteins can be stained using specific amino acid tests, e.g. Millon's test for tyrosine. These are more selective in that only proteins or free amino acids will react, but since free amino acids are usually lost during processing the method is quite selective; however, it is more difficult than simple staining.

5. The enzyme may be lost by simply diffusing out of the section into the substrate solution or the activity may be diminished by fixatives or other chemicals.

6. Cyanide is used in dehydrogenase reactions to poison the electron transport enzymes in the mitochondria. This should prevent any reduced NAD being processed by the electron transport chain where the hydrogen would be converted to water. In the absence of the electron transport chain reactions, all of the hydrogen generated by the dehydrogenase reaction will then be available to reduce the tetrazolium salt and give a visible reaction.

7. The α-naphthol method can be used to demonstrate any hydrolase enzyme provided an artificial substrate can be prepared that the enzyme can hydrolyse to release α-naphthol. The α-naphthol can then be trapped and visualized using an azo-coupling reaction.

8. Biochemical measurement is easier to control. For histochemical measurement, the section thickness is crucial for accurate measurement but is very difficult to control or measure accurately. The reaction may not follow the Beer–Lambert relationship in sections and may give a false quantification.

Chapter 10

1. Carbon, asbestos and silica.
2. Lipofuscin gradually accumulates with age.
3. Haemosiderin and asbestos bodies are both positive with the Perls technique.
4. Melanin, lipofuscin, argentaffin and chromaffin pigments are all positive with the Schmorl's technique.
5. APUD cells secrete amine neurotransmitters and peptide hormones.

6. Haemosiderosis and haemochromatosis.
7. Insoluble calcium salts can be demonstrated using alizarin red. The von Kossa technique will demonstrate mineralization, which usually involves mainly insoluble calcium salts. Xylenol orange, tetracycline and alizarin can be used *in vitro* to show the deposition of calcium during bone growth and remodelling. Soluble calcium salts can only be demonstrated in intact living cells using ion-selective dyes.
8. Copper accumulates in Wilson's disease and can be demonstrated using rubeanic acid.
9. Ratiometric fluorescence measures the concentration of an ion by comparing the intensity of fluorescence at two wavelengths. Ratiometric measurements are independent of the concentration of the dye. Single-wavelength measurements use the change in intensity at a single wavelength to measure the concentration and are dependent on dye concentration. The single-wavelength methods use an ionophore and ionic buffer to calibrate the amount of dye present within the cell.

Chapter 11

1. Specimens for microbiology are living bacteria that allow cultures to be grown and biochemical and sensitivity tests can be carried out using these cultures. Paraffin wax sections only contain dead cells, which limits the investigation to staining and serology. Cultural and biochemical tests are more flexible and can identify a wider range of organisms.
2. Staining methods, e.g. Gram staining, can be used to identify the organism as Gram-positive or -negative. Immunological methods and *in situ* hybridization can be used to confirm the type of bacteria. Careful observation of the morphological changes in the tissue may help to identify the causative organisms.
3. Viruses (e.g. herpes virus), bacteria (e.g. *Staphylococcus aureus*), fungi (e.g. *Aspergillus fumigatus*) and protozoa (e.g. *Entamoeba histolytica*) can all appear in sections.
4. Amyloid is an abnormal protein with a β-pleated sheet conformation that accumulates in certain pathological conditions. Amyloid is resistant to most proteolytic enzymes because of the unusual β-pleated sheet conformation.
5. See table below.

Amyloid type	Diseases with which the amyloid is associated
AL Amyloid light chain immune-associated amyloid	Multiple myeloma, Waldenström's disease
AA Reactive amyloid	Rheumatoid arthritis, tuberculosis, Hodgkin's disease, familial Mediterranean fever
AE Endocrine-related amyloid	Insulinomas, medullary carcinoma of the thyroid
β-Amyloid	Alzheimer's disease
AS Senile amyloid	Old age (80+)
AF Familial amyloid	Familial amyloidosis
AD Dermal amyloid	Lichen amyloidosis
AH Haemodialysis-associated amyloid	Renal failure with haemodialysis

Chapter 12

1. An antibody is a soluble protein produced as part of the immune response to the presence in the body of a foreign material. An antigen is a material capable of eliciting the production of an antibody when introduced into the body. Antibodies are good histological reagents as they can be highly specific and very sensitive in the detection of their specific antigen.

2. Fluorescein isothiocyanate (FITC) is a derivative of fluorescein and can react with proteins to form a conjugate. This conjugate is stable and will not dissociate spontaneously from the dye. The labelling process involves a simple reaction between FITC and the protein followed by purification of the conjugate to remove unreacted FITC.

3. Immunofluorescence techniques suffer from several disadvantages compared with enzyme-labelling methods. Immunofluorescence requires a fluorescent microscope and not just a routine light microscope. The fluorescence can be quite faint and needs a darkened room to allow the eyes to become dark-adjusted. The fluorescence fades quite rapidly, so specimens can only be considered temporary. It is difficult to counterstain the specimen effectively without interfering with the fluorescence of the fluorescein.

4. Alkaline phosphatase and horseradish peroxidase are two enzymes used in immunocytochemistry.

5. For the peroxidase–anti-peroxidase (PAP) method, the following steps are needed:
   (a) Bring the test sections to water.
   (b) Treat with diluted primary antibody.
   (c) Wash well to remove excess antibody.
   (d) Treat with bridging antibody.
   (e) Wash well to remove excess bridging antibody.
   (f) Treat with PAP reagent.
   (g) Wash well to remove excess PAP.
   (h) Incubate in enzyme substrate solution to detect sites of binding.
   (i) Counterstain.
   (j) Mount the sections.
   In addition to these steps, it may be necessary to destroy any endogenous peroxidase activity

and to prevent non-specific binding by blocking with normal (non-immune) sera. It may also be necessary to unmask antigens. These three extra steps would normally be done before step (b).

6. The final product of the PAP technique is the brown deposit from diaminobenzidene and this can be treated with osmium tetroxide, which makes it visible with the electron microscope.

7. Antigen retrieval is the treatment of sections to make the antigens available to the antibody. This may be needed after some fixation and processing schedules. Antigen retrieval can be done using proteolytic enzymes, microwave heating or heating in a pressure cooker.

8. Lectins can be labelled using fluorescein, enzymes, gold particles, etc. in the same way as antibodies. Lectins will bind to specific carbohydrates in tissue and can be visualized in the same way as for antibodies.

9. It is mainly used in electron microscopy where the colloidal gold particles are easily recognizable.

10. Avidin is a high-affinity binding agent for biotin. Avidin can be used to form a bridge between biotin-labelled antibodies and biotin-labelled enzymes.

Chapter 13

1. Any method used for screening must not have any serious risk for the patient, it must not be painful for the patient and it must be reasonably inexpensive. Exfoliative cytology uses only cells being shed naturally from the body and so does not seriously damage the patient. Cervical cytology sampling should not be painful although patients may consider it uncomfortable or embarrassing. It is not too expensive as it can be performed by GPs or nurses without the need for a hospital appointment or full surgery.

2. The main method of staining gynaecological smears is the Papanicolaou method. This method gives good nuclear detail, transparent cytoplasm and good cytoplasmic differentiation between the different cell

layers. Other methods do not have this range of properties although they may be useful in more specific cases.

3. (a) Superficial cells are normal constituents of smears. (b) Parabasal cells are unusual but not abnormal constituents of smears and are more common when oestrogen levels are lowered. (c) *Trichomonas vaginalis* is a pathogenic organism although in some patients it may not cause any symptoms. (d) Dyskaryotic cells are abnormal although mild dyskaryosis may revert to normal. (e) Koilocytes are indicative of viral infection and are not normal. (f) Erythrocytes are normal at around menstruation but at other times they indicate a haemorrhage. (g) Plasma cells are uncommon in normal smears and usually indicate a chronic inflammation.

4. A navicular cell is elongated with curling edges giving a boat-shaped cell with an eccentric nucleus. This is usually found in pregnancy.

5. The best stages at which to take smears are around ovulation, i.e. late proliferative phase, ovulatory phase and early secretory phase. Smears taken around menstruation are unsuitable.

6. Acute inflammation will typically show a large increase in the number of neutrophils present, possibly with more mucin and sometimes the presence of fibrin and red blood cells from a haemorrhage. If there is an infection, characteristic organisms may be seen and the cells may show signs of degeneration. Chronic inflammation will typically show an increase in the number of macrophages and lymphocytes rather than neutrophil leukocytosis.

7. *Trichomonas vaginalis, Candida albicans,* herpes virus and human papilloma virus.

8. Cervical intraepithelial neoplasia.

9. Too few cells in the smear, the presence of red cells obscuring detail, the presence of pus obscuring detail, and the presence of artefacts caused by air-drying or inadequate fixation.

Chapter 14

1. Centrifugation uses high G forces to compact the cells which are heavier than the fluid. Simple centrifugation compacts the cells at the bottom of the centrifuge tube and allows most of the fluid to be decanted off. The pelleted cells can then be spread as a smear. The cytocentrifuge uses a special cuvette which spins the cells directly onto the slide as a discrete circular pellet. The excess fluid is soaked up with an absorbent pad. The cytocentrifuge can only deal with small fluid samples and has a limited range of cell concentrations at which it is an effective method. Filtration uses a membrane with precise small holes which allows the fluid to pass through but retains the cells on the filter. The cells are stained *in situ* without the need for further smearing. Filtration is one of the techniques used in automated liquid-based cytology but otherwise has limited application.

2. Superficial and basal urothelial cells, superficial intermediate squamous epithelial cells, glandular cells, spermatozoa, red blood cells, neutrophils and macrophages (histiocytes) are all found in normal urine. Transitional cell carcinoma (bladder cancer) is a very common type of malignancy (fourth commonest in men) and is caused by chemical carcinogens that become concentrated in the urine. Smoking is the most common source of such carcinogens. Many industrial chemicals can also cause bladder cancer so chemical workers are a high risk group.

3. (a) Bacterial infections can cause inflammation of the bladder (cystitis) with *E. coli* being a common causative organism. Inflammation of the urethra (urethritis) can be caused by chlamydia and gonococcus. (b) Viral infections include herpes, HPV and cytomegalovirus. (c) Fungal infections can be due to candida. (d) Parasitic infections of the urinary tract include trichomonas and schistosoma.

4. (a) Sputum samples can be obtained by collecting cough specimens and the sputum can contain cells from many parts of the respiratory tract. (b) Bronchoscopy in which a bronchoscope with a sampling device can obtain cells from the upper respiratory tract and directle sample suspicious lesions. (c) Washing the walls of the upper respiratory tract can dislodge cells into the irrigation fluid and cells can be reovered from the fluid using standard fluid cytology techniques.

5. Squamous epithelial cells (superficial, intermediate types), ciliated epithelial cells, mucus-secreting goblet cells, red blood cells, neutrophils and macrophages can all be found in respiratory samples.

6. Asbestos fibres indicate exposure to asbestos and are the cause of the industrial disease asbestosis. Asbestos can be carcinogenic, causing mesothelioma and increasing the risk of lung cancer. It can be recognized as elongated fibres which are brown with brightfield microscopy and are birefringent in polarized microscopy. In the body they become protein coated, forming distinctive beaded 'asbestos bodies' which show a Perls-positive reaction.

7. Effusions are accumulations of fluid in spaces or potential spaces in the body. They can occur in the pleura, peritoneum and pericardial spaces.

8. Fine needle aspirates are used to collect fluids from structures which can be deep inside solid organs. Typical sources of samples are cysts.

9. CSF samples are frequently seen in haematology, clinical chemistry and microbiology.

Chapter 15

1. Cytogenetics is the study of changes in chromosome number or structure that can be seen by microscopy.

2. Cytogenetics can diagnose syndromes associated with chromosomal changes before and after birth. Although diagnosis in children and adults cannot lead to the disease being cured (the abnormalities are currently not curable), the recognition of a genetic abnormality will allow treatment of the symptoms and save the need for further diagnostic investigation. Prenatal diagnosis allows the possibility of termination of the pregnancy to avoid the birth of a severely affected child. Cytogenetics can also be of help in diagnosing somatic diseases such as leukaemia where there may be characteristic chromosomal changes in the diseased cells.

3. Cells are taken from the patient (blood cells, chorionic villus cells or amniocentesis cell samples) and grown in culture. Cells may need to be stimulated to divide in some cases and cells may be treated with colchicine to accumulate mitotic cells. Cells are then treated with hypotonic solutions, fixed and spread on to a slide and finally stained to show banding. The hypotonic solution causes the cells to absorb water by osmosis and this inflates the cells and causes the chromosomes to separate.

4. The banding patterns can be stained using fluorescent dyes such as quinacrine or by Giemsa staining following digestion of the cells.

5. A normal female has the constitution 46XX. 47 XY, +21 would indicate a male Down's syndrome patient.

6. At conception it is estimated that up to 1 in 12 (~8%) human conceptions are chromosomally abnormal and this falls to about 1 in 200 at birth (0.5%) due to spontaneous abortions.

7. The Y chromosome is small and contains only genes needed to produce a male foetus. Thus, the absence of a Y chromosome simply results in the female phenotype being produced. The presence of an extra Y chromosome results in only a few extra genes and seems to cause only minor physiological consequences. The cells have a mechanism to switch off any X chromosomes in excess of the single X found in males to maintain the chromosome balance between the different sexes. This means that supernumerary X chromosomes do not cause the severe abnormalities seen with comparably sized autosomes.

8. In translocation Down's syndrome, there is a single large chromosome, which corresponds to the long arm of chromosome 21 attached to the long arm of another chromosome (commonly chromosome 14). This translocation chromosome can be seen in chromosome spreads. Translocation Down's syndrome-affected patients have only 46 chromosomes, not the 47 found in the commoner form of Down's syndrome. Translocation Down's syndrome shows a familial pattern of inheritance with the presence of carriers who are themselves unaffected by Down's syndrome but who have an increased chance of passing it to their children. Translocation Down's syndrome is independent of maternal age.

9. A pericentric inversion occurs when there is an inverted piece of chromosome that includes the centromere. The inversion is symptomless until meiosis occurs, when it can result in the death of the cells if a crossover occurs in the inverted region.

10. Hydatidiform moles or molar pregnancies involve a placental growth with no foetus present. This is due to an abnormal conceptus with an abnormal euploid set of chromosomes. An incomplete molar pregnancy has a triploid chromosome complement, mostly due to a double fertilization. A complete molar pregnancy has a diploid chromosome complement but usually the chromosomes all come from one parent.

Chapter 16

1. *In situ* hybridization uses a specific labelled DNA probe that will hybridize with complementary DNA sequences in the fixed tissues and cells by base pairing. Identification of the site of the label used then indicates the site of the target DNA and identifies this as being complementary to the DNA probe.

2. The target DNA must be dissociated into two single strands using a high temperature (called the melting temperature). The labelled probe is then added to the section and allowed to anneal (hybridize) with the native target DNA at a lower temperature. The probe can then be visualized. The method of visualization depends on the nature of the labelling used to produce the probe.

3. Radioisotopes, fluorescent dyes, haptens (e.g. biotin) and colloidal gold can be used for DNA probes. Radioisotopes produce results only slowly and have safety complications but can be very sensitive. Fluorescent probes give a quicker result but are less sensitive and require a fluorescent microscope which limits counterstaining potential. Haptens allow amplification so can be more sensitive but can give more background staining; but, they are more rapid, do not need special microscopes and have a wide range of counterstains available to show the general tissue structures.

Colloidal gold is most applicable to electron microscopy.

4. PCR uses a short primer complementary to part of the target DNA. The two strands of the target DNA are separated into single strands (denatured) using heat and the primer can then bind to the target DNA. DNA polymerase then adds deoxynucleotides to form two new duplex DNA strands. Repeating the process doubles the number of new strands and by repeating the process many times the number of DNA molecules will be doubled. This results in a rapid exponential increase in the number of copies of the original target DNA.

5. Tissue for ISH can be prepared by buffered formaldehyde fixation followed by routine paraffin wax embedding. Frozen sections can be used but these still require formalin fixation to immobilize the DNA in the tissue. Following sectioning the main requirement is cleanliness, to avoid contamination with exogenous DNA. Prior to hybridization the target DNA may need to be unmasked and DNases which could destroy the target or probe DNA need to be inactivated.

6. The main applications of ISH are research, identification of infective organisms, tumour identification, therapy strategies, detection of abnormal genes and assisting in karyotyping.

7. Autoradiography uses chemicals labelled with radioisotopes. The position of the radioisotopes can be determined using a photographic emulsion. When the radioisotopes decay, the ionizing radiations affect the film. After development of the film, the areas of the emulsion exposed to the ionizing radiations show black and this locates the radioisotopes.

8. The prostaglandin will be removed by hydrophobic solvents; however, since it will be bound to a receptor and has a low solubility in water, it is likely to withstand freezing and thawing without being displaced. Therefore the tissue can be cut as unfixed cryostat sections which can be air dried. The dry sections can then be coated in emulsion. Following successful exposure the slide could be stained with a standard H&E.

9. (a) There are two possibilities: ^{14}C ($^{14}C_2H_5OH$) or ^3H ($C_2{}^3H_5OH$). (b) The labelling could be

on the sulphur: ^{35}S ($H^{35}SCH_2CH(NH_2)CO_2H$), or any of the carbons could be labelled with ^{14}C, or tritium could be used to label the three hydrogens directly attached to a carbon ($HSC^3H_2C^3H(NH_2)CO_2H$). (c) Labelling here is not very reliable with any atoms. Oxygen has no suitable radioisotopes and labelling the H with tritium is not likely to be stable, so autoradiography is possibly not the best technique. (d) Only the phosphorus is suitable ($H_3{}^{32}PO_4$).

Chapter 17

1. The numerical aperture is related to the resolving power of the lens, which is one of the most important characteristics of any microscope lens. Numerical aperture is also related to the depth of field of the lens and determines how much depth of specimen is in focus at any one time.
2. Empty magnification is where the image is enlarged to give a bigger image but with no increase in detail. Empty magnification may even result in less detail being observable than at a lower but useful magnification. Useful magnification is approximately 1000 times the numerical aperture of the objective.
3. An achromatic lens is a lens that is partially corrected for chromatic aberration. The colour fringing (a spectrum of light produced by the lens) is greatly reduced, but not totally eliminated.
4. The darkground microscope uses oblique illumination in the form of a hollow cone of light. None of the direct light can enter the objective so the field of view appears dark. Objects which can bend the light by diffraction or refraction will appear bright against this dark background.
5. In the polarizing microscope two polarizing filters are used, with the first filter between the light source and the specimen and the second between the specimen and the eye. The first filter produces plane polarized light; the second polarizing filter is rotated or 'crossed' so that in the absence of a specimen no direct illumination is seen. Urea crystals are birefringent and rotate the plane of polarization, allowing some

light to pass through the second polarizing filter. Thus, the crystals appear bright against a dark background.
6. Two filters are used in a conventional fluorescent microscope. The first filter is placed between the light source and the specimen and this allows only short wavelength light to pass. The second filter is placed between the specimen and the eye and allows only long wavelength light to pass. Only fluorescent light which has been emitted by the specimen will be seen with this arrangement.
7. Confocal microscopy uses a very narrow aperture (pinhole) at the conjugate plane of the focal point of the lens to restrict the fluorescent light to a single plane. Only a single point can be imaged effectively at one time, so the illuminating light needs to be scanned across the specimen to build up a complete image; this is electronically visualized using a television-style monitor. This is most effectively achieved using a laser as the illumination source in an incident light microscopy arrangement; the laser is then scanned across the specimen by deflection mirrors.

Chapter 18

1. The human eye can be fooled by optical illusions; apparent size is a subjective assessment by the eye and the brain and not an absolute measurement.
2. 0.55 or 55% (area proportion is 110/200 = 0.55).
3. The density is not necessarily homogeneous. The amount of light passing through the lightly stained areas will be disproportionate to the amount of light coming through the darker areas and the amount of material will be underestimated.
4. A digital camera body can be attached to the eyepiece, or an equivalent camera port in a trinocular system, to project the microscope image onto the light sensor chip in the camera. The camera captures the image as three monochrome digital images: one image for each of the three primary colours. This usually involves a Bayer grid with each colour channel being the output from four adjacent pixels

(typically two green, one red and one blue sensitive pixels). The image is then stored as a standard file format (e.g. JPG or TIFF) which can be processed by a computer to reconstruct the original three-colour image.

5. Image processing and image analysis both involve mathematical algorithms to process images. An algorithm for image processing will produce a new image, using the output from the calculations, which has been enhanced in some specific way (e.g. contrast enhancement) to help interpretation of the image. Image analysis produces a numerical output which represents a specific attribute of the image (e.g. optical density). Both may be employed in a single analysis where image processing is used first to simplify the image and allow easier image analysis.

6. Digitally based microscopy can be used to identify those specimens which are clearly normal; this removes the need for human screening. It can identify clearly pathological specimens which can be passed on for immediate diagnosis without the need for further screening. Any specimens identified as indeterminate can be passed on for further human screening. Digital analysis can also find the most abnormal cells in a specimen and mark the slide to facilitate human examination of a smear. Digital systems can be faster than equivalent human screening. They are also more consistent than human screening and can work continuously for long periods without becoming fatigued. The method currently cannot reliably be used for full diagnosis nor for reliable interpretation of the borderline cases.

7. Telepathology allows images and data to be shared or analysed at distant sites without the need to exchange physical specimens. This allows pooling of expertise, sharing of information and the ability for staff to access data when away from the department.

Chapter 19

1. CTEM differs from light microscopy by using electrons instead of light. It has a better resolving ability since electrons can have much shorter wavelengths than light. The lenses in the electron microscope are electromagnetic, not glass, and magnification is changed by altering the magnetic strength of the lens rather than changing lenses as in the light microscope. The electron microscope must work in a vacuum and this prevents the examination of living specimens. The electron microscope needs thinner sections and they need better support than in light microscopy, so specimens are mounted on copper grids not glass slides. The image cannot be viewed directly in the electron microscope but must use either a phosphorescent screen or a camera to take a photograph. The image in the electron microscope is always monochrome, whilst the light microscope image can be multicoloured. Contrast is achieved in the electron microscope using heavy metal salts rather than the dyes used in light microscopy.

2. Resin or plastic embedding is needed to give better support to the specimen and allow thinner sections to be cut. The plastics are also more stable than wax when subjected to a high vacuum and an electron beam, so they support the tissue better during examination.

3. There are no lenses used in SEM to image the specimen so there are no lens aberrations. The condenser lenses used to focus the electron beam on to the specimen play no direct part in producing the image.

4. Critical point drying is the use of a liquid that has a critical temperature and pressure at which there is no distinction between the gaseous phase and the liquid phase. At this critical point, the gas and liquid have exactly the same density. One such liquid is CO_2. By dehydrating the tissues and then replacing the water with liquid CO_2, it is possible to remove the CO_2 without evaporation simply by warming it to above its critical temperature. The liquid becomes a gas without boiling or evaporating and can be allowed to escape gently without damaging the tissue.

5. It is necessary to coat SEM specimens to make them electrically conductive and gold is the best metal for this purpose. The electrical conductivity is needed to earth any static electrical charge, which would otherwise interfere with the image.

6. The positive identification of a material in the light microscope is easier because there is a wide range of available dyes, which can be quite selective in their staining and which have a wide range of contrasting colours. In electron microscopy, there are only a few heavy metal ions available for use. These are less selective in their staining properties and will react with a wider range of materials. Finally, the electron microscope image is only in shades of grey so there is less distinction between materials.

7. The X-rays produced in the electron microscope will have wavelengths characteristic of the elements within the specimen, so examining the spectrum of X-rays allows determination of which elements are present in the specimen.

Index

Bold indicates main entry

439